Cosmogony and Ethical Order

Contents

Preface

The present collection of essays represents the most tangible result produced by a series of three conferences on "Cosmogony and Ethical Order" held at the Divinity School of the University of Chicago in March 1981, April 1982, and October 1982. Sponsored jointly by the Institute for the Advanced Study of Religion at the Divinity School of the University of Chicago and the Harvard/Graduate Theological Union project on "Values in Comparative Perspective," these conferences received supplementary funding from both the National Endowment for the Humanities and the Luce Foundation.

As organizers of the conferences, and as editors of the volume, we have written the introductory essay. The fourteen papers that constitute the collection include thirteen that were presented at the conferences and one (Burkhalter's study of cosmogony and ethics in Islam) that was written by a participant to fill a lacuna in the range of traditions discussed within the conference setting itself.

We are very appreciative of the cooperation we have received from our colleagues. Taking all three conferences into account, more than thirty scholars have been involved in the project. In addition to those who have contributed essays to this volume, the following attended at least two of the three conferences, and made major contributions to the discussion: Joseph Kitagawa and Wendy Griswold from the

University of Chicago; John Reeder from Brown University who, in addition, read the editors' introduction and made many invaluable suggestions; Ronald Green from Dartmouth College; and David Little from the University of Virginia. We are also grateful to James Gustafson of the University of Chicago, Kenelm Burridge of the University of British Columbia, Frederick Carney of Southern Methodist University, and Sumner Twiss of Brown University (a conference participant), each of whom presented a paper; to Charles Reynolds of the University of Tennessee, whose insightful critiques and commentaries concluded each of the sessions; to Bruce Grelle and Paul Williams, who injected intellectual contributions far more extensive and substantive than those usually associated with conference "recorders", and to Jeffrey Stout of Princeton University, who read the introduction and collected papers with great care and made many creative and useful suggestions.

Many other persons have helped in a variety of crucial ways. Dean Franklin I. Gamwell of the Divinity School of the University of Chicago, Professor John Carman, Director of the Center for the Study of World Religions at Harvard University (who was also a conference participant), and Professor Mark Jurgensmeyer of the Graduate Theological Union in Berkeley provided strong encouragement and essential administrative support. Thomas Terrell, Janet Summers, and Martha Morrow-Vojacek of the Divinity School staff ably assisted with matters of organization and practical detail.

Our intention in organizing the "Cosmogony and Ethical Order" project was to advance the fledgling discipline of comparative religious ethics by bringing religious ethicists (as well as interested philosophers) to join with historians of religion (as well as interested anthropologists and biblical scholars) to explore a common theme that would be accessible to both groups. We believe that our experiment has been a success, and trust that our readers will agree.

Robin W. Lovin
Frank E. Reynolds

1 In the Beginning

Robin W. Lovin and Frank E. Reynolds

1. Cosmogony and Ethics

In ancient China, the scholar-bureaucrats who were responsible for the internal order of the warring states conceived of their task as a "rectification of names," an ordering of human relationships according to a system of ranks and duties that some schools believed were established by a permanent, natural order (Gernet 1982, 87, 96–97). Thomas Aquinas could not account for the freedom of human action without explaining how it escapes the control of the stars and planets whose positions determine most material events (Wildiers 1982, 54–55, 65). In the nineteenth century, Marx and Engels sought an interpretation of ethics that would explain moral imperatives in terms of material facts and relate conflicting moral positions to the dialectical processes of history (Giddens 1971, 18–34).

Such diverse ways of explaining and justifying moral imperatives may be taken as indications of a very general human effort to relate the changing requirements of action to a permanent and unchanging order of things. Familiar concepts in Christian theological ethics, ideas of "natural law" or of the "orders of creation" can be interpreted as systematic expressions of a far more widespread belief that truly significant actions recapitulate the primordial cosmogenesis or participate in a pattern established outside of the flux of ordinary events.

Eliade (1959) has demonstrated the pervasive effort of traditional cultures to relate the order of present action to cosmogony. From the perspective of the historian of religions, many of the ethical systems of the West—both religious and philosophical—appear to be linked to that effort.

Natural order provides the foundation for much of everyday moral thinking, and the origins of natural order can be invoked to justify the choice of one way of acting over another. We are not surprised by such arguments in the Homeric literature (Adkins, "Cosmogony and Order in Ancient Greece"), nor by reports that link cosmogony and ethics in the traditional cultures of the Andes and highland Guatemala (Sullivan, "Above, Below, or Far Away"; Warren, "Creation Narratives and the Moral Order").[1] We are, however, less prepared to see the impact of cosmogonic beliefs on the moral systems of our own society, both in popular thought and in theoretical reflection.

Consider, for example, the legal battle of "scientific creationism" that made headlines early in this decade. The creationism position was often dismissed as a bizarre survival of premodern ways of thinking, and on narrow legal grounds, the creationist idea had little to recommend it. What was at issue legally was the question of whether the belief that the world and the species of life in it originated as the direct result of divine creation can be defended as a "scientific" view of origins, a view comparable to evolutionary theory as that has been worked out since Darwin. On that legal question, the opinions of the experts and the decisions of the judges have uniformly gone against the creationists.

Set the legal issue in a broader cultural context, however, and the winners and losers are less clear. Many religious conservatives want to make creation "scientific" to bolster the warrants for a definitive set of moral beliefs. Most evolutionists insist that a proper scientific theory has no moral content, and that the real issue is the freedom to pursue and to teach scientific truth. The distinction between scientific truth and moral import, however, is not so easily maintained. A modern scientific cosmogony that sees the evolution of life as continuous with the evolution of matter challenges us to formulate ways of living in a vast universe that seems largely indifferent to our existence (Gustafson 1981). Freedom becomes an existential given as well as a methodological requirement.

The impulse to link moral orientations to accounts of cosmic origins does not end, then, with the rise of modern science, and while we would join the experts and the courts in rejecting "scientific creationism" as a way of making that connection, the underlying notion that there is a relationship between origins and ethics may be more astute. Modern scientific cosmogony is not simply a report of the results of an investigation. Our culture appropriates it in a larger view of the human place in the universe, and contempory cosmogony thus becomes a part of that older enterprise through which people identify the truly significant actions that are somehow distinguished from self-interested choices and everyday events. "The Creation, the Apocalypse, the Foundations of Morality, the Justification of Virtue: these are problems of perennial interest, and our contemporary scientific myths are only one more installment in the series of attempted solutions" (Toulmin 1982, 84).

One reason why we often miss the relationship between origins and ethics in modern thought is that it has been neglected in twentieth-century philosophy. Many moral philosophers in the Anglo-American tradition have accepted some form of G. E. Moore's dictum that moral terms cannot be defined in terms of nonmoral realities. Having wrestled unsuccessfully with the problems of securing agreement on an account of reality that would justify particular moral choices, the philosophers have sought instead to render morality as far as possible logically independent of conflicting beliefs about the world.[2] Religious thought has often accommodated itself to this tendency in philosophy, even when it has affected scorn for the philosophy itself. Thus Karl Barth insists on direct reliance on the Word of God, rather than on any investigation of the created order, as the source of moral imperatives (Barth 1957, 509–732). More recently, other authors have sought to take religious ethics seriously by accepting the philosophical premise that there is a distinctive moral rationality and arguing that the traditions of religious ethics are based, in part at least, on just that way of reasoning (Donagan 1977). While not every author who employs this neo-Kantian method makes cross-cultural claims for the findings, Ronald Green (1978) in particular has made Kantianism a powerful comparative tool by claiming to demonstrate the existence of a common pattern of moral rationality in religious systems which obviously have quite different views about the nature of nonmoral reality.

Presently, however, this understanding of reason itself is under attack in philosophy (Rorty 1979; Stout 1981; MacIntyre 1984). The important consequence of this philosophical development for the study of comparative ethics is that we are led back to an approach that treats a system of beliefs as a whole and refuses to isolate moral propositions for analysis apart from propositions about how things are in the world and how they came to be that way.

The moral philosopher will recognize this method as "naturalistic" in that it does not sharply distinguish the ways that people identify and test their moral choices from the ways they identify and test their beliefs about facts.[3] Once largely rejected in Anglo-American philosophy, naturalism in this sense now receives extensive, if controversial, reconsideration by moral theorists who seek an account of moral reasoning that would be closer to the patterns of thinking that persons actually employ in resolving questions about action (White 1981; Flanagan 1982).

What is controversial for the moral theorist is, however, the normal way for an anthropologist or historian of religion to proceed. Confronted with the merit-oriented moralities of Theravada Buddhists (Reynolds, "Multiple Cosmogonies and Ethics") or the divination rituals that Andean shamans use to give advice on moral choices (Sullivan, "Above, Below, or Far Away"), one seeks to understand them not as isolated systems of moral reasoning but as activities integrated into a complex cultural whole that includes both moral beliefs and beliefs about reality. This approach to the description of a moral system is the starting point for many methods of comparative study. Anthropologists, cultural historians, and historians of religion make a tradition available for comparative study by detailing the modes in which its moral beliefs may be expressed (legal codes, wisdom literature, ritual actions, etc.), relating these modes of expression to other aspects of the tradition's view of the world and identifying the typical moves by which problems of choice and action are resolved in the context of the whole system of factual and moral beliefs (Reynolds 1980).

Any approach to comparative religious ethics thus runs quickly into controversial issues in ethical theory. The essays in this volume are primarily descriptive studies in ethics. They deal with the ways that different cultures and traditions have resolved questions about action, both in widely shared cultural presuppositions and in conscious moral choices. For the most part, the authors of the essays do not make judgments about what a moral system should be or should include. Yet the very choice of a method to describe the ethics of a culture or a tradition implies a judgment about what moral thinking is. A method that begins with the particulars of a moral system and tries to understand them in relation to other elements in the culture is congenial to a naturalistic understanding of ethics. It presumes a close connection between moral ideas and other ideas about what persons find good and what they think their world is like. By contrast, a formalist understanding of ethics usually leads to a descriptive method that first articulates and clarifies systems of moral reasoning, only later attempting to identify examples of this reasoning in particular cultural contexts.

Our task in this introduction, then, is to consider these alternative ethical theories in a way that will provide a framework for understanding the specialized studies that make up this volume. We begin in section 2 by identifying more carefully the idea of *cosmogony* that links the essays and provides the starting point for thinking about origins and ethics. Even a brief foray into the varieties of cosmogonic speculation raises the question of relativism. In section 3, we trace the problem of relativism in comparative ethics back to alternative solutions formulated by Hume and Kant, and in sections 4 and 5 we show how those early theoretical alternatives are recapitulated in recent proposals for the study of comparative religious ethics. Our own proposal is outlined in section 6, where we set out an empirical approach that rejects the primacy of a formalistic study of moral reasoning and takes seriously the interaction between moral ideas and beliefs about reality. Section 7 provides an extensive review of the essays that appear in this volume. Finally, in section 8, we venture to suggest that our empirical approach to descriptive comparative ethics also provides an appropriate starting point for an ethical theory which could guide normative discussions in cross-cultural settings.

2. *Identifying Cosmogonies*

A cosmogony, as the etymology of the term suggests, is usually an account of the "generation or creation of the existing world order" (*Oxford English Dictionary* 1971, 568). The term includes the connotation that this account has implications for understanding the present human condition, and so it applies well to those classical and biblical accounts of the origins of the physical world that seventeenth- and eighteenth-century English writers first classified together as "cosmogonia." Ralph Cudworth (1617–88), for example, tried to demonstrate a convergence of Christian and pagan views among the ancients on the point that the constituents of the world as we know it have fixed natures, ordered toward discernible ends (Cudworth 1820, 316ff.) The comparative study of cosmogonies thus provided an antidote both to atheism and to the theological voluntarism that made all reality directly dependent on the will of God.

So understood, both the study of cosmogonies and the cosmogonic myths themselves are attempts to find a pattern for human choice and action that stands outside the flux of change and yet within the bounds of human knowing. What unites the cosmogonies is this underlying function of bestowing on certain actions a significance that is not proportioned to their empirical effects or to the individual goals of their agents, but derives from their relation to an order of the world that begins with the beginning of the world as we know it. The philosopher's insistence on "an artificial, orderly, and methodical nature" (Cudworth 1820, 316) makes explicit what Eliade discerns in the traditional civilizations that interpret the flux of history by "perpetually finding transhistorical models and archetypes for it" (Eliade 1959, 141). Interest in cosmogony marks a hope that despite the perpetual changes evident to traditional cultures and the diversity of beliefs apparent to moderns, some realities may be permanent and some things may be dependable enough to build a way of life on them.

Once this function of cosmogony is clear, Eliade notes, its scope can no longer be sharply restricted to those myths that describe the origin of a primordial totality from which nature and human society later develop. The beginnings of the order of rivers, fields, and mountains, the origins of human beings and their societies, and the establishment of hierarchies among both human beings and the gods must all be included among the foundational events that give meaning to present action (Eliade 1969, 80–87).

So cosmogony encompasses a wider range of myths than the etymology of the term might lead us to expect. The foundational events may be located in historical, as well as mythic, time, and where monotheistic faiths emphasize the world's dependence on a creator, the cosmogonic event may be extended indefinitely through time by an affirmation that creation goes on continuously, sustained at each moment by divine will. The sense of precariousness that attaches to any reality that is so utterly dependent on a continuing act of will is perhaps at odds with the permanence sought in cosmogonic speculation, but this juxtaposition of ideas does occur when Christians try to integrate the natural pattern of events into a system of providential care (Lovin, "Cosmogony, Contrivance, and Ethical Order") and when Muslims insist that God intends to perfect the course of history in a way that corresponds to his original perfection of nature (Burkhalter, "Completion in Continuity"). Moreover, the acts of religious founders, though they occur in historical time, can acquire such significance that they mark the beginning of a new aeon, in which the world itself is recreated. So a recent Christian liturgy affirms that in the Resurrection, Jesus "destroyed death, and made the whole creation new" (*Book of Common Prayer* 1977, 374). This assimilation of historical events to cosmogonic foundations may also be seen in the Buddhist case, where the institution of *dhammic* order becomes not merely an epochal event in the story of the Gotama Buddha but a pattern which all Buddhas, past and future, are said to follow (Reynolds, "Multiple Cosmogonies and Ethics").

These few examples should significantly broaden our notion of the mythic, legendary, and liturgical literature which is relevant to an inquiry into cosmogony and ethical order. Even quite contemporary writings, which deliberately set aside traditional religious cosmogonies in favor of a more scientific account of origins may themselves be cosmogonies in this sense: by identifying inescapable features of reality which result from the circumstances of its beginnings, we discover what amount to necessary constraints on our own choices and possibilities. Darwin, Marx, and Freud may have replaced divine purposes to which we might conform by a materialism in which all subsequent events follow with causal necessity from an initial state of affairs that had no meaning or purpose hidden within it; but what is thus ultimately contingent is nonetheless proximately necessary. To that extent the practical application of Marxian or Freudian materialism often duplicates the action-guiding function of cosmogonic myth.

It may seem that a concept of cosmogony that includes virtually all myths of origin *and* the many attempts to provide systematic historical, scientific, or metaphysical alternatives to traditional myths is too broad to be of use in the study of actual cultural and religious systems. David Little, for example, argues that when we use terms in comparative studies, we must be careful to limit their meanings. We must be "more, rather than less sensitive and attentive to the various jobs those terms do for us" (1981, 221). That warning is well taken, but one of the jobs words like "cosmogony" can do is to provide a generic concept under which a wide variety of narratives and explanations can be subsumed. The concept is, of course, shaped by the intellectual interests of our own culture, but once we understand those interests we can see that nothing is to be gained by attempting narrowly to restrict the application of this term or others like it.[4] In fact, it is this breadth that makes cosmogony a useful comparative concept. No approach to cosmogony and ethical order that claims comparative validity can, for example, simply apply the Protestant scheme of the "orders of creation" to a variety of cultural contexts. The problem of finding genuinely comparable units of organization is too great, and even if that were possible, the meaning and importance that each culture assigns to the social system in question may vary dramatically. So cosmogony, if it is to be a useful comparative concept, cannot be limited to the study of a set of origins in which we happen to be interested. Those who use the term must take the culture's understanding of origins as a whole, not demanding in advance that certain questions be answered but viewing the traditional cosmogonic system—with its accompanying anthropogony, sociogony, and divine or human hierarchies—as clues to the questions the culture finds important. In the same way, it is inappropriate to truncate the inquiry by restricting it to *myths* of origin, as though there were no other way for a culture to deal with the problems of origins or to solve the problems that may arise from ambiguities and contradictions in the mythic traditions.

The comparative study of cosmogony must, then, be broadly construed as a study of the ways in which cultures and individuals relate their basic notions of the origins of the reality in which they live their lives to the patterns of action that they consider to be dependable and worthy of choice. The authors in this volume understand cosmogony in this broad sense, and they structure their questions and hypotheses accordingly. The essays themselves are seldom comparative in a direct way, since the authors focus their work within the context of a single tradition or survey the impact of conflicting traditions on a common culture. Nevertheless, the broad meaning of cosmogony and ethics which the authors employ allows them to delineate significant patterns of cosmogonic thinking and significant modes of relating cosmogonic motifs to ethical norms and issues. This, in turn, permits comparisons that still respect the authors' grasp of the unity and integrity of the cultures they have studied. What we must do now is to think more explicitly about the method for doing comparative ethics which these procedures imply and embody.

3. *Rationality and Ethical Diversity*

While comparative ethics is a rather new academic enterprise, the methods employed in that study have their origins early in the history of modern philosophy. In the seventeenth century, Ralph Cudworth could maintain a Platonic doctrine that the standard of morality is rational, immutable, and universal. This standard could, he thought, be discerned in the best thought of pagan philosophers and Hebrew prophets, as well as in its final revelation in Christianity. By the middle of the eighteenth century, however, this ethical universalism could not be asserted in so straightforward a fashion. Whatever effects the discovery of other cultures and civilizations may have had on European consciousness, the beginnings of the modern period were also marked by new awareness of the diversity and change in European civilization itself. Montesquieu (1699–1755) observed the variations in law and culture between England and the Continent (Montesquieu 1975), and David Hume (1711–76) pointed out the differences between the morals of ancient Greece and those of Christian Europe. "The Athenians surely, were a civilized, intelligent people, if ever there were one; and yet their man of merit might, in this age, be held in horror and execration" (Hume 1975, 333).

These empirical observations about the diversity of human conceptions of the good life seemed at the outset to threaten the ordered moral world with the menace of ethical relativism. If basic moral beliefs differ so widely, might it not turn out that they have nothing in common at all, so that there is no way to adjudicate conflict between the choice of one act and another?

Hume was concerned to answer this question, and to answer it within the framework of the empirical philosophy which had raised the issue in the first place. Although Hume is well known for his insistence that "ought" and "is" must be distinguished (Hume 1973, 469), he does not isolate the "ought" statements in a special realm of rational or intuited truths, different from our knowledge of the reality available to the senses. Moral evaluations are intimately connected with our affective responses to ordinary events and our usual assumptions about what is, in fact, the case. Only closer attention to the morals people espouse in the full context of the rest of their beliefs and values will dissolve the problem that confronts us when we first begin to attend to the diversity of morals.

In his dialogue on ancient and modern morals, Hume has his interlocutor pose the relativist's question as sharply as possible: "What wide difference, therefore, in the sentiments of morals, must be found between civilized nations and barbarians, or between nations whose characters have little in common? How shall we pretend to fix a standard for judgments of this nature?" Hume then provides the outline of an answer: "By tracing matters, replied I, a little higher, and examining the first principles, which each nation establishes, of blame or censure. The Rhine flows north, the Rhone south; yet both spring from the *same* mountain, and also are actuated, in their opposite directions, by the *same* principle of gravity" (Hume 1975, 333).

Hume suggests that if we attend not to the specific acts which are praised or blamed, but to the basis on which a culture decides how to apply praise and blame, much of the apparent diversity in morals will disappear. Quite opposite courses of action can be approved as expressions of the same moral sentiments: "Had you asked a parent at Athens, why he bereaved his child of that life, which he had so lately given it. It is because I love it, he would reply; and regard the poverty which it must inherit from me, as a greater evil than death, which it is not capable of dreading, feeling or resenting" (Hume 1975, 334).

Hume's solution to the problem of ethical relativity anticipates that of later anthropologists and ethicists (Duncker 1939; Dyck 1977). So, too, his interest in the empirical study of cultures and peoples gives rise to the more exact methods of history of religions, sociology, and anthropology which provide most of the data for contemporary comparative religious ethics. Later social scientists do not always share Hume's confidence that common *moral* principles can be found at the source of diverse moral systems, but the eighteenth-century study of society was a starting point for attempts to understand diverse cultures in terms of common social dynamics (Schneider 1967). In this broad sense, Weber's comparative studies of religious responses to the demands of social and economic rationalization (Weber 1964, 207–22) or Durkheim's identification of a functional role for religion in sustaining human community (Durkheim 1965, 474–79) are Humean inquiries. They accept the appearance of moral diversity as the starting point for study, and the patterns which they identify emerge from the interaction of morality with other beliefs and with the inescapable constraints of social life. If the studies in this volume show the immediate methodological influences of Weber, Troeltsch, Wach, and Durkheim, their approach to the variety of moral systems often continues the main lines of Hume's empirical studies of morals.

From the beginning, however, there were those who rejected Hume's assurance that the empirical approach would reveal common features that allow us to compare different moral systems and assess their responses to similar social forces. Explaining moral diversity is not a matter of understanding moral systems in social contexts, these critics insisted, but of understanding the formal features of moral reasoning. Morality cannot be vitiated by relativism because it has a rational structure that is fundamentally identical in all times and places. Hume's empiricism thus provoked the formalism of Immanuel Kant (1724–1804). Kant rejected any attempt to base moral choices on what people actually choose and desire. Even if we reinterpret their immediate responses on the basis of their long-term interests or the needs of society in general, we cannot thereby establish what they *ought* to do (Kant 1964, 61–62). The categorical demand which a moral argument imposes rests on a structure of practical reason which establishes that to deny the moral requirement involves one in self-contradiction. To say that an act is morally obligatory is to say that I could will it not only as a solution to my own immediate problems but as a universal rule for action. Keeping solemn promises qualifies as a moral obligation, according to Kant, because it passes this formal test. Breaking solemn promises is immoral, not because we find it distasteful, but because making *that* behavior into a universal rule would mean that promise-making would be pointless in the first place. That self-contradiction, rather than our feelings or our assessment of the consequences, is what makes an act immoral (Kant 1964, 88–94). Kant thus moves the scrutiny of moral claims from the diverse, disordered, and confusing substance of what persons want and value into a formal, logical test that concerns only the structure of their moral claims.

Kant did not think that most people understood their moral choices in this way. Most were motivated by fear of punishment or by more immediate feelings of delight or disgust for a particular act. What a philosopher does is not to describe the moral life of persons as they experience it, but to provide a rational account of the structures of thought which make that experience possible and which alone provide valid reasons for the moral choice. Kant did believe, however, that a *complete* account of the structures of practical reason would show that this respect for moral law is compatible with the search for happiness that motivates much of humanity's nonmoral choice and action (Kant 1956, 117–36).

Kant's transcendental philosophy thus provides a rigorous defense against the moral relativism that seemed to threaten Western philosophy and religion when it confronted the fact that moral standards do not converge, even between classical and Christian Westerners. If a single structure of reason underlies all valid moral thought, we need not be puzzled by its apparent diversity. Instead, the appearance of diversity is a philosophical challenge, inviting us to demonstrate the underlying unity in moral reasoning. If one's chief objective is to insure that moral judgments retain some universal authority, Kant's proposal has great appeal, but it does introduce a certain distance between moral philosophy and ordinary moral experience, and in so doing it tends to separate moral beliefs from other sorts of beliefs and conclusions.

For comparative ethics, then, one basic methodological choice was formulated before the turn of the nineteenth century in the controversy between Hume's empirical study of moral affections and Kant's formal explanation of valid moral reasoning. The students of particular cultures have, as we noted, often been unwitting disciples of Hume. They have tried to understand ethics as part of a whole cultural system, and they have tried to keep their descriptions of moral systems close to the way that the participants actually understood them. By contrast, many students of ethics have consciously chosen to follow Kant. They begin by asking how a particular culture exemplifies universal features of moral reasoning, even though this analysis may be rather far removed from the terms in which the participants explain their own morality.

4. Comparative Religious Ethics A Formalist Approach

Ronald Green's *Religious Reason* (1978) is an outstanding contemporary example of the Kantian approach. Green insists that we must establish the structures of moral rationality independently of empirical generalizations about what people happen to approve morally. When this is done properly, with exclusive attention to the moral logic of religion and a careful elimination of the particularities of Western theism, Green suggests, we will find a set of "requirements of pure religious reason" (1978, 109). These requirements provide a structure within which we can understand the moral teaching of different religious traditions, either as exemplifications of the structure or as characteristic maneuvers to correct divergences from it.

The explanatory power of Green's method appears most clearly in his treatment of two religions of Indian origin, Hinduism and Theravada Buddhism (Green 1978, 201–46). In contrast to the common assertion that the religions of India are the faiths of mystics who have no use for ethics and, indeed, no belief in the substantial reality of the human person as moral agent (see, for example, Weber 1958, 180–230), Green argues that Hinduism and Theravada Buddhism conform precisely to the structure of pure religious reason which he has outlined. In the idea of *karma*, he finds the insistence on the significance of intentional actions which is essential to an ethical system and which is, indeed, developed in the religions of India far more rigorously than in the moral philosophies and theologies of the West. The iron law that connects will, act, and consequence in India obviates much of the Western discussion of whether virtue is rewarded. Indeed, Green notes, the certainty that right action *will* be rewarded would, taken by itself, reduce morality to a simple matter of prudence and obscure the distinction between duty and desire that Kant found so important. What keeps that reduction from happening is precisely the doctrine that the ultimate good is an annihilation of the self (*moksha* or *nirvana*) and that this annihilation of the self ends the karmic cycle of deed and recompense. Only by this teaching, which seems at first glance antithetical to Western ideas of moral responsibility, can a religion which believes in *karma* retain a moral rationality which distinguishes action that respects the moral law from action that merely seeks the long-term good of the agent. For Green, the methodological conclusion that follows from this reading of Hinduism and Theravada Buddhism is clear: empirical attention to the details of a culture's morality may lead us to miss "a more basic moral and religious logic that can help further explain the seeming inconsistencies" that the empirical method inevitably turns up (1981, 112).

Other observers see other possibilities. Frank Reynolds sees moral life in the Theravada tradition as a complex pattern of action devoted both to the alleviation of suffering in the world of cause-and-effect events and to the cessation of suffering in the attainment of *nibbana*. The choice of a particular line of action, however, is justified less by direct reference to the goal that is sought than by how that action is understood in the context of a worldview shaped by one or more of the canonical cosmogonies. An action may be taken because it conforms to rules that sustain the monastic or social order and thus stays the course of cosmic devolution that ends in the destruction of the present world. An action may be taken because it cultivates virtue, conforming the individual's life to the *dhamma* and helping to order the world according to it. Examples could be multiplied, but the point should already be clear: When we begin our investigation with the question of how moral choices are understood in Theravada Buddhism, we find that they are not always connected to any direct line of practical reasoning that ends in *nibbana* or depends upon *nibbana* to make sense of the choice.

Green recognizes this variety in Buddhist ethics. Speaking of Indian religion generally, he writes, "Lower caste individuals might be sufficiently attracted by the goal of a higher birth to be content with a world dominated by *karma*, and even today it is the simple promise of higher birth that principally motivates the Hindu lower castes and the Buddhist laity" (Green 1978, 216). From his formalist theoretical perspective, however, this widespread pattern of moral thought and action makes sense only when it is set in a context dominated by the *nibbanic* ethic, which conforms to his pattern of religious reason. The forms of Buddhist ethics which concentrate on rooting out personal vices or increasing personal merit may have sociological or historical interest, but the key to Theravada Buddhism as a system of religious reason lies in its pursuit of *nibbana*. The question may be asked, of course, whether this emphasis on basic moral and religious logic has truly resolved "inconsistencies" in Buddhist thought, or whether it has merely drawn our attention away from its genuine diversity.

The diversity of moral systems may be developed to an unusual degree in Theravada Buddhism, but more than half of the essays in this collection study traditions that include multiple cosmogonic myths, each with specific moral implications that develop from it. (See the essays by Knight, Burkhalter, Warren, Yearley, and Sturm.) To insist in advance that this apparent moral diversity within the traditions is best understood in terms of a single, underlying moral and religious rationality obviously sets the formalist explanation at odds with the moral experience of persons in some of these traditions. For it is precisely the diversity of moral values and cosmogonic warrants that characterizes the experience. One may always solve the problem by identifying one or another of the moral systems as primary, as Green identifies the *nibbanic* ethic in Buddhism, because it appears to bring the whole complex tradition into line with the formal requirements of religious reason. That approach, however, purchases comparative scope at the price of descriptive adequacy. It suggests the uncontroversial principle that if you go looking for things that are the same, the things you find will be very much alike.

The problems which a formalist approach has in dealing with the diversity within religious traditions reappear in a striking way in comparisons between traditions. Consider, for example, the positions taken by Alan Donagan (1977, 28–29, 32–36) and by Green. Donagan, who employs a Kantian, formalist approach combines the premise that there is a universal structure of moral rationality with the belief that only some cultures and traditions have rules that apply to persons generally without regard to status or station. He concludes that while principles of morality are universal in the sense that they explain moral requirements where they occur, only certain societies do, in fact, have true moral requirements. Thus the scope of comparative religious ethics is severely restricted. Green, on the other hand, argues that the formalist pattern of religious reason he elaborates actually does appear in all of the very diverse religious traditions he has examined. He develops this argument by employing a "diachronic" method (1978, 121) which allows him to count as conformity to rational principles not only straighforward exemplifications of them but also the characteristic moves made in traditions to correct deviations from these principles. Given this method, it is difficult to know what sort of evidence might count against the theorist's presumption that moral rationality is at work in any culture under consideration.

All such formalist approaches which seek to illuminate the commonality and diversity of religio-ethical traditions must incorporate, at a secondary level, a description of the beliefs, wants, and value of persons in various cultures. However, these formalistic approaches, by virtue of their a priori character, inhibit adequate descriptive accounts. Donagan's formalist understanding of ethics, in combination with his historical understanding of various religions and cultures, leads to a dichotomy between religions and cultures which have ethics in a strict sense and those which do not. In practice this turns out to be a distinction between Western, Judeo-Christian cultures on the one hand, and primitive and Eastern cultures on the other. In the case of Green, a very similar formalistic understanding of ethics leads to the construction of historical interpretations designed to justify the contrary assumption that the requirements of religious reason are, in fact, manifest universally. These interpretations, though they are certainly less presumptuous in tone than those of Donagan, are, for the empirically oriented historian, equally questionable.

5. Comparative Religious Ethics A Semiformal Approach

The attention to actual moral reflection and actual human communities that necessarily concludes Green's formalist approach to comparative religious ethics provides the starting point for David Little and Sumner B. Twiss (1978). In *Comparative Religious Ethics: A New Method*, they begin with the social problems of cooperation and scarce resources, and they ask how societies formulate action guides in response to these facts about the human condition (1978, 27). What emerges in their investigation is, however, a structure of practical reason which shares many of the characteristics of Green's "religious reason," though it is not, of course, derived from a priori analysis of the requirements of moral rationality. In terms of the fundamental methodological alternatives for comparative ethics, Little and Twiss attempt to occupy an intermediate position between the formalist and the empirical approaches.

Little and Twiss suggest that in any system of action-guides, we will find basic norms and the rules and principles governing their application. Also as part of the system, there will be procedures for the "validation" of these norms by divine command, authoritative interpretation, and so on, and for the "vindication" of the validated norms by generally held beliefs or special ways of knowing. No claim is made that the same normative content will appear in every system brought under examination, but Little and Twiss discern a way of thinking about how principles are applied and justified for which they do claim a cross-cultural validity.

For example, a basic norm of not injuring others might be applied through a set of rules that require rigid observance of the norm with respect to innocent bystanders but allow exceptions for purposes of self-defense or restraint of malefactors. These applications may proceed through a quite rigorous casuistic logic. A less formal reasoning, however, usually governs the validation of the norms. One may find a proscription against inflicting injury in sacred scripture, for example, or one may claim that it simply belongs in a list of self-evident truths. The vindication of the norm would then depend on the coherence of the validating claim with more general ideas about reality, such as the idea that scripture originates as the word of God, or the Socratic claim that all persons have a set of moral ideas in their minds that need only to be awakened by conscious reflection. Cosmogonic beliefs, then, are chiefly relevant to the last-named "vindication" element in the structure of practical reasoning. The same structure applies, according to Little and Twiss, whether the action-guides in question are moral, legal, or religious.

Clearly, this method is distinguished from the formalist procedures by the way in which the structure of practical reason is established. Green settles his in advance and uses it to guide empirical investigations of culture. Little and Twiss allow the structure to emerge in the investigation. The diverse ways of thinking that mark application, validation, and vindication are more clearly recognized, but it is correspondingly less clear why these, and just these, should be called "rational." The claim here is not Kant's proposition that we must think according to certain laws if we are to think at all, but rather that there is in fact a very general agreement on the ways in which it is appropriate to argue for one norm or another.

The classics of comparative method shed little light on this notion of rationality. Indeed, social theorists often suggest that what makes a practice socially rational will conflict with any account we can give of why individual persons might find it worth choosing. Weber, for example, seems to regard rationalization primarily as a matter of bringing individual habits and practices into line with the requirements of the social and economic system as a whole. This tendency to reduce reason to instrumental rationality, while at the same time subordinating human purposes to social necessity, may make it impossible to reconcile Weber's rationality with the Kantian account of reason (cf. Lukes, 1970, 207). Weber certainly introduces a "struggle in principle between ethical rationalizations and the process of rationalization in the domain of economics" (Weber 1964, 216). With Durkheim, the distance from the ethicist's usual notion of practical reason is even more pronounced, since he gives up the "struggle in principle" and equates rationality strictly with the requirements of the social system. "Since morality expresses the nature of society and since this nature is no more directly apprehended by us than the nature of the physical world, individual reason can no more be the lawmaker for the moral world than that of the physical world" (Durkheim 1961, 116).

The attempt to establish by empirical investigation a structure of practical reason that closely matches the structure of persons' actual deliberations represents a genuinely new inquiry which Little and

Twiss have brought to the more established comparative enterprises of the anthropologists and sociologists of religion. What is sought is not the social-functional rationality that subordinates human wants to the requirements of an economic and cultural system, but a way of explicating what persons find worthy of choice in terms of what they believe to be lasting and important in the world in which they live.

The structure of practical reason Little and Twiss describe (1978, 99) certainly encompasses the elements of that explication, but the rigidity of the structure itself may arbitrarily limit the relationships counted as rational. The structure hangs suspended, as it were, connected to the rest of the world only through the informal logic of "vindication," which is "an effort to persuade others of the acceptability of the recommended validating norms and procedures" (1978, 114). It is in such a connection that appeals to ideas of human nature or the cosmic order are appropriate. This raises at once the question of whether the upward movement of the structure—application, validation, vindication—does not ignore important dialectical interactions between the levels (Stout 1980, 292). Little (1981, 219) concedes that he and Twiss have not emphasized such interactions, but he says they did not mean to imply that they do not happen. However, even when it is understood dynamically, a structure which implies that basic norms are subject to question and reformulation by cosmogonic beliefs or natural realities only in relation to their validation by higher-level norms retains echoes of the separation of "ought" from "is" that has marked ethical theories in Anglo-American philosophy. A more thoroughgoing empirical approach, by contrast, suggests that a norm called into question is susceptible to confirmation or rejection in terms of its relationship to any of the other beliefs the inquirer or inquirers have. Beliefs about the way things are in reality impinge upon action-guides not only in the vindication of a whole system of action-guides but directly in the formulation of particular norms. This seems apparent in the case of the Andean *mesa* and other forms of divination (Sullivan, "Above, Below, and Far Away"), but it is also true for the Hesiodic landowner who receives from the gods not only a sanction for his efforts to attain prosperity but quite specific suggestions about how to do it (Adkins, "Cosmogony and Order in Ancient Greece"). It is even true for William Paley, the theological utilitarian of the eighteenth century, who finds the will of God clearly indicated in such specific things as the structure of the eye or the function of the teeth (Lovin, "Cosmogony, Contrivance, and Ethical Order"). We may regard these ways of connecting order and action as mistaken, outmoded, unusable, or merely incomprehensible, but it is hard to say why we should call them irrational, unless we have begun the empirical investigation with a few formalist assumptions of our own.

6. Comparative Religious Ethics An Empirical Approach

What we need is not a universal pattern of religious reason or a single structure of practical reason, but an explication of rationality adequate to the real complexity of the relations between what is and what ought to be done. If that is not to be found in the social-scientific approaches that reduce moral rationality to a form of social utility, it is also unlikely to be provided by a reassertion of the formal criteria of moral reason that guide Green's investigations and, as we have seen, intrude upon the empirical intentions of Little and Twiss.

Dissatisfaction with Kantian and Cartesian systems that begin reasoning from necessary conditions of thought or self-evident truths has led recent philosophers to seek models of rationality that seem closer to the actual processes of scientific investigation and discovery (Toulmin 1972; Putnam 1975, 196–214). What is more, some philosophical ethicists have already begun the task of applying these models of rationality to the study of ethics. (Putnam 1981; White 1981). These scholars attempt to keep their accounts of rationality close to the experience of persons whose reasoning they study, and they discern rationality in the various procedures by which accounts of anomalous and unexpected experiences are formulated and tested against a larger body of beliefs. In this investigation, the requirement of reason is precisely that no element in the inquiry be immune to revision. Accounts of new experiences may be rejected, or reformulated to cohere with the main body of beliefs, or the main body may be revised to accommodate the new experience. Rationality on this account is, of course, no longer a matter of applying unchanging moral principles to particular circumstances, and moral judgments become complex equilibrations of fact, principle, and feeling. What we think should be done cannot be neatly separated from what we believe to be the case. Moral principles may guide responses to particular experiences, but particular experiences, if we feel strongly enough about them, may also revise our principles or send us in search of additional facts. "The moral judge builds conceptual bridges that not only get him from some sensory experiences to other sensory experiences, but also from sensory experiences to his moral feelings" (White 1981, 37).

Applied to the study of religious ethics, this account of rationality supports our interest in the relationship between cosmogonic beliefs and moral order. An account of how all things began is, as we suggested in section 2, an indication of a tradition's most general ideas about reality. A cosmogony may guide choice and action in quite specific ways, but it is also true that strong convictions about specific actions may send one in search of a worldview in which that action makes sense. An empirical approach to comparative ethics directs our attention especially to points in the history of religions where mythic accounts of the world's order intersect with human action. It is in the many ways that conceptions of world order and human action are correlated that we find some of the most interesting expressions of moral and religious rationality.

To be sure, this rationality presents itself quite differently in different cultural contexts. In most instances it appears less in the arguments persons make than in what they say they want. A Greek in the time of Homer or Hesiod defines success purely in terms of material prosperity. The gods share his craving for honors and riches. There is no higher aspiration to call the satisfactions of luxury into question and no consolation if these satisfactions are not forthcoming. A Theravada Buddhist, by contrast, is mindful that the world of appearances is finally an illusion, and will be careful to distinguish the conditions for immediate material prosperity from the requirements of a karmic law that is even more ironclad than the laws of economics. In some cases the moral problem is a problem of discerning and conforming to a given cosmic reality. One may do that as the Andean Indians do, divining the appropriateness of an action by determining its location in space and time; or as the Taoists do, by easing out of the grip of imposed orders and into undifferentiated natural reality. In other cases, the moral problem is a more constructive task, the creation in human history of an order that conforms to guidelines laid down in natural creation, or the tentative construction of a way to cope with a universe that may be indifferent or hostile to our wants and needs. Paley's cheerful formulation of an order of law that reflects his knowledge that God intends to maximize human happiness is an example of the former sort of construction. Wendy O'Flaherty's summary of the ethics of the *Jaiminiya Brahmana*—"feed the gods, feed the priests, feed the survivors of the inevitable debacle"—suggests that Hindus have at least at times seen ethics as the second kind of construction.

Traditions may, moreover, present the problem of ethics in more than one way. The essays in this collection suggest that for Confucianism, especially, but also for Islam, ethics is both a conformity to the created order and the construction of a human order within it. Controversies arise as different interpreters decide how much weight should be given to each task.

The studies of cosmogony and ethical order that make up this collection were not written to illustrate a methodological point, but we think the essays present a variety of ways of relating moral choice and world order that is given better theoretical order by the empirical understanding of rationality as an interactive testing of moral possibilities against factual limits than by a formal pattern of moral rationality. Taken together, the essays lend themselves to a comparative method that would be attentive to the patterns of moral thinking that actually go on in various cultural settings. Such a method might also suggest some normative guidelines concerning the way in which conflicts between moral systems might be resolved. Our immediate task is to review the essays as studies in descriptive ethics, reserving the consideration of normative implications for ethical theory for section 8.

7. *The Essays*

The patterns that link cosmogony and ethical order could be organized in several ways. If one begins, as Little and Twiss do, by identifying patterns of moral reasoning, the comparative task is to ask whether a mode of moral reasoning relates in some significant way to a belief about cosmogony. Do result-oriented ethics, for example, generally presume a benign power that originally ordered things for human benefit? The theological utilitarianism of William Paley does, in fact, display that enthusiasm for God's creation. Or is the emphasis on consequences, considered across a range of traditions, equally compatible with a negative attitude toward the world of experience? A Buddhist who acknowledges that the world as a whole grows worse with the passing ages may nonetheless cultivate virtues with a view to making the world no worse than it need be.

Another possibility would be to begin not with moral reasoning but with the moral feelings that traditions evince toward their cosmogonies. On this basis, the Buddhist cosmogony of dependent co-origination would be situated at the "negative" pole. Both the process and its results are found wanting. The Marxian cosmogony, on the other hand, would be at the "positive" pole. Both the process and the results receive an approval that is explicit and enthusiastic. Many of the other cosmogonies in this volume could be placed at other points on this continuum.

For an overview of the essays, however, we find it most useful to focus on the cosmogonies themselves. Of course, the judgments a culture makes about the reality its cosmogony presents react on the cosmogony in important ways that may change the cosmogony itself, and will certainly change its correlation with ethics. The cosmogony or cosmogonies nevertheless seem to us to provide an appropriate starting point for comparative studies, if only because their places in the traditions under study can be rather easily typed and classified. In the survey which follows, we identify five such types and briefly discuss the studies that seem to exemplify each of them.

Single Cosmogonies and Ethical Order

In the most straightforward cases, a single cosmogonic myth or unified complex of mythic elements is taken as the focus of interest and provides an explanation of the attitudes and patterns of life that the tradition approves. Often a way of life is simply implied in a narrative, in the sorts of deeds the tradition recalls and the sorts of persons it finds worth talking about. Justifying arguments and disputes about cases may come later, but for a particular narrative or a particular school of thought, consideration of a single cosmogony momentarily excludes the alternatives, and the possibilities for life in *that* world are explored in detail.

Arthur Adkins's paper, "Cosmogony and Order in Ancient Greece," reminds us forcefully that the terminology of moral philosophy in the West originates in just such a specific system of divine and human relationships. The *arete* of the Greek hero, what we should call his "virtue" or "excellence," was, in fact, his ability to maintain himself and his dependents in a particular world. This world was one in which social struggles for honors and possessions mirrored a primordial struggle in which Olympian Zeus had established his supremacy over the gods who preceded him, and had established the norms for ethical relationships and reciprocities that prevailed under his rule.

Norman Girardot's study of Taoism indicates that a tradition may retain its cosmogony even when the world the cosmogony presents is judged an unhappy or unsatisfactory place. Chinese culture offered a variety of cosmogonic myths, but Taoist writers focused on the devolutionary myth of Hun-tun, the primordial emperor who was literally "bored to death" by the agents of advancing civilization. The Taoist cosmogony is clearly devolutionary, but unlike familiar Buddhist and Hindu cosmogonies which describe the world's decline in straightforward terms of material deprivation and social disorder, Taoism suggests that human society brings an increase of order. The increase is clearly recognized as part of a continuous cosmogonic process, and it is just as clearly deplored. What one should do, then, is not dictated by conformity to the order the myth relates, but by a self-directed return to the primordial state that seems more congenial than the orderly, constraining present.

Lawrence Sullivan's essay on Andean Indian traditions explains the use of a very complex but unified cosmogony to orient human actions and relationships. In this case, the cosmogony directs action by indicating the spatial and temporal framework in which human life must be lived and to which specific personal choices should conform. Andean ritual serves to reveal the order of reality, and ritual experience locates the participant in that order. This is by no means simply a collective ritual affirmation of the cosmogony. Individuals facing difficult personal choices seek the advice of shamanic diviners, who resolve the problems by a mode of imagistic argument based on a cosmogram. Here, as in the case of ancient Greece, a unified cosmogony provides guides to conduct which may be ambiguous or difficult to follow, but which indicate obligations that admit no sharp distinctions between prudence, piety, and moral rectitude.

Cosmogonic Traditions and Ethical Order

Where we can trace the relationship of cosmogony and ethics through literary works over a long period of time, we find that the relationship does not remain static, even when the fundamental mythic elements remain unchanged. The three essays in this section trace the development of cosmogony and ethics in two different traditions. Douglas Knight and Hans Dieter Betz consider important elements in the tradition of the Jewish and Christian scriptures. Knight provides a general survey of cosmogonic themes and their ethical implications in Genesis and in other parts of the Hebrew Bible. Betz describes a new kind of argument from cosmogonic imagery to ethical expectations put forward much later in the Sermon on the Mount. According to Betz, the Sermon on the Mount is addressed to a religious situation in which an older confidence in divine creation that permeates the Hebrew scriptures had broken down. In this new situation, the metaphor of God the Father reasserts God's cosmogonic activity in a new and compelling way, and clarifies the implications of that divine activity for human moral practice.

Wendy Doniger O'Flaherty traces a parallel tradition that was developing in a very different setting during roughly the same period of time. O'Flaherty focuses attention on a series of Hindu myths which depict the primordial cosmogonic act as the separation of heaven and earth. The earliest myths, from the Rg Veda, have little to do with ethical responsibility or prescriptions. The myths that appear in the first half of the first millennium B.C.E. in the Brahmanas clearly have more ethical relevance. In the later myths associated with the rise of devotional religion in India and dating roughly from the centuries around the beginnings of Christianity in the West, the correlation that is made between the character of the cosmogonic processes and the norms for ethical behavior is very close indeed.

Multiple Cosmogonies and Ethical Order

Some of the most interesting issues in relating cosmogony to ethical order arise when a religious tradition includes more than one cosmogonic myth. Cosmogonic thinking in the Hebrew tradition, for example, tries to integrate two myths in the early chapters of Genesis, and Douglas Knight identifies a total of six different types of Hebrew cosmogonies. Philosophical and scientific reflection, too, may include more than one interpretation of origins. Yearley considers carefully the relationship between the cosmogony of the "heavenly mandate" and that of the "sagely rule" in the thought of Mencius. Yearley also mentions a problem of multiple cosmogonies in his discussion of Freud, and Douglas Sturm notes it in his study of Marx. The problem arises most directly, however, in traditional systems where a variety of myths are apparently in competition. This issue is explored most fully in the essays by Reynolds, Burkhalter, and Warren.

Reynolds' article, "Multiple Cosmogonies and Ethics," highlights the presence within Theravada Buddhism of two ways to relate cosmogonies and ethics within a single tradition. The first pattern involves the presence of two cosmogonies, each with its own ethical implications, which correspond in each case to a different level of reality. In the Theravada case, one of these is formulated in the Buddhist doctrine of dependent co-origination of all elements that constitute the world of experience. The other is the cosmogony that describes the devolution of the particular natural and social world in which we actually live. Within the Theravada tradition, there is often a tension between these two cosmogonies and their ethical correlates. Because they operate at different levels of practical concern, however, the cosmogonies are able to retain their own distinctive identities.

The second pattern involves two cosmogonies which are related to one another by the fact that the one articulates a basic problem of religion and ethics to which the other offers a solution. In the Theravada case, the problem is posed by the cosmogony that presents the doctrine of dependent co-origination, describing the generation of the world in terms of ignorance, desire, and suffering. The problem-solving cosmogony, then, is the soteriological cosmogony which recounts the Buddha's establishment of a new world for himself and his followers, a world in which ignorance is overcome by wisdom, desire is replaced by compassion, and suffering gives way to salvation.

Sheryl Burkhalter suggests another form of problem-posing/problem-solving relationship in her treatment of cosmogony in Islam. Here we are presented with one Muslim myth which recounts the original cosmogony, and with another which recounts the "completion" of the creation in the context of human history. The Muslim myth which tells of the original creation points to the coexistence of a natural world that is complete and perfect with another element of creation that is clearly imperfect and incomplete, namely, humanity itself. The second Muslim myth recounts Allah's further creative activity, focusing on the mission of the prophet Muhammad and his establishment at Medina of a human community that, like nature itself, is truly in accord with the divine will. The Muslim account of the primordial creation is, of course, very closely related to the cosmogonies in the book of Genesis. However, the ethical problems posed by human corruptibility and the resolution of those problems by another creative act are perhaps clearer in Islam than in the Jewish and Christian traditions, which share essentially the same original elements of the creation story.

Kay Warren's essay, which focuses on Indian traditions in the highlands of Guatemala, presents another variant of the problem-posing/problem-solving pattern, one that combines elements of continuity and juxtaposition. One of the two coexisting cosmogonies which Warren describes poses the central problem of religious and moral life in the village. The other recounts the establishment of an alternative world in which a resolution of that problem is possible. The myth which poses the problem describes a cosmogonic process that moves from the primordial creation by God the Father, through the subsequent activity of Jesucristo through the transmission of Christianity to the Spaniards. It then recounts the imposition of Spanish rule in Guatemala in a way that results in the establishment of Christian moral demands without any provision for the freedom that the Indians need to meet those demands. The coexisting myth, which recounts the establishment of an alternative world in which the problem can be resolved, describes a parallel cosmogonic process that moves from the same primordial creation by God the Father, through the same secondary activity of Jesucristo, but then diverges from the first account. In this second myth, the creative and ordering activity of Jesucristo is mediated directly by apostolic saints who establish an indigenous Indian community and cultus through which the religious and moral life can be pursued in a way that is both distinctively Indian and eminently practical.

Myth, Philosophy, and Ethics

A major transformation takes place when cultural developments lead to new ways of thinking that supersede the traditional cosmogonic myths. When the prevailing cosmogony no longer answers the questions that social life raises, a quite different way of thinking about action may emerge.

This breakdown may occur either because the myth itself is no longer believed or, more commonly, because the way of life it authorizes is no longer satisfying. Adkins's "Cosmogony and Order in Ancient Greece" points to the problem of the justice of the gods that arises already in the archaic period. By the classical period that problem had, for some reflective Greeks, become intolerable. Adkins's second essay, "Ethics and the Breakdown of Cosmogony in Ancient Greece," details the development of classical Greek ethics following the general intellectual rejection of the order of life lived by the Homeric gods and heroes. The traditional cosmogony "could not withstand the strains put upon it by those who demanded just deities reliably punishing individuals in their own person in this life. It was not that kind of cosmogony" (293).

We must for the present set aside the interesting question of what ways of thinking lead to the rejection of a prevailing moral system on apparently moral grounds. What is important to note is that in cultures as unrelated and different as those of ancient Greece and ancient China, the displacement of traditional cosmogonies is accompanied by the emergence of ways of thinking that are recognizably philosophical.

In ancient China, the emergent moral philosophies involved two quite different evaluations of rule, rituals, and courtly formalities. Girardot, in his paper, associates these two different evaluations with the Taoist tradition on the one hand and the Confucian tradition on the other. However, Lee Yearley, in his study of Mencius, shows that a similar distinction existed in Confucianism itself. Yearley's study suggests that this philosophical tension can be traced to the two quite different cosmogonies which Confucianism attempts to systematize. The principle of the "Mandate of Heaven" and the principle of "sagely rule" are closely related, but systematic treatment cannot entirely reconcile the underlying differences between discovering an order that heaven has already established and wisely fashioning an order for human affairs that fits within the constraints of reality but is not itself a part of those constraints.

The development of new forms of moral philosophy after the breakdown of a traditional cosmogony can also be seen in the Christian West, beginning as early as the eighteenth century. Even before the effects of Darwin's work called Christianity's cosmogonic myth into question, utilitarian moral philosophy had broken the notion of a natural moral order and had begun to treat moral systems as human contrivances. Theology attempted to reclaim the connection between creation and duty, but Lovin's "Cosmogony, Contrivance, and Ethical Order" shows that this was possible only by rethinking the original creation along lines closely analogous to the human creativity that marked the age of invention. So it is true for Archdeacon Paley, as for the Vedic poet, that "We must do what the gods did in the beginning."[5] The human project, however, is now not the re-creation of the cosmos but the creation within it of a system that insures the general happiness.

Science, Cosmogony, and Ethics

This moral creativity became more important as scientific discoveries displaced traditional cosmogonic myths in the modern cultures of the nineteenth and twentieth centuries. Here it was no longer a matter of elaborating a moral philosophy to cover the ethical inadequacies of the mythic system. The myth itself was discredited, and the principal architects of the new systems of conduct saw themselves as providing norms for action which had scientific foundations that all earlier religious and philosophical ethics had lacked. Marx and Freud affirm, quite vehemently at times, that the old religious cosmogonies which had previously identified the principles and archtypal patterns for human conduct are a form of false consciousness. Having done this, they then summon us, each in his own way, to a new form of praxis that reestablishes the connection between what really is and what ought to be done.

Lee Yearley and Douglas Sturm provide a significant conclusion to this inquiry into cosmogony and ethical order by showing the persistent and fundamental influence of cosmogonic thinking on the ethics of Marx and Freud. For Marx, and even more for the Marxian tradition mediated through Engels, the generative importance of the *act* in a material world of endless causal connections is decisive for the human condition and the scope of human possibilities. Just how much freedom persons have to initiate these acts remains a vexed question in the interpretation of Marx and Engels, but the primacy of acts and events over ideas in the determination of reality has never been in doubt in Marxist materialism. This means that whatever teleology, whatever vision of the end of history the Marxian tradition may have, that conclusion can only be grounded scientifically in an account of origins. The Marxian orientation is, in fact, grounded in an account of the nature of material reality and the way things contingently fell out to set in motion successive self-destructions of economic systems and the final reunion of the workers with their means of production. Sturm not only details the view of history that Marx and Engels regarded as scientific. He shows us that the undeniable eschatology of that history depends on an initially less apparent cosmogony.

In Freud, the emergence of a new cosmogony is more obvious. His theory of the primal horde, in which the sons murder their father to gain access to his women, provides an explanation with truly mythic qualities of the incest taboo and the fundamental social structures that accompany it. Lee Yearley takes us beyond this familiar account of the origins of society to examine the cosmogony that sets the life instinct against the death instinct and so provides an account of even more basic forces that shape the subsequent development of society and provide the interpretative key to the tensions and problems of individual lives.

Freud dispels the comfortable illusions about our situation that are embodied in some traditional cosmogonies, but the result of that critical task is not the displacement of myth by philosophy that we find in Plato, the Confucians, or even in Paley's theological utilitarianism. Freud's contribution is a new cosmogonic myth, founded in "science" rather than in tradition. Yearley reminds us that Freud himself was ambivalent about the mythic qualities of some of his explanations. Freud warns us not to hypostatize the psychological functions he describes, and he seems aware at times that the behaviors he attributes to the death instinct are susceptible to other, less dramatic explanations. Still, the conflict of the two primordial and powerful forces has an appeal to our imaginations and an explanatory power that goes beyond any strictly scientific evidence that can be presented on its behalf. An alternative which would reduce all human actions to illustrations of the pleasure principle might have more theoretical elegance in terms a physicist could appreciate, but it would not have the power of Freud's cosmogony of death versus life. In the end, Freud's cosmogony is not unlike the myths Plato weaves in the *Republic*. They are, strictly speaking, more than what reason requires to provide the foundations for ethics (Adkins, "Ethics and the Breakdown of Cosmogony," 306). One is not sure that the mythmaker himself believes them, or that he believes them all the time. But we are sure that he does not want to do without them.

8. Relativism and Rationality

The essays included in this collection mark the beginning of an inquiry, not its conclusion. The patterns of relationship between cosmogony and ethics which they document confront us in rich and not yet fully systematized variety. The links between prudence, piety, and morality in traditional cosmogonies, the coexistence of different cosmogonic myths in a single culture, and the ways that philosophy and scientific inquiry take on the characteristics of earlier cosmogonic myths—all these could be explored in greater depth and in many more traditions.

For the reader who is more interested in formulating a normative ethics than in historical and comparative studies it may seem, however, that this rich mix of religious and moral traditions merely recapitulates the philosophical problem that the discovery of moral diversity introduced in the first place. One reason why Western philosophers turned from an examination of ways of life that could be deemed good to an examination of ways of reasoning about them was the hope that the relativistic implications of this kind of diversity could be overcome.

The Enlightenment philosophers hoped that by arriving at a formal standard of rationality they could eliminate many of the alternatives as irrational and could choose one on the basis of logical requirements, without recourse to the seemingly interminable debates about human nature or about what persons really want. A more liberal formulation of the same strategy was to establish the minimal universalizable features of a good life that had to be maintained if people were to live together at all, and thus to reduce the alternatives to a range of options that were equally worthy insofar as they all met the criteria of moral rationality.

By denying that all traditions somehow employ a common structure of practical reason and insisting instead on the variety that marks their understandings of the good life, our empirical approach could perhaps be accused of introducing a relativism that turns comparative ethics into a cabinet of curiosities and undercuts any notion that moral traditions are about a common task on which there might, ultimately, be some common agreement. The problem must be taken seriously, and we cannot conclude without giving some consideration to its solution.

If this new empirical approach points the way toward a solution to the problem of relativism, it cannot, of course, be the solution offered by its Humean ancestor. There seems to be no common principle that determines morality throughout history in the way that the principle of gravity determines the flow of the Rhine and the Rhone. Indeed, as we have suggested, we cannot make even the more limited claim of Little and Twiss that there is at least some agreement on how one would structure a good argument for such a principle.

In the absence of suitable candidates for the moral equivalent of Newton's law of universal gravitation, those who take an empirical approach are able to focus instead on the hopeful fact that when moral systems do come into conflict, the parties involved are able to establish communication on the issues that are at stake, and may eventually arrive at an agreement that incorporates some of the elements present in each of their ways of looking at reality and evaluating action. Even without an apparently neutral first principle to which all parties can appeal, we find that "however different our images of knowledge and conceptions of rationality, we share a huge fund of assumptions and beliefs with even the most bizarre culture we can succeed in interpreting at all" (Putnam 1981, 119).

To be sure, this pragmatic confidence in conversation and dialogue is not altogether satisfactory from the formalist perspective. There is the evidence of experience that it can work, but there is no guarantee that it always will work. Some contemporary philosophers simply accept this uncertainty with a sort of sangfroid. "The pragmatists tell us that the conversation which it is our moral duty to continue is *merely* our project, the European intellectual's form of life. It has no metaphysical or epistemological guarantee of success. Further (and this is the crucial point) *we do not know what 'success' would mean except simply 'continuance'*" (Rorty 1982, 172).

The pragmatists are right, of course, if they mean that ethics is "*merely* our project" in the sense that there is no single form of practical reason in terms of which we might require everyone else to do ethics our way, or not at all. History of religions, however, suggests a slightly different approach. By drawing attention to the fact that different patterns of choice and action are related to different myths of origin and understandings of reality *even within a single tradition,* an empirical, history-of-religions approach calls into question the formalist proposition that rationality in morals can be reduced to a single pattern which will be independent of humanity's substantive preferences and diverse beliefs about the facts. But such an approach also shows that the diversity which does, in fact, exist within cultures is not necessarily destructive or debilitating.

For example, it appears that the traditions of Theravada Buddhism and of the Trixano community in Guatemala are able to sustain personal integrity and human community in part *because* they provide a variety of ways to think about moral choices. Perhaps what we need is not a single pattern of rationality to explain those inconsistencies, but an appreciation of the relationships between the cosmogonies and the ethics they support that is dialogical or "conversational." Insofar as some traditions have successfully maintained these differences, they suggest the possibility of a similar "conversation" between traditions that are separate or even competing (see, for example, the suggestions of Tracy 1983).

The sort of ethical theory that this conversation requires is a form of ethical naturalism; that is to say, an ethical theory that rejects the separation of moral reasoning from beliefs about the facts and, indeed, suggests that most reasoned moral discussions are precisely discussions that try to justify particular choices and value judgments by relating them to a more comprehensive view of reality. Contemporary moral philosophers such as White and Flanagan have, as we have seen, begun to develop this kind of theory and the methods appropriate to it. But much more remains to be done. Beliefs about the way the world is may relate in many different ways to beliefs about what we ought to do, and each of these requires careful explication.

A historian of religion will add the caution, too, that these explications must also take account of the range of ways in which the beliefs may be expressed. Not all correlations of fact and value are formulated as straightforward verbal propositions. When the Marxist asserts "From each according to his ability; to each according to his need" as a workable law of social life, the Christian or Jew may say, "Cast thy bread upon the waters: for thou shalt find it after many days" (Eccles. 11:1). They may mean it, too, either as an affirmation of trust in God's providence or as prudent investment advice. The Theravadin may speak to the same subject in a discourse on merit or may, more likely, say nothing at all but express the point in a ritual offering to the monks.

Where our partner in the conversation is a whole culture or tradition, we must not limit our attention to worldviews and moral beliefs expressed in verbal propositions, nor, as we have seen, can we assume that the investigation will turn up only one worldview and one way of moral reasoning per culture. A naturalistic approach to comparative ethics seeks to describe the relationships between worldviews and norms in ways that accurately reflect the tensions and controversies in a community's experience, in ways that reproduce the complexity of a tradition and allow the identification and meaningful comparison of the most crucial elements within it. This kind of description is the primary result of the essays in this collection; but the similarities between patterns that are discerned in quite different settings already suggest the possibility of further cross-cultural comparisons, and of investigations that would explore their normative implications.

The ways that people seek to live well within the limits reality imposes are multiple, perhaps irreducibly so, and yet the constraints and the choices themselves are sufficiently similar that, over time, the effective ways of relating them will tend to reappear. A naturalist approach to comparative ethics thus holds out the prospect that, when actions and beliefs of various human cultures are properly understood, the "European intellectual's form of life" may not be as lonely as it seems. The study of the very diverse ways of identifying, justifying, and living a good life that have been worked out in the course of human history may become an invaluable resource for our effort to identify, justify, and live a good live in the new global environment presently confronting us.

That is not to say that the strategies of Enlightenment rationalism should be abandoned. In a pluralistic, individualistic society such as the one we inhabit in the North Atlantic community, the test of universalizability, for example, may be a very important part of moral reasoning. In a situation where we cannot practically scrutinize our own plans from every relevant point of view, a principle of universalizability helps us to separate our real moral claims from our most powerful wants. It helps us to formulate our values in public terms that can be understood by others. These are important ways of locating our own lives within the constraints imposed by the real world; but we should not be surprised to find that the same thing is accomplished in traditional cultures by reciting an epic poem or by examining the entrails of a sacrifice. Nor should we assume that in our own culture other strategies should be disparaged or neglected.

We want, obviously, to avoid conflating Kant's *Groundwork of the Metaphysic of Morals* with Hesiod's *Works and Days*, or either of them with the admonitory tales of the *Jatakas*. Rather, we take the variety of cosmogonic myths, the cosmogonies of the philosophers, and even the scientific cosmogonies of Marx, Freud, and the modern-day cosmologists to be meaningful attempts to say what is really there in the human encounter with nature and social reality, what is fundamental in the changing configurations of experience. And we take the historical variety of ethical norms, values, and prescriptions to be attempts to formulate how, in the context of these realities, life should be lived. If there is a high point from which all the Rhines and Rhones of our moral systems flow, it will not be found in a common set of abstract principles, nor even in an agreed-upon structure of practical reason, but in the continual discovery and rediscovery of what we must do to live as we would in a world that is as it is.

This process of discovery and rediscovery is one in which our conversation partners, who are often the subjects of our historical and comparative research, have been engaged in the past. And it is a process which we, through our involvement in the conversation (and in many other ways as well), must enter into and continue. But what is involved is not, as the pragmatists have phrased it, "merely continuance." What is involved is the renewal, day after day and generation after generation, of the effort to delineate and actualize the fullest potentialities and richness that the human condition allows.

Notes

1. References in the introduction to papers in this volume are by author and title of the essay. References to other published works are by author and date of publication. Complete bibliographical information will be found at the end of the introduction.

2. See Moore's account of the "naturalistic fallacy" in *Principia Ethica* (Moore 1922, 9–15).

3. It need not, of course, be naturalistic in the reductive sense that it defines moral good in terms of some single factual criterion, e.g., maximized feelings of pleasure or optimal evolutionary advantage (see White 1981, 104–5). Neither need it be naturalistic in the sense of applying a restrictive criterion of empirical verifiability to the experiences that will be allowed to count in the testing of moral choices and factual beliefs. Moral affections and religious experiences appropriately figure in the testing of both moral and factual beliefs.

4. "Eschatology," "legalism," and "ontology" come to mind as terms that are similar in this respect.

5. *Satapatha Brahmana*, VIII, 2, 1, 4. Cited in Eliade (1959, 21).

References

Barth, Karl
1957 *Church Dogmatics*, II/2. Edinburgh: T. & T. Clark.

The Book of Common Prayer
1977 New York: Seabury Press.

Cudworth, Ralph
1820 *The True Intellectual System of the Universe*. London: Richard Priestly. [First published 1678.]

Donagan, Alan
1977 *The Theory of Morality*. Chicago: University of Chicago Press.

Duncker, Karl
1939 "Ethical Relativity." *Mind* 48:39–57.

Durkheim, Emile
1961 *Moral Education*. New York: The Free Press.
1965 *The Elementary Forms of the Religious Life*. New York: The Free Press.

Dyck, Arthur
1977 *On Human Care*. Nashville: Abingdon Press.

Eliade, Mircea
1959 *Cosmos and History*. New York: Harper and Row.
1969 *The Quest: History and Meaning in Religion*. Chicago: University of Chicago Press.

Flanagan, Owen J.
1982 "Quinean Ethics." *Ethics* 93:56–74.

Gernet, Jacques
1982 *A History of Chinese Civilization*. Cambridge: Cambridge University Press.

Giddens, Anthony
1971 *Capitalism and Modern Social Theory*. Cambridge: Cambridge University Press.

Green, Ronald
1978 *Religious Reason*. New York: Oxford University Press.
1981 Review of David Little and Sumner B. Twiss, *Comparative Religious Ethics*, in *Journal of Religion* 61: 111–13.

Gustafson, James
1981 *Ethics from a Theocentric Perspective*, Vol. 1. Chicago: University of Chicago Press.

Hume, David
1973 *A Treatise of Human Nature*. Oxford: Clarendon Press. [First published 1739.]
1975 *Enquiries*. Oxford: Oxford University Press. [First published 1748.]

Kant,Immanuel
1956 *Critique of Practical Reason*. Indianapolis: Bobbs-Merrill. [First published 1788.]
1964 *Groundwork of the Metaphysic of Morals*. New York: Harper and Row. [First published 1785.]

Little, David
1981 "The Present State of the Study of Comparative Religious Ethics." *Journal of Religious Ethics* 9:186–98.

Little, David, and Sumner B. Twiss
1978 *Comparative Religious Ethics: A New Method*. San Francisco: Harper and Row.

Lukes, Steven
1970 "Some Problems about Rationality." In Bryan R. Wilson, ed., *Rationality*. Oxford: Basil Blackwell.

MacIntyre, Alasdair
1984 *After Virtue*. 2d ed. Notre Dame, Ind.: University of Notre Dame Press.

Montesquieu, Charles Baron de
1975 *The Spirit of the Laws*. New York: Hafner. [First published 1748.]

Moore, G. E.
1922 *Principia Ethica*. Cambridge: Cambridge University Press.

Oxford English Dictionary
1971 Oxford: Oxford University Press.

Putnam, Hilary
1975 *Mind, Language, and Reality*. Cambridge: Cambridge University Press.
1981 *Reason, Truth, and History*. Cambridge: Cambridge University Press.

Reynolds, Frank
1980 "Contrasting Modes of Action: A Comparative Study of Buddhist and Christian Ethics." *History of Religions* 20: 128–46.

Rorty, Richard
1979 *Philosophy and the Mirror of Nature*. Princeton: Princeton University Press.
1982 *Consequences of Pragmatism*. Minneapolis: University of Minnesota Press.

Schneider, Louis
1967 *The Scottish Moralists*. Chicago: University of Chicago Press.

Stout, Jeffrey
1980 "Weber's Progeny, Once Removed." *Religious Studies Review* 6:289–95.
1981 *The Flight from Authority*. Notre Dame, Ind.: University of Notre Dame Press.

Toulmin, Stephen
1972 *Human Understanding*. Princeton: Princeton University Press.
1982 *The Return to Cosmology*. Berkeley: University of California Press.

Tracy, David
1983 "Religion and Human Rights in the Public Realm." *Daedalus* 112:237–54.

Weber, Max
1958 *The Religion of India*. New York: The Free Press.
1964 *The Sociology of Religion*. Boston: Beacon Press.

White, Morton
1981 *What Is and What Ought to Be Done*. New York: Oxford University Press.

Wildiers, Max
1982 *The Theologian and His Universe*. New York: Seabury.

Part I

Single Cosmogonies and Ethical Order

2 Cosmogony and Order in Ancient Greece

Arthur W. H. Adkins

There are no canonical documents in ancient Greek religion; and there were about a dozen prose or verse theogonies or cosmogonies ascribed to Greeks of the archaic period (West 1966, 12–16). However, the *Theogony* of Hesiod alone is extant; and that poem, together with Hesiod's *Works and Days,* will be the subject of this essay. I shall take the whole of both poems into account. Were we in these discussions employing a narrow definition of cosmogony, only a small part of the *Theogony,* and none of the *Works and Days,* would be relevant; for most of the *Theogony* is, as the title suggests, theogonic rather than cosmogonic, and much of the remainder consists of aetiological myth; while the *Works and Days,* apart from digressions to explain the unhappy lot of mankind, is for the most part wisdom literature. For the present purpose, however, the whole of the *Theogony* and *Works and Days* may be regarded as cosmogonic; for from the poems may be derived a quite coherent early Greek view of the origin of the basic order of the world in which the Greeks lived; of the origin of the gods and of humanity; of the origin of the basic social structures of the gods and, to some extent, of human beings; and of an ordering principle or complex of principles which accounts for the—perhaps limited—intelligibility and other characteristics of the cosmic order. The poems are also linked with the religious ethics of early Greece.

The *Theogony* and *Works and Days* together comprise some 1,850 hexameter lines. The figure is a maximum: most scholars believe that some passages are interpolations (West 1966, 1978, notes passim). The question is not important for the present purpose; all of the lines seem to reflect archaic belief, and throw light on the links between cosmogony and order. Even if some passages were dated rather later than the archaic, their presence in the poem presumably indicates the existence of a period when all of the beliefs were held; and this suffices for the present discussion.

I shall treat the *Theogony* and *Works and Days* as one myth here. They in fact contain an indefinitely large set of myths. Some of these, like the "Five Races of Men" (*W*109–201) or "Zeus, Prometheus and the Origin of Sacrifice" (*T*535–612) are easily demarcated (though *T*535–612 contain units which may once have existed separately); but the general myth of the generations and relationships of the gods may be viewed as the result, in part, of assembling a large number of individual myths, some at least of which may have been narrated more fully elsewhere before Hesiod composed his poems.

I shall not discuss the question of Hesiod's sources, direct or indirect. Like the Homeric poems, the *Theogony* and *Works and Days* bear the marks of relationship to an oral tradition. How much of the genealogies in the *Theogony* is Hesiod's own work we cannot be certain; but it seems unlikely that so complex a system is the work of one poet. There are non-Greek influences present: resemblances with theogonies from further east render it virtually impossible that the sequence of chief deities (Ouranos, Kronos, Zeus) is Hesiod's own invention. This fact argues for Greek predecessors too, for it seems unlikely that the material reached Hesiod directly from outside Greece (West 1966, 18–31). My task here, however, is not to try to reconstruct the prehistory of the Hesiodic poems, but to analyze the structure of fact and value presented by the cosmogony and theogony, and to relate it to Greek religious ethics and other values relevant to behavior.

I begin with the cosmogony, or rather with that part of the theogony which most evidently overlaps with cosmogony: *T*116–38. Any Greek who accepted Hesiod's account believed that the first generation consisted of three or four members (*T*116–22):[1] "First of all Chaos[2] came to be, and then broad-bosomed Gaia [Earth] . . . , and misty Tartaros . . . and Eros [Sexual Desire]." The second generation was produced from these (*T*123–132): "From Chaos, Erebos [Darkness] and black Night came to be. . . . Earth first bore strong Ouranos [Heaven] equal in extent to herself, so that he might cover her all over, and there might be a safe seat for the gods for ever. And she bore the great

Mountains . . . , and she bore . . . Pontos (Sea), without the joys of love."

The third generation was produced in a variety of ways from the first two. Night (II)[3] conceived by her brother Erebos (II) and bore Day and Aither, the bright sky (*T*124–25); and Earth (I) "having lain with [her son] Ouranos (II) bore Okeanos with his deep eddies, and Koios, Kreios, Hyperion, Iapetos, Theia, Rheia, Themis [Custom-law], Mnemosyne [Memory], Phoebe with her golden crown, and lovely Tethys; and after them youngest of all was born Kronos of crafty counsels, and he hated his lusty father."

Despite "last of all," the *Theogony* as we have it immediately (*T*139–55) mentions two other groups of three gods: Brontes, Steropes, and Arges, the Kyklopes, makers of thunder and lightning; and the huge Kottos, Briareus, and Gyes, each with fifty heads and a hundred arms.

Now Ouranos would not let his children—or possibly only the last six—be born.[4] Consequently, all his children and Earth, his wife and their—and Ouranos's—mother, hated him. Kronos volunteered to take revenge, and with the sickle furnished by Earth castrated Ouranos (Heaven) when he came at nightfall and stretched himself over Earth to make love. Kronos threw the members away. Drops of blood fell to Earth, who conceived from them and bore the Furies, the Giants (Earth-born) and the Nymphs of the Ash-Trees. The members fell into the sea, and from them Aphrodite, Goddess of love, was born.

In addition to producing Day and Aither sexually, Night (II) also asexually produced (*T*211–25) Moros, black Ker, the Hesperides, the Moirai, the Fates, Nemesis, Deception, Philotes (Friendly Cooperation), Old Age, and Strife.

Earth (I) also conceived by her son Pontos (II) and bore Nereus and a number of other sea deities.

The first two generations come closer to simple cosmogony than does anything else in Hesiod. But the myth is not merely an account of the origin of the physical universe. Indeed, it will become apparent that it is misleading to use the term "physical universe" at this period. Virtually every feature mentioned is also treated as a person.

Tartaros is a doubtful case for scholars who reject the Typhoeus-episode (*T*820–80), since Tartaros has then no progeny.[5] Nor can one be certain about Chaos, from which—or from whom—Erebos and black Night "came to be"; but "came to be" *(gignesthai)* is used of being born from a mother in sexual or asexual reproduction,[6] so that Chaos may be viewed personally. Earth is certainly personalized,[7] since she reproduces sexually also; and since Earth's other asexually produced offspring sexually generate other offspring, and must be personalized, presumably the mountains too are personalized here.

Elsewhere in Hesiod Atlas, later a mountain in North Africa, is personalized (*T*509,517), and is the father of the Pleiades (*W*383); but it is not clear that he is thought of as a mountain at this date.

But all the other aspects of the physical universe are evidently treated also as persons, both in the passage quoted and elsewhere in the *Theogony*. That Eros (Sexual Desire) is treated as the moving principle of creativity is not surprising. For other examples of personalization see *T*243–64 (Nereids) and *T*346–66 (Okeanids). One might argue that neither the Okeanids nor Nereids mentioned by Hesiod are closely linked with particular pieces of water, while the Rivers (*T*337–45), which bear the names of well-known rivers, are not really personalized at all. They receive no adjectives that are appropriately used only of persons, as do the Nereids and the Okeanids; but the Rivers, like the Okeanids, are children of Okeanos and Tethys, and are referred to (*T*368) as "the sons of Okeanos, whom lady Tethys bore." In general, parts of the physical universe are also personalized; for example, Dawn, the Sun, and the Moon (*T*372).

The third generation contributes more aspects of the universe. The first two have provided earth, sea, mountains, and—apparently—darkness but no light. Ouranos is termed "starry," which suggests night. Night certainly already exists; but even starlight is not yet available. (The stars are not created until *T*382.) It is only in the third generation that Day and Aither (Bright Sky) are born (*T*124). Hyperion may be the personalized sun, Phoebe the personalized moon (*T*134,136). ("He who goes over" and "the bright one" are probable meanings of their names.) If so, they are doublets of Helios and Selene, children of Hyperion and Theia (*T*371). Okeanos is evidently personalized Ocean, and Tethys possibly some part or aspect of the sea.

However, such deities as Koios, Kreios, Iapetos, Theia, Rheia, Kronos, and the Hundred-Handers seem not to be personalizations of anything.[8] The *Theogony* is not a quasi-scientific account of the universe expressed in figurative language.

The Kyklopes (*T*139–46) should perhaps be listed with the personalized aspects of the universe. Their names—Brontes, Steropes, and Arges—are transparently formations from words meaning Thunder, Lightning, and Bright Flash, a fact which strongly suggests that they are personalized phenomena rather than the deities who manufacture the phenomena. But they are more than either, for they are—or supply—the power by means of which, in the generation of Zeus (IV), Zeus maintains his power over the world: these deities are also instruments of order.

A hint of order is present even in the first generation. The importance of Eros derives from general personalization and from procreation as the means of generating the world. Procreation produces not merely individuals but families in which individuals have relationships; and these relationships, together with the values and behavior deemed appropriate or usual in the culture for persons so related, may be invoked as explanations of order—or disorder—in the myths which follow.

The third generation furnishes other deities of order. Among the children of Ouranos and Earth are Themis (Custom Law) and Mnemosyne (Memory). The functioning of any organized community depends on the existence of law: in a nonliterate society, necessarily a custom law remembered by the appropriate members of society. The Hundred-Handers and the Kyklopes represent power and strength, which might be used either to enforce the order which that law helps to maintain, or to subvert it. It is interesting that Ouranos is portrayed as trying to prevent the emergence of the power into the world.

The asexually generated Children of Night have an analogous role. *Moira, nemesis,* and transgression are central concepts of social, moral, and political order in early Greece. The Moirai as deities are personalized *moirai,* the share in existence and goods in terms of which, in the value-system of early Greece, one's position, status, and claims were evaluated. Both gods and men were deemed to have a *moira.* To "go beyond one's share" was to commit a serious transgression; for one's share was one's due share, the share which, in some sense of "ought," one ought to have. So (*T*220) the Moirai "pursue the transgressions of men and gods, nor do they cease from their terrible wrath until the wrongdoer gives them requital." How *moira* fits into the value-system of early Greece as a whole will become apparent later. It is not the most powerful value-term in early Greek (Adkins 1960a, 1972a, 1975).

Genealogies are traced from many of the third-generation deities to produce a fourth generation. Some of these are overtly personalized aspects of the physical universe; for example, the Rivers (*T*337–45), children of Okeanos (III) and Tethys (III), or Sun, Moon, and Dawn (*T*371), children of Hyperion (III) and Theia (III); but many are not. Interspersed in the narrative are accounts of the descendants of the fourth generation, some of which also are personalized aspects of the physical universe: the children of Dawn and Astraios (*T*378) were Zephyros (the West Wind), Boreas (the North Wind) and Notos (the

South Wind). But in this generation too the majority of deities are not of this kind.

In the third generation lines of descent are traced from Tethys and Okeanos (*T*337–70), Theia and Hyperion (*T*371–74), Kreios and Eurybie (*T*375–77), Phoebe and Koios (*T*404–52), Rheia and Kronos (*T*453–58), and Iapetos and an Okeanid of the fourth generation (*T*507–14). The most important line of descent is that of Rheia and Kronos, for Zeus was among their children; but the other lines should not be forgotten. The children of Rheia and Kronos were Hestia, the hearth; another important social-religious concept (Vernant, 97–143), Demeter, goddess of agriculture, Hera, Hades, principal god of the underworld, Ennosigaios, god of earthquakes, usually identified with Poseidon, and Zeus. (In the light of Vernant's illuminating discussion of the conceptual-spatial relationship of Hestia and Hermes, it is interesting that while Hestia is the child of Kronos, Hermes is the child of Zeus.)

Kronos swallowed his children when they were born, since Ouranos and Earth had told him that it was ordained for him to be overthrown by one of his sons (*T*459–65). But Ouranos and Earth are not merely partisans of Kronos; they also advised Rheia to go to Crete, where Earth hid Zeus, and gave Kronos, as a substitute for Zeus, a stone, which Kronos swallowed. Zeus grew in safety, and subsequently, for reasons at which Hesiod only hints, Kronos vomited up all his remaining offspring.

The battle between Zeus and his followers and the Titans, Kronos and his followers, is imminent. It is not simply a battle between the generations. Ouranos or Kronos had imprisoned some of the children of Ouranos. ("Father" in *T*502 might refer to Zeus's father Kronos or to the father of the imprisoned children of Ouranos.) They are unnamed, but must be at least Brontes, Steropes, and Arges; for when Zeus released them (*T*503) "they remembered his benefits gratefully and gave him the thunder and the smoky thunderbolt and the lightning flash, which huge Earth had hidden before. Relying on these he rules over mortals and immortals."

Here the poet gives another hint of the foundation of Zeus's power and the manner in which he obtained his support. This too contributes to Hesiod's picture of the universe and the powers and values it contains.

During the general genealogy, Hesiod pursues the descent of the Okeanids beyond the fourth generation (*T*383–96): "And Styx [*IV*], daughter of Okeanos, having lain with Pallas [*IV*]—son of Eurybie and Koios [*T*375–76]—bore Zelos [Rivalry] and Victory of the beautiful ankles, and Kratos [Strength] and Bie [Force], her famous children. And these have no abode away from Zeus, nor any road save that on which Zeus leads them. Always they sit by Zeus the loud thunderer. For so did Styx the immortal daughter of Okeanos plan it on the day when the Olympian god of the lightning summoned all the immortal gods to lofty Olympus, and said that he would never deprive of their privileges any of the gods who fought with him against the Titans, but that each one should have the privileges he enjoyed before among the immortal gods; and that any god who was without *time* and privilege should attain these, according to what is right *[themis]*."

Styx came with her children, who were duly enrolled among the allies of Zeus; and he consequently has Rivalry, Victory,Strength, and Force as his supporters. "And he himself has great power and rules" (*T*403).

Now Ouranos had imprisoned the Hundred-Handers Kottos, Briareus, and Gyes under the earth, or, one might say, within the goddess Earth (*T*617–20). She now (*T*626) told the gods that victory would be won with the help of the three; and Zeus freed them and fed them on ambrosia and nectar. Zeus then asked them for help in return for his *philotes* (friendly acts) to them. (For the importance of *philotes* in early Greek society, see Finley 99–103; Adkins 1963; and for *philotes* and gods, Adkins 1972; 1975.) The three acknowledged their obligation; and their strength, and the thunderbolts of Zeus, settled the outcome.

The Hundred-Handers drove the Titans into Tartaros, "as far underground as earth is distant from heaven," where they guard them to prevent their escape. There the Sun and the Olympians never come, but only Night, Sleep and Death, the Children of Night, Hades, Persephone, and the mortal dead. The power of the Hundred-Handers was vital in giving Zeus his victory, whose continuance it now helps to assure.

After the battle (*T*881–85) "when the blessed gods had completed their toil and had successfully contended with the Titans for their *timai*, then on the advice of Earth they bade Olympian Zeus to be King and ruler of the immortals; and he well shared out their *timai* for them."

There follows a list of those goddesses and mortals by whom Zeus had children; Metis (Wisdom, Skill, Cunning), Themis (Custom Law), Eurynome (an Okeanid), Demeter, Mnemosyne (Memory), Leto, and "lastly" Hera (*T*886–923). Despite "lastly," Maia (*T*937) and the mortals Semele and Alkmene (*T*940–44) appear later. They bore to Zeus Hermes, Dionysus, and Herakles respectively.

For the present purpose, Zeus's marriage with Metis is the most important one, for the son to be born of that union would have overthrown his father. But Zeus swallowed Metis (*T*899–900), "so that the goddess might counsel him as to what was beneficial and what was harmful." He thus both prevented the birth of the son and acquired practical wisdom.

Attempts are being made to bring the sequence of divine rulers to an end, to ensure that Zeus is not overthrown in his turn. Zeus had already in the *Theogony* (*T*820–80) destroyed Typhoeus, offspring of Earth and Tartarus. Typhoeus would have been ruler of gods and mortals (*T*837) had not Zeus destroyed him with a thunderbolt, presumably before he had come to full strength. Again, the sea-nymph Thetis was married not to Zeus but to the mortal Peleus, because her son would have been mightier than his father. (The wedding, but not its motive, is mentioned, *T*1006–7.)

The absorption of Metis not merely eliminated a possible usurper, but endowed Zeus with a nonviolent safeguard against overthrow. Some of his other marriages account for the existence of personal deities not yet fitted into the genealogy: Persephone (*T*913), Ares and Eileithuia (*T*922). Zeus also produced Athena from his own head (*T*924). Other marriages produced deities of a different kind. To Zeus Themis (Custom Law) bore the Horai, whose names here[9] are Eunomie, Dike, and Eirene (Peace). All are personalized abstracts. I shall discuss Eunomie and Dike later. All three are said to "watch over the cultivated land [*erga*] of mortal men." Themis also bore to Zeus the Moirai, Klotho, Lachesis, and Atropos, "to whom Zeus gave the greatest *time*" and "who give benefit and woe to mortal men" (*T*904–6). The list of Zeus's children also includes the Charites (Graces), whose mother was Eurynome (IV), an Okeanid (*T*907–9), and the Muses, whose mother was Mnemosyne (Memory, III, *T*915–17). The names of the Muses are not given here; the Graces' are Aglaia (Joy, Splendor), Euphrosyne (Festivity), and Thalia (Abundance).

It will be useful at this point to consider ways in which this genealogical account might be linked to a value-system applicable to human beings. There are at least three, which arise from the personalization of every aspect of the world; from the inclusion of personalized concepts of value and power, together with other social phenomena, as an integral part of the genealogies; and from the organization, values, and behavior of the gods themselves, including the personalizations.

The personalization of what we regard as the inanimate world will be discussed in the context of the *Works and Days*.

Personalized abstracts in the third generation include Themis (Custom Law) and Mnemosyne (Memory), daughter of Ouranos and Earth;

the Erinyes (Furies, Curses) sprung from the castrated male organ of Ouranos; and the fatherless children of Night, who include the Moirai (Apportioners), the Fates, Nemesis (Righteous Indignation), Philotes (Helpful Cooperation, Friendship), and Strife. In the fourth generation the offspring of Strife include Toil, Forgetfulness, Famine, Battles, Homicides, Quarrels, Lies, Dysnomie (the opposite of Eunomie), Ate (Blind Ruin), and Oath "who does most damage to mortal men, whenever anyone wilfully breaks his oath" (*T*231–32). In the fourth generation are Hestia (Hearth), an Olympian, daughter of Kronos and Rheia, and Metis (Practical Wisdom), an Okeanid. In the fifth generation are Rivalry, Victory, Power, and Might, the children of Styx, greatest of the Okeanids; and also the Moirai, in the alternative account by which they, together with Eunomie, Dike, and Eirene, are the children of Zeus (IV) and Themis (III).

Some of these concepts, and their role in archaic and early Greek society, require explanation. A *moira* is a share: one's share both in life in the sense of its length and in livelihood, in this world's goods, notably the family farm and the goods and chattels thereon, the quantity and quality of which, in the agricultural society of archaic Greece, sharply demarcated and delimited one's opportunities in life. (The range of usage of the Greek word *bios* spans both "life" and "livelihood.") The goods might be referred to as *time,* status-producing possessions and the status that accompanied their possession; one's position in society could be expressed—was expressed, in a society little given to abstraction—as one's *moira* of *time,* whose quantity and quality differentiated the life of a ruler from that of a small farmer, and both from that of a craftsman or a beggar. One's *moira* is one's due *moira* and one should not be deprived of it; and within one's *moira*—on one's own land—one is virtually autonomous.

When Zeus apportioned the *timai,* the status and functions of his supporters, after the battle with the Titans (*T*885), each of them acquired a *moira* of *time,* or was confirmed in the *moira* of *time* already possessed: the apportionment of functions to deities existed before. The social-political system of the gods, like that of men, may be expressed in terms of *moirai* of *time.*

Eunomie also requires some explanation. The word is usually derived directly from *eu* ("well") and *nomos* ("law"), and is often said to imply not so much having good laws as having laws which are obeyed. This was certainly the view of Aristotle (*Politics* 1294a3ff.); but Aristotle was living between three and four hundred years after Hesiod, and his knowledge of the archaic period was not much greater than ours. It is worth observing that *eunomie* by the usual rules of Greek word-formation may also be developed from *eu* + **nem* ("allot," "apportion"): *nomos* itself is from **nem*. Such a derivation would endow *eunomie* with the—additional—sense of "a situation in which the apportionments [*moirai*] are correct," in which everyone possesses his land in security.

Dike, which at this time may mean "claim (to one's own)" as well as "justice,"[10] is a desirable prerequisite to ensure the possession of one's land, one's share, one's *moira*. *Eirene* (peace) is equally desirable.

If the archaic Greek read or heard Hesiod's *Theogony* he would discover all the important social and political concepts there as gods or goddesses: Custom Law, Share, Good Distribution, Justice (Claim), Peace, Righteous Indignation, Memory, Cooperation. But he would also discover Bad Distribution, Quarrels, Lies, Deception, Battles, Curses, Forgetfulness, and Blind Ruin. The archaic Greek readily personalized the abstractions which had important good or bad effects on his life. The baneful powers are as persistently present as the good. Bad Distribution, indeed, with the other fatherless children of Night, is two generations older than Good Distribution, Justice, and Peace.

True, Zeus has linked the benign powers to himself and his rule as closely as possible. Custom Law and Memory are his consorts, the Apportioners, Good Distribution, Righteous Claim, and Peace are his daughters, together with Splendor, Festivity, and Abundance. Hearth is his full sister. Zeus made Styx and her children, Rival, Victory, Power, and Might, his allies by guaranteeing their *time* under his rule.

All the methods of establishing cooperative bonds known to early Greek society—marriage, parentage, blood-kinship, alliance (*philotes*)—are used by Zeus to link him with these important benign deities of the third, fourth, and fifth generations, and of course with many personal deities as well. The Erinyes and the fatherless children of Night are not so linked; but the agreement of the gods on the winning side, after the defeat of the Titans, that Zeus should apportion *time* to all presumably guarantees to all of them, including the Erinyes and the children of Night, their traditional shares, *moirai*, of *time*. That an intelligent Greek could thus think about his gods many generations after Hesiod is not mere inference; the problem of Orestes in Aeschylus's *Eumenides* derives precisely from the irreconcilable claims upon him of Apollo and the Erinyes, expressed in terms of *moirai* of *time*. (Adkins 1982a, 229–32).

If the archaic or later Greek considered the *Theogony* merely as a static pattern, he might draw the foregoing conclusions. But the *Theogony* is not merely a static pattern; it describes the establishment of a divine world-order by personal and personalized deities, and such an order must rest on power of some kind.

The deities of order and the personal deities have (physical or quasi-physical) power of their own, some more, some less; but in the *Theogony* the greatest power is evidently furnished by the Kyklopes, as manufacturers of the thunderbolt, and the Hundred-Handers (*T*139–53). Now Hesiod's account of the role of Ouranos and Kronos is sketchy; but Ouranos would not let (at least) the Kyklopes and the Hundred-Handers come into the world; and on one interpretation he treated Custom Law and Memory in the same way. Ouranos fell. At the end of the reign of Kronos, the Kyklopes and the Hundred-Handers were imprisoned. Zeus made them his allies *(philoi)* by his act of *philotes* in releasing them. Kronos fell, for Zeus's power, thus augmented, was too great for him.

Hesiod's universe partly is, and partly is under the governance of, a polytheistic pantheon whose principle of organization is a socio-political order superimposed on a familial, genealogical order.[11] Zeus's administration of the universe is a political one, based on *moirai* of *time*, apportionments of functions and privileges, not on justice. The administration is made possible by the possession of power; not omnipotence, but a superiority of power, not all possessed by Zeus himself, to any that can be mustered against it. That Zeus might fall was not unthinkable; indeed, several myths recounted narrow escapes that he had had in the past. Consequently, Zeus and the other, less powerful, gods are likely to be very sensitive to any challenge to their power, or to any slight inflicted upon it, whether by other gods or by men.

Two other points should be borne in mind. First, Greek gods and men are evaluated on the same scale. The difference between them is said in Homer to be that gods have more *arete, time,* and *bie* (strength, force) than men (*Iliad* 9.498). (Gods are also immortal, of course.) Second, a god or goddess can mate with a mortal and produce offspring. *Theogony* 963–1018 records a list of children born to goddesses whose consorts were human, and the *Catalogue of Women*, or *Eoiai*, of which only fragments survive, recorded mortal women who bore children to gods. It seems a reasonable inference that by a superabundance of *arete* a human being—especially but not necessarily only a human being with one divine parent—might become a god. Herakles, the once mortal offspring of Zeus and the mortal Alkmene, was widely believed to have done so. Hesiod records the belief (*T*950–55).

Now *arete* is the most powerful noun of commendation in Homeric and later Greek. It denotes and commends the qualities which a man or a god needs to defend his *moira* of *time* and, if possible, acquire more. *Arete* is very competitive, and draws a god or a man anywhere on the scale of *arete* and *time* to try to rise higher. It may thus be disruptive of the orderly arrangement of *moirai* of *time*, since in most cases one can acquire more *time* only by taking it from the *moira* of someone else; and to increase one's *arete* and *time* is, if one can succeed, more important than merely behaving in accordance with *moira*

(Adkins 1960a, 30–60; 1971, 13–14; 1972b,19). A mortal of great *arete* might well aspire to become a god, or at all events he might claim that he could achieve any mortal success that he wished to achieve without the assistance of any god, and even against the wishes of any god.[12] Nonomnipotent gods may be expected to be particularly jealous of their status and prerogatives, and prompt to strike down the presumptuous mortal, in the manner of human aristocrats who strike down a presumptuous social inferior to prevent him from encroaching on their privileges.

It is within the framework of value and belief that Hesiod's myth of the Five Races of Men and his myths of Prometheus and Pandora should be located. I shall discuss the values in more detail later.

In Hesiod, the large-scale structure of "history" and man's place in it is supplied by the myth of the Races *(Gene)* of Men (*W*109–201). Five races are described; the Golden, the Silver, the Bronze, the Race of the Heroes, and the Iron. As this list shows, there is a blend of the schematic and the historical; the Race of the Heroes, that of the Theban and Trojan Wars, regarded by Hesiod and by most Greeks throughout antiquity as historical and a comparatively good time to have been alive, interrupts both the sequence of metals and the picture of continuous deterioration. A further impression of historicity is given by the fact that iron was in use by the Race of Iron, whereas the metal used for tools and weapons by the Race of Bronze was bronze; but there is no suggestion that the Golden and Silver Races used tools or weapons of these materials—which presumably Hesiod would have known to be technically impossible. (On the other hand, Hesiod does not say that either the Golden or the Silver Race used tools or weapons at all; and this could be represented as part of their good fortune.)

The presentation is in other respects mythological. The races are not five stages of the human race, but five separate creations, the first by "the immortals" in the time of Kronos, the other four by Zeus.

The Golden Race (*W*109–26) was blessed. They neither fought wars nor tilled the land. No moral excellences are explicitly ascribed to them, but they must have cooperated peacefully with one another. The end of their race is not deserved; it merely occurs. They are now blessed spirits who bring good and avert evil as they roam the earth. This is their "royal privilege" (*W*126). It is interesting that this race is assigned to the reign of Kronos, who himself suppressed the deities of power and violence, and was deposed in consequence. This too suggests that the reign of Zeus is not the ideal, but a practical "political" solution.

The Silver Race was inferior in every way to the Golden (*W*127–42). Its members took a hundred years to reach maturity. When they did so, they did not live much longer, as a result of their folly. They manifested *hubris* against each other, and would not worship nor sacrifice to the gods, as men should. So Zeus in anger "hid them, because they would not give *timai* [worship and sacrifice] to the gods who possess Olympus" (*W*139). They are now blessed spirits below the earth, however, inferior to the Golden Race, "but nevertheless *time* comes to them too" (*W*142).

The Bronze Race (*W*143–155) was terrible and mighty. Its members fought wars and engaged in acts of *hubris*, destroyed each other and went down to Hades, nameless.

The Race of the Heroes (*W*156–173) was "more just and better" (*W*158), and fought at Thebes and Troy, where some of its members died. Zeus placed those who did not die in the Islands of the Blest in the Ocean, where they enjoy three harvests each year.

The Race of Iron is the worst and, as the race of which Hesiod is a member, is treated at greatest length. But though the situation is bad now, Hesiod expects matters to become worse, until every vestige of justice and morality has vanished from the earth (*W*180–201).

The picture is bleak, and does not encourage a belief in divine justice. The Silver and Bronze Races deserved to be brought to an end, as will the Iron Race in its turn. The Golden Race and the Race of the Heroes did not. The Silver Race, as described, seems inferior not only to the Golden but also the Race of the Heroes; yet its reward, its status, its *time* seems greater.[13] Again, the successive races do not deserve to be inferior to their predecessors. They cannot, since they are new creations. It is simply a fact that Zeus successively creates races inferior to the Golden, and to their predecessors.

The gradual accumulation of the woes of humanity is not further explained in the Five Races myth. It is the inscrutable will of Zeus. Hesiod's two Prometheus myths (*W*42–105, *T*535–616) explain why work, disease and other woes came to be the lot of mankind.

The *Theogony* account begins with an aetiological myth to explain Greek sacrificial practices (535–57). Prometheus, who is always on the side of mankind, divided a great ox into two portions, "trying to deceive Zeus." In the one portion he included the flesh and offal, concealing them in the stomach of the ox; in the other he placed the bones, wrapping them in the fat. In other words, he arranged the portions as in a Greek sacrifice to the Olympian gods.[14] Prometheus then invited Zeus to choose one portion. Zeus is not usually represented as omniscient in early Greek belief. He is not so represented here; but he recognized the trick. Nevertheless, he chose the inferior portion; and "from that point the tribes of men burn white bones to the gods on the smoky altars" (*W*556–57). In consequence, Zeus refused to let mortals have fire, presumably to prevent them from cooking meat for themselves; but Prometheus stole fire for them in a fennel-stalk, at which Zeus was again very angry, and "devised a woe for mankind in exchange for [the good of] fire." The woe proved to be Pandora, the first woman (*T*590). The version of the *Works and Days* introduces Pandora's Jar, rather allusively, as the source of the woes of mankind.

The role of this myth as an explanation of Greek sacrifice is evident. It also gives some indication of the relationship between gods and mortals. Hesiod's train of thought is something like the following. Gods are more powerful than human beings, and the more powerful usually get the best of what is available; human rulers fare better than human peasants. The mortal of higher status becomes angry if an inferior attempts to usurp the privileges of his betters; and the gods may be expected to behave in the same way. Human beings have many woes; but they have fire, which philosophers as late as the Stoics continued to regard as divine, as an unexpected good, and they also receive the better portion when sacrifice is made to the Olympians.

From these materials Hesiod—or his source—constructs his explanation. He informs us also (*W*42–49) that Prometheus's actions explain why men have to till the soil and sail in ships; had Zeus not in anger hidden away man's livelihood, one day's work would have furnished enough sustenance for a year.

We may however ask about the nature of the sustenance. The answer is not clear, for Hesiod does not tell us what apportionment of sacrificial meats would have seemed equitable to Zeus. Had gods and men not met at Mekone at all (*T*536), presumably men would have neither fire nor meat. At the sacrifice at Mekone, had Prometheus made the same apportionments and had Zeus chosen the other one, men would have received the bones and fat while the gods received the meat; a state of affairs presumably pleasing to the gods. Alternatively, Prometheus might have divided the meat into two equal portions, or into a larger and smaller (or better and worse) portion. Since the *arete/time* scale is essentially hierarchic, and the individual (or group) with more *arete* expects to receive more *time,* evidently Zeus and the other gods would not have been satisfied with merely an equal share. But provided that the gods were willing to let mankind have any meat at all, men might have had some meat, the use of fire, and no Pandora, and have continued to work one day each year for their livelihood.

The last paragraph is entirely speculative, for Hesiod's myth is constructed to explain an existing state of affairs, and he had no inducement to consider what the human condition would have been like otherwise. I include the speculation to emphasize (a) that the myth is concerned with shares, *moirai,* and indeed with *moirai* of *time,* for to sacrifice is to give *time* to the gods, and (b) that the gods' attitude to men is in general grudging, and that it rests with Zeus to determine what size of human *moira* is acceptable to the gods. (In cult and general belief, of course, the gods were believed to be pleased by their portion of the sacrifice, and to be angry if they did not receive it.)

Hesiod draws other morals from his Prometheus myths. Prometheus had advised Epimetheus (Afterthought) never to accept a gift from Zeus, lest it should prove to be a woe for mankind; but Epimetheus forgot, which was foolish of him (W85–88). There is no way of avoiding what Zeus has intended (W105). Not even Prometheus could do so; and he was terribly punished for his attempt.

It is against this background that the wisdom literature of the *Works and Days* should be read. It should not be regarded as the other panel of a diptych, produced by Hesiod with the conscious goal of relating human life and behavior in as detailed a manner as possible to all the possible demands of deity, and the problems which these demands might create. I have discussed elsewhere other works which illustrate these demands and problems (Adkins 1982a); there is no need to suppose that any thinker of the eighth or seventh centuries was capable of producing works of that kind. Nor is it appropriate to demand complete point-by-point correspondence between the *Works and Days* and the relevant aspect of the *Theogony;* not even the Five Races myth and the contiguous Prometheus myth (W42–105, 109–201) achieve such correspondence. Nevertheless, the *Works and Days* is written in the context of the beliefs and values of the *Theogony.*

Suppose that an early Greek believes Hesiod's account of the genealogies of the gods, of the successive ruling gods, and of the "political" dispensation under the rule of Zeus, and suppose he also accepts the Five Races myth of successive and unmotivated creation of human races each inferior to its predecessors, and the not entirely compatible Prometheus-Pandora myths with their implied evaluation of the relationships between god and men. The Greek, like anyone else, wishes to prosper as well as he can in this world, which has not merely a multiplicity of governing deities but has parts which are themselves deities; and the Greek has a value-system which explicitly commends upon him success as his goal. How is he best to achieve success?

The *Works and Days* is a complex poem. Lines 1–10 are a proem, 11–41 a discussion of the two strifes which Hesiod finds in the world, and an exhortation to his brother Perses to engage in the good one which leads to prosperity. Lines 42–105 (the Prometheus myth) explain why human prosperity is attainable only by striving for it. Lines 106–285, the myth of the Five Races, give an alternative account of the need for work in this, the Race of Iron. Lines 286–319 give agricultural advice with the explicit goal of achieving *arete*, prosperity, and a more leisured existence. Lines 320–82 furnish a variety of moral, religious,and social/prudential advice, in every case urging that following Hesiod's advice will bring prosperity; ignoring it, disaster. Lines 383–617 are an agricultural calendar, diligent observance of which in one's farming is the most reliable route to a good harvest. Lines 618–94 give similar advice to the seafarer, with autobiographical digressions. Lines 695–705 advise on the age of marriage and the kind of wife to choose. Lines 706–64 contain a mixture of rationally prudential, magical, and religious advice, and 765–825 a list of lucky and unlucky days.

No account of the *Works and Days* will render its structure simple. But it becomes more comprehensible if it is remembered that justice in early Greece is not an end in itself, any more than is any other of the individual topics of the poem. The goal is prosperity, for the prosperous man is *agathos*, has *arete* and leisure, and can lead the most desirable kind of life; and the *Works and Days* discusses the different means necessary to ensure prosperity, or rather increase one's chance of gaining or retaining it.

At *W*285 Hesiod explicitly advises Perses how to avoid *kakotes* (poverty) and reach *arete* (prosperity, leisure, and social status); the good Strife—hard work in a spirit of emulation—is good because such strife leads to wealth (*W*20–24); and prosperity as a reward, or disaster as a punishment, is a constant theme. The gods reward by granting prosperity in the form of good harvests, fertility of field and flock, successful voyages, victory in war, health and general well-being; they punish by denying these desirables. It is urgent to know how the gods may be placated.

If justice is to be worth choosing, the just man must be seen to prosper; and Hesiod supplies as many arguments as he can muster, not necessarily all compatible with one another, to convince others, especially the rulers, that justice does pay; for Hesiod's own prosperity evidently depends in part on the justice of those more powerful than he.

There are two manifestations of justice: that of the man—or god—set in authority to administer justice to others, and that of persons cooperating justly with one another under the law, written or unwritten.

Much of Hesiod's treatment of justice is concerned with just and unjust judges. "Oath runs straightway together with crooked judgments [and punishment follows]" (*W*219). When judges give straight judgments to strangers and to their own people, the city and its people flourish: there is peace and prosperity, and fertility of field, flock, and man. *Hubris* and harsh deeds produce the opposite result (*W*225–39). There are thirty thousand spirits who invisibly watch over judgments and harsh deeds (*W*252–54). Justice is personalized and goes to tell Zeus when she is maltreated, "so that the people may pay for the misdeeds of their rulers who . . . twist the course of justice by crooked decisions" (*W*256–62). A variety of different divine agents is invoked: perhaps one account may persuade the unjust rulers. Hesiod even seems to treat the matter naturalistically also, saying that the person who does harm harms himself most of all (*W*265–66).

Hesiod's train of thought may begin naturalistically also in *W*214–15, when he proclaims that *hubris* is bad for a poor man *(deilos)*, and that not even a man of wealth and position *(esthlos)* can easily sustain his *hubris*. Since the gods can bring any mortal low, the gods seem not to be in Hesiod's mind in writing the line; but he passes on to an explanation in terms of deities. Here the poor man's *hubris* may bring him disaster. In *W*240 Hesiod maintains that a whole city may suffer for the misdeeds of a *kakos*. At this time *kakos*, like *deilos* above, is likely to be a social term, and denote and decry a poor man. If so, Hesiod is warning that not only the injustice of those set in authority may bring disaster to the whole community: it is threatened that Zeus will send famine, plague, and barrenness of women (*W*242–44). Hesiod is concerned with the profitability of justice for the individual, and for his brother Perses in particular, again in *W*274–85, where he says both that Zeus has endowed men, alone among living creatures, with justice, and also that Zeus gives prosperity *(olbos)* to the just and truthful man and his family, disaster to the liar and foresworn and his family.

(It is a datum of archaic Greek thought that the group may be punished by heaven for the misdeeds of the individual. If one believes that disaster is divinely sent punishment, the datum is indeed empirically verifiable, since disasters strike whole families and cities in circumstances when one cannot suppose that all the members are at fault. Whether such divine behavior appears just or unjust depends upon the relative importance accorded to the group and the individual).

But after all his exhortations and warnings, Hesiod adds two significant reflexions. "The eye of Zeus which sees and notes all sees these things too, if he wishes, and it does not escape his notice what kind of justice this city contains within it" (*W*267–69). King Zeus (who is not omniscient, omnipotent, or reliably just) may take note of these misdeeds and punish them *if he wishes.* But whether justice is worth choosing depends upon whether it is profitable. Hesiod immediately adds (*W*270–72) "Now may neither I nor my son be just among men, since it is a bad thing for a man to be just, if the more unjust is to come off better." True, the next line is "But I do not think that . . . Zeus will yet bring this to pass;" but the apparent confidence is belied by the tone of much of the work, including the immediately preceding lines; and that justice is only contingently desirable is undeniable (Adkins 1960a, 67–70; 1972b, 43–44).

When the early Greek read or heard the agricultural calendar (*W*383–617), he was aware of the personalized divinity of the earth, the winds, the rivers, the sun, and of the rain as the gift of Zeus, the crops as the gift of Demeter, and grapes and wine as the gift of Dionysus (*W*415–16, 488, 393, 465, 597, 805, 614). At least some of the constellations were divine and personalized. Hesiod mentions surprisingly few constellations. (Indeed, the calendar as a whole is remarkably sketchy.) The Pleiades, as daughters of Atlas, are certainly goddesses (*W*383). Orion (*W*598, 609, 619) was once mortal; but as a constellation he must be immortal, and "when the Pleiades flee the might of Orion and fall into the misty sea" (*W*619–20), indicates that Orion is thought of, like the Pleiades, as a living being. The same conclusion may be drawn with respect to the other constellations mentioned. Even in Aristotle regular movement indicates intelligence, not mindless mechanism.

The Greek farmer—and most ancient Greeks were farmers—was surrounded by gods and goddesses, on some of whom he depended for prosperity and indeed life, while others marked the regular cycle of his appointed tasks. (The belief that the very earth he worked was a goddess, and one of the oldest generation, presumably in itself affected his whole attitude to farming and the cycle of the year. See Sophocles, *Antigone* 337–39.) The account of justice might suggest that just behavior, particularly on the part of rulers, would by itself ensure good harvests and victory—if Zeus chose. No community could survive with such beliefs. Justice might—possibly—be necessary, but it is not sufficient. Deities must be satisfied in other ways too, and then—possibly—prosperity will come.

The farmer must do everything when the constellations decree, in due season; and the Horai, the Seasons, are goddesses too. Begin the harvest when the Pleiades are rising (*W*383); cut wood when mighty Zeus brings on the rains of autumn (*W*415–16); when Zeus has completed the period of sixty more winter days, after the Sun—a god—turns in his course, prune the vines (*W*564–65); and the vines are the gift of Dionysus (W614). When Orion first appears, bid your slaves winnow the grain of Demeter (W597–98); when Orion and Sirius are due south at dawn, and Dawn of the rosy fingers beholds (the star) Arktouros the Bear-watcher, prepare and press your grapes, the gift of Dionysus (W609–14).

And the farmer must work hard. One must strip for ploughing, sowing, and reaping if one wishes to attend to all the tasks of Demeter (W391–93). One must perform the tasks which the gods appointed for men to perform, for otherwise one may go hungry and be forced to beg from one's neighbors (W397–400).

Work, pray, and do sacrifice. Pray to Zeus of the earth and to Demeter, plough early and well, and have a small boy to follow you and cover over the seed which you sow (W465–71). Work, and Famine will hate you and avoid you, and revered Demeter will be your friend and fill your barn with livelihood (W299–301).

But the gods are inscrutable. If one ploughs and sows early with due prayer, one will have a good crop, *if* Zeus grants a prosperous outcome (W473–74). If one ploughs late, there will be a scant harvest. "Yet the mind of aegis-bearing Zeus is changeable, and difficult for mortal men to understand" (W483–84). If Zeus rains for three days about the time of the first cuckoo, the late planter might fare as well as the early (W485–90).

No working farmer could suppose that the earth, the winds, the sun, the rain—the gods—always give a good harvest to the diligent, or even that they invariably give a better harvest to the prompt than to the dilatory, any more than he could believe that the just always prosper at harvest time. It is in the fertility of field and flock that the rewards of the gods are seen in an agricultural community. The gods' failure to reward the just and the diligent may at any time be apparent.

The situation on the sea is similar: "For fifty days after the turning-point of the Sun [the summer solstice] . . . there is a reasonable time for mortals to sail. At that time you would not break up your ship nor would the sea drown the men, unless . . . Poseidon or Zeus king of the gods eagerly desires to destroy it; for in their hands are all outcomes, successful or unsuccessful" (W663–69).

There is no suggestion that it is only the unjust whose prudent summer voyages end in disaster, or the diligent but unjust farmer whose crops fail. After all his exhortations to justice, diligent farming, and careful seamanship as routes to prosperity, Hesiod can write (W717–18) "Do not be so hard-hearted as to mock a man with baneful poverty that devours the spirit: poverty is the gift of the immortal gods." When all is said and done, the race is not necessarily to the just, the hardworking, and the careful: the gods give good and ill as they choose.

It is now possible to draw some conclusions about the relationship of early Greek religious ethics (if that is an appropriate term) to the cosmogonic myth of Hesiod. To do so, it is necessary to put oneself in the position of the archaic Greeks, and consider their religious beliefs as an attempt to orient themselves in a baffling world, filled with mysterious and possibly life-threatening powers.

To those powers the earliest Greeks of whose beliefs we have evidence assigned personality and plurality. It goes beyond the scope of this paper to inquire why the members of some societies personalize the powers they encounter in nature and culture (the latter being the "abstracts"), or whether it would be possible for monotheism rather than polytheism to develop at this stage in such societies. The earliest Greeks also ascribed human motives and values to the powers.[15] It is difficult to see what else they could have done: if the motives and values of deities are not similar to those of human beings,[16] how is one to discover what they are? And it is essential to discover what they are, since one's well-being or disaster depend on doing nothing to offend the deities.

If the powers have personality, plurality, and human motives and values, everything else follows. The deities are the most powerful inhabitants of the universe. It seems not unreasonable to suppose that their values as individuals will resemble those of the most powerful beings to whose motives and values the archaic Greek had direct access, the human *agathoi;* and since, like the *agathoi,* they are numerous, it seems not unreasonable to suppose that their organization, and the values manifested in that organization, will resemble those of human society.[17]

Reasonable or not, these were the conclusions drawn by the archaic Greeks. I have spoken of the solution of the *Theogony,* the arrangement on which Zeus's rule rests, as a political one. That solution and its values are not those which Hesiod or his sources chose from an indefinitely large number available and known to them. The only organization and values which they knew well were those of archaic Greek society, known to us from the Homeric poems and other early Greek texts (Adkins 1972a, 1982a); and these they employed as an explanatory tool to account for the benign and malign behavior of their deities. This situation affects both the kind of behavior to be expected from the Greek gods, whether to one another or to human beings, and also the role which the fear or hope of divine intervention plays in modifying the behavior of the early Greek.

Space does not permit a detailed account of archaic Greek values to be given here: I refer my readers to what I have written elsewhere

(Adkins 1960a, 1960b, 1963, 1972a, 1975). As head of *oikos* (landed household) or as leader of army-contingent, the Homeric *agathos* is virtually autonomous. In Homer's Ithaca Odysseus is a kind of paramount chief, and it is he who leads the Ithacan contingent at the Trojan War. He is the most important *agathos* in Ithaca. But during the twenty years of Odysseus's absence, no assemblies are held in Ithaca, not because only he could call an assembly (*Odyssey* 2.28–29), but because there was no public business to warrant calling one. The absence of Odysseus caused great hardship to his own *oikos,* but there is not a word to suggest that there were any problems in Ithaca in general.

Each *agathos* concerned himself with the well-being of his own *oikos,* his *moira* (share) of *time* (status-producing material goods, and the status that accompanies their possession). Similarly, in the Greek army before Troy the other leaders acknowledged Agamemnon as in some sense their commander-in-chief; but unless each leader received his due *moira* of *time,* in this case war booty, he could withdraw from the fighting without any imputation of treachery, or of desertion in the face of the enemy. The individual *agathos* is concerned with increasing his own *moira* of *time,* or at all events preventing its diminution.

Among the *agathoi* the kings *(basileis)* are spoken of as receiving special privileges, including a *temenos,* enclosure, of cultivable land (*Iliad* 12, 310–21). These too are *timai,* accorded to superiors by inferiors. We have seen that the gods are the super-*agathoi* of the Greek world, that they have most *arete* and *time,* that they have a *moira* of *time* that marks off one god's domain from another's. They receive *time* also from human beings, their inferiors. *Time* and the verbs *timan* and *tiein,* "to give *time,*" are routinely used of offering sacrifice to deities; and deities too have a *temenos,* an enclosure of land assigned to them. When a Homeric character is promised that others "will give him *time* like a god," the emphasis is on receiving material goods, and the analogy with the situation of the deity is close.

The *agathos's* means to attaining the goal of keeping and if possible increasing his *time* are his courage and strong right arm, his *arete,* his weapons and his associates by blood, marriage, or guest-friend alliance, his *philoi;* and the pursuit of prosperity, and the claims of blood, marriage, or alliance, may require of the *agathos* behavior which neither we nor the early Greeks would regard as just. Furthermore, the pursuit of prosperity and *arete* is more important than being just, as Hesiod himself says (*W*270–73); and it is also more important than remaining within one's *moira* (Adkins 1960a, 30–60; 1972a; 1972b, 19; 1975, 1982a).

In the world of men, there is no strong centralized system of enforcing justice which can be invoked against a powerful individual or individuals: Achilles cannot go to court against Agamemnon, nor Odysseus against the suitors. (Indeed, even in fifth- and fourth-century democractic Athens, much of what we take for granted in a judicial system is absent.) The weaker members of society may go to court and cry for justice from the kings in their role as judges. To judge from Hesiod's words, they may well cry in vain; and there is no redress on earth to be obtained from the unjust judge in such a society. Why then do the weaker not cry to heaven for justice at the hands of a being more powerful than any mortal, and all-just?

Possibly because the Greeks of this period had never conceived the possibility of such a being. Hesiod, or his sources in the *Theogony,* does not portray a Zeus who became ruler of the gods because he was just and Kronos was not, or because he was more just than Kronos. He defeated Kronos because he gathered allies by promising to respect the *moirai* of *time* of any god or goddess who would support him, and offered *time* to those who lacked it under Kronos's rule; and in particular because, having made the Kyklopes and the Hundred-Handers his *philoi,* he had greater power than Kronos and the Titans. Having led the other gods to victory, he was acknowledged as their ruler. The analogy with a human seeker for power is evident; as is the analogy with a human political solution in the allotment of *moirai.*

By analogy with the human situation, Zeus as ruler will take his place on the judicial bench to settle the complaints of the weaker, in this case mortals; but, again by analogy with the human situation, why should such a ruler thus come to power not be unjust to the weaker when he is not on the bench, or when he is on the bench if there are other claims upon him more powerful than justice—or even out of caprice? Hesiod himself does not expect reliable justice from Zeus (*W*270–72). Certainly Justice is Zeus's daughter and an important adviser at Zeus's court, and the Moirai too are his daughters (*T*902, *W*256–62); but rulers do not always heed their advisers,[18] and the requirements of justice and *moira* may conflict.[19]

Cooperative relationships *(philotes)* with human superiors may be created by doing services and giving *time.* One may have similar relationships of *philotes* with gods. Through industry (*W*300) and prayer (*W*465) the farmer becomes Demeter's *philos;* but that will not guarantee a good harvest. Sacrifice, too, is expected by Greek deities, and such giving of *time* creates a bond of *philotes;* but *time* received may be ignored by the god, if he so chooses, as does Zeus in *Iliad* 4.25–67 (Adkins 1972a, 16–17). No reasons are given in the *Works and Days* for the possible caprice of Zeus, Poseidon, Demeter, or any other god; but the archaic Greek knew that his human superiors would not necessarily discommode themselves on his behalf, whether in response to the claims of justice or to those of *philotes*, which may of course conflict. Again, how does the motivation of *time*, sacrifice, given to the gods accord with the requirements of strict justice? The gods, like Hesiod's human judges, receive gifts which are intended to influence their behavior.[20]

Again, divine reward and punishment are an empirical matter in archaic Greece. There is no reward or punishment after death for ordinary human misbehavior (Adkins 1960a, 67); reward consists in well-being in this life, punishment in disaster. Hesiod is aware that the just and diligent farmer may not receive his reward, if Zeus wills otherwise (W574, above, p. 57).

The Five Races myth and the Prometheus-Pandora myths portray a divine order whose attitude to mankind is indifferent or grudging. Any benefit attained by man must be at least counterbalanced by an evil. Those for whom Hesiod is writing seem themselves hardly likely to attract the envy of heaven by excessive prosperity; but the belief, common enough in early Greece, suits Hesiod's general worldview, and the Prometheus-Pandora myths in particular.

In short, Hesiod's super-*agathoi*, the gods, are as unreliable in behavior to their inferiors as are the human *agathoi* to theirs, and for very much the same reasons.

To understand Hesiod's answer, it is necessary to understand his question, and the premises from which he begins. The *Theogony* and the *Works and Days* do not constitute a botched attempt at a theodicy: Hesiod is not trying to prove that the gods, or the world, are reliably just. He is trying to explain why the world is as it is, and the gods are as they are. Like that of the human society he knew, the divine order of the world rests on power. The power is devoted to maintaining a system in which justice may sometimes be found but is not of paramount importance to the powerful, whose values give precedence to other qualities, or indeed to the weak, if justice does not lead to prosperity, or to more prosperity than injustice. But there is a system of values, a hierarchic system which reaches from Zeus to the humblest beggar, in terms of which everything that happens can be explained. The archaic Greek universe is instinct with value and pullulating with divine purposes, frequently in conflict with each other.

The *Works and Days* is precisely the kind of handbook that one would expect to find in such a world of belief and value. The goal is to become *agathos*, prosperous, to attain to *arete* (or to remain *agathos* if one is so already). These are the most powerful terms of commendation in Greek society. All activities are instruments to their attainment, and though none is fully reliable, the importance of each is measured by its effectiveness. Seen in this light, Hesiod's advice, which seems to us to cover such disparate fields as prudence, ethics, agriculture, seamanship, magic, and religion, marks out parallel routes to the same goal. In these circumstances it becomes difficult to identify clearly a field of ethics in a familiar sense; and certainly the qualities emphasized by the value-system are not those brought to mind first when the word "ethics" is used; but there is a value-system, and it is related to the deities in which these early Greeks believed, and to the values of those deities.

My own recommendation is to use "ethics" to denote the totality of the system of general value-terms in a culture: here *agathos, arete, time, moira, philotes, dike,* and a number of associated terms. Even if one does not include the competitive excellences under "ethics," it is possible to acknowledge the links in a culture between the behavior of those value-terms which are deemed to be ethical and those which are not; but the usage which I favor draws attention to the links in any culture that is examined; and the study of such totalities will be much more fruitful, both in the study of particular cultures and in cross-cultural research.

Notes

1. Some scholars reject 119, which contains "Tartarus." See West 1966, ad loc. The "coming-to-be" of Tartarus should be mentioned somewhere; and this is the appropriate place.

2. Chaos means not "confusion," but "chasm" or "gap." For the different views of Chaos corresponding to "chasm" and "gap" see West (1966, 116n.) and the works there cited.

3. Roman numerals after the name indicate the generation to which a deity belongs. For the most part deities procreate with other deities of the same generation. There are useful genealogical tables in Lattimore, 222–26.

4. There is something seriously wrong with the narrative. West (1966, ad loc.) argues plausibly that 139–53, the lines concerned with the Kyklopes and the Hundred-Handers, were added by Hesiod when he reached the Titanomachy (*T*617–725), and realized that the existence of these deities had not been accounted for. But West also supposes Hesiod to mean that all eighteen of Ouranos's children were still in Gaia's womb, prevented from egress by the continuous copulation of Ouranos. This seems much more difficult. Allowance must doubtless be made for the confused imagery of cosmogonic myth; but

to suppose that Gaia asked for volunteers, Kronos volunteered, and Gaia conveyed a sickle to him in these circumstances seems unlikely; and it would require remarkable agility for a child in the womb to castrate his father while in the act of copulation. We might dispense with rationality and treat the passage as variations on Freudian themes; but *T*176–77 make it quite clear that Okeanos was not engaged in continuous copulation with Gaia. For the present purpose the important feature of the narrative is that the separation of Heaven and Earth is treated in personal terms.

5. For my purpose, the Typhoeus episode is relevant, whether by Hesiod or not; but West's arguments (1966, ad loc.) for Hesiodic authorship are persuasive. Even if *T*820–80 were rejected, the case against Tartarus being treated personally is doubtful; Eros has no children in the *Theogony,* but seems to be treated as a person at *T*201.

6. I use "asexual" to cover what a modern biologist would distinguish as asexual (by budding or cell division) and parthenogenetic reproduction. Hephaestus was evidently born parthenogenetically from Hera (*T*927–28); but, as West points out (1966, 132) Earth's production of sea, sky, and mountains might be asexual reproduction in the other sense. In either case, the offspring might be thought of in personal terms, as Erebos and Night certainly are, since they subsequently procreated sexually.

7. I use "personalized," rather than "personified," since "personification" denotes a self-conscious literary device. See Lesky (1963, 98). Lesky's discussion of Hesiod employs some premises very similar to those of the present paper, but he does not draw all the conclusions which seem to me to be appropriate.

8. With the exception of Kronos, they are also obscure deities who had no cult, so far as is known. "So far as is known" is, however, a very important qualification in any generalization about Greek religion. The only constant is variety, and our ignorance is far more extensive than our knowledge.

9. *Hora* means "season" and the Horai are usually the personalized seasons. In Attica they received cult, and had the names Thallo and Karpo, names associated with growth and harvest.

10. Consequently, for this and other reasons, "justice" is an inadequate rendering of *dike,* "just" of *dikaios*. But both, even in Hesiod, are at least sometimes value-terms used to commend cooperative behavior. Gagarin (1973, 1974) denies this; but see Dickie, 91–101. "just" and "justice" will serve for my present purpose.

11. *T*885 (Zeus's distribution of *timai* and *T*390–96 (his promise not to reduce the *time* of any ally) would permit Zeus to improve the

status of any of his allies in reward for merit. Hesiod—or his source—may have found the device useful to account for what otherwise might appear anomalous; the status of Greek deities is not closely correlated with generation or with any principle of succession found on earth. In a mortal family the older members possess greater status, and a Greek by analogy might have expected the deities of an earlier generation to possess greater status than those of a later. The preexisting status of Greek deities rendered such a solution impossible; Athena (V) and Apollo (V) were far greater deities than any of the three thousand Okeanids (IV), and pre-Hesiodic Greek belief cannot have been so fluid as to enable Apollo and Athena to be placed in an earlier generation than the Okeanids. In any case Zeus (IV) was now the chief deity; and he inevitably acquired more status than any of his seniors as a result. To say of the existing distribution of *timai* that it is so because Zeus, without reducing the *time* of any ally, decreed that it should be so, solves that problem.

12. To make such a claim is to commit *hubris,* and divine reprisals are to be expected. See, e.g., Aeschylus, *Seven against Thebes,* 375–416 (Tydeus), 422–51 (Kapaneus); Sophocles, *Antigone,* 117–40.

13. "Second" W242, seems intended thus to rank them. In any case, they receive great *time* though they are portrayed as violent and impious boobies. The Olympians frequently overlook violence among mortals; but a failure to offer sacrifice is usually represented as evoking reprisals.

14. Sacrifices to chthonic powers were holocausts. The Greek would not share a meal with the dead, or with the gods of the dead.

15. To do so does not entail that they are necessarily anthropomorphized in the literal sense. An divine owl or a sacred snake might have the same values and motives as a human being.

16. "Similar" would not exclude "but nobler, finer, more exalted"; but the motives of Greek gods do not surpass those of Greek mortals in the archaic periods (or later, in most cases), though philosophers—e.g. Xenophanes, D-K, B11 and 12 (vol. 1, p. 132)—begin to complain about 500 B.C.

17. There are of course no poor among the gods; all are *agathoi,* but some are more *agathoi* than others.

18. Where personalization of abstracts is so vivid, e.g., W256–62, such a comment seems justified, as it would not be in the case of personification as a literary device.

19. The distinction is difficult to draw in the archaic period; but it is crucial later, in Aeschylus's *Eumenides* (Adkins 1982a). See the argument between Zeus and Hera in *Iliad* 4. 25–67 (discussed in Adkins 1972a, 16–17).

20. There is no clear statement that the gods are influenced by *time* only when not on the bench; and even in Aeschylus, the *Seven against Thebes* would suggest the contrary (Adkins 1982b).

References

Adkins, A. W. H.

1960a *Merit and Responsibility: A Study in Greek Values.* Oxford: Clarendon Press.

1960b " 'Honor' and 'Punishment' in the Homeric Poems." *BICS* 7:23–32.

1963 " 'Friendship' and 'Self-Sufficiency' in Homer and Aristotle." *CQ* 13: 30–45.

1971 "Homeric Values and Homeric Society," *JHS* 91: 1–14.

1972a "Homeric Gods and the Values of Homeric Society." *JHS* 92: 1–19.

1972b *Moral Values and Political Behaviour in Ancient Greece.* London: Chatto and Windus; Toronto: Clarke, Irwin.

1975 "Art, Beliefs and Values in the Later Books of the *Iliad.*" *CP* 70: 239–54.

1982a "Laws versus Claims in Early Greek Religion." *HR* 21: 222–39.

1982b "Divine and Human Values in Aeschylus' *Seven against Thebes.*" *Antike und Abendland* 28: 32–68.

Dickie, Matthew

1978 "*Dike* as a Moral Term in Homer and Hesiod." *CP* 73: 91–101.

D-K

1951 Hermann Diels and Walther Kranz. *Die Fragmente der Vorsokratiker.* 6th ed. Berlin: Weidmannsche Buchhandlung.

Finley, Moses

1978 *The World of Odysseus.* 2d ed., New York: Viking Press.

Gagarin, Michael

1973 "*Dike* in the Works and Days." *CP* 68: 81–94.

1974 "*Dike* in Archaic Greek Thought." *CP* 69: 186–97.

Lattimore, Richmond, trans.
1978 *Hesiod*. Ann Arbor: University of Michigan Press.

Lesky, Albin
1966 *A History of Greek Literature*. Translated by James Willis and Cornelis de Heer. London: Methuen.

T
1966 Hesiod. *Theogony* (text in West 1966).

Vernant, Jean-Pierre
1965 *Mythe et Pensée chez les Grecs*. Paris: Maspero.

West, M.L.
1966 *Hesiod: Theogony*. Oxford: Clarendon Press.
1978 *Hesiod: Works and Days*. Oxford: Clarendon Press.

W
1978 Hesiod. *Works and Days* (text in West 1978).

3 Behaving Cosmogonically in Early Taoism

Norman J. Girardot

Hold on to the Tao of old to master present existence.
The ability to know the primal beginnings (of the world),
Is called the principle of Tao.
Lao Tzu, chap. 14

Therefore the sage . . .
Assists the spontaneous development [*tzu-jan*] of all things,
And does not act presumptuously.[1]
Lao Tzu, chap. 64

Myth and Cosmogony in Early China

It has been a commonplace observation that early China, in comparison with other ancient civilizations, is strikingly lacking in any complete or coherent myths, especially creation myths. Thus in nineteenth-century Western scholarship China was sometimes portrayed as a welcome anomaly in the history of world cultures since in early China, unlike in other more "superstitious" traditions, there was never any real indulgence in mythological speculation. The absence of myth and cosmogony was in sympathy with the Confucian humanistic admonition to keep the spirits at a distance and seemed to insure that ancient China was "singularly pure" when contrasted with all other archaic civilizations more blatantly caught up in the irrational throes of religion and myth (see Girardot 1976). At the very least and less judgmentally, some more recent scholarly opinion (Mote 1972) would contend that the absence of a Chinese tradition of creation mythology is the essential basis for the overall "cosmological gulf" between Chinese and Western patterns of thought and behavior.

There are, of course, important differences in the ancient Chinese worldview in comparison with other traditions. However, the notion that the "cosmological gulf" is a result of the nonexistence of Chinese creation myths, or any form of mythic or cosmogonic thought, is without foundation. It is more historically accurate, and more heuristically significant, to say that much of the cosmological, religious, philosophical, and ethical distinctiveness of ancient Chinese tradition actually depends on the kind and use of mythological and ritual materials. Rather than an almost incredible lack of mythological expression or some peculiar imaginative poverty, the issue more realistically concerns the manner in which different types of myths and mythic images have been interpreted and transformed in early Chinese tradition.

Furthermore, with regard to the particular case of cosmogonic myth, the issue is not the complete lack of creation myths in early Chinese tradition but a question of the kind of creation myths, the type of cosmogonic images, and the logic of creation variously emphasized in the early texts. The heart of the matter "is the manner, rather than the fact, of cosmic creation" in ancient China (Plaks 1976, 18). As this essay will endeavor to show, the question of a Taoist ethical perspective is clearly related to the concern in all of the early Taoist texts for a certain kind of mythological understanding of the creation and the creative life of the world and man. If the moral vitality of a religion is "conceived to lie in the fidelity with which it expresses the fundamental nature of reality" (Geertz 1973, 126), then it may be said that in early Taoism "fidelity" refers most profoundly to behavior that emulates the spontaneous action and impartial interaction of creation. To know what "ought to be" implies most of all a renewed awareness of what "is" as the continuous manifestation of what "was" at the beginning.

The work of Chang Kwang-chih (1976) convincingly demonstrates that during the Eastern Chou period (ca. seventh to second centuries B.C.E.—i.e., embracing the crucial foundational period of the "one hundred schools": Confucius, Lao Tzu, Chuang Tzu, Mencius, Mo Tzu, Hsun Tzu, etc.) there was a rich traditional mythological lore that affected all levels of early Chinese tradition. There is also sufficient literary evidence to argue for the presence of mythological systems of "cosmogonic formations and construction" that in different ways influenced all of the major text traditions or "schools." Moreover, amid the congeries of mythology reconstructed by Chang (1976, 157), there is finally the central and shared cosmogonic theme that in the beginning the "cosmos was . . . a chaos *hun-tun*, which was dark and without bounds and structure."

The word *hun-tun* in its Chinese use for a kind of embryonic monad is, above all, an excellent example of what Lewis Carroll's Humpty-Dumpty called a "portmanteau" word, that is, a word "packed up" with several meanings. Unpacking these meanings and reconstructing the thematic order of their relationship can, therefore, tell something of the intent of the wayfarer. Indeed, the specific conceptual content of *hun-tun* as a word is not so important as the fact that it serves as a reference for an underlying cosmogonic form, theme, system, structure, pattern, or shape in the early Chinese texts, especially the early Taoist materials.

The difficulty here is that *hun-tun* as a term for chaos conveys the idea of the relatively "undifferentiated" or "shapeless" form of the primal condition in contrast to the cosmetic form of the cosmos. Mythologically and cosmogonically, however, both Humpty-Dumpty and Hun-tun have a "body" that is imaginatively shaped like a large egg, dumpling, wonton, or gourd and, as stories, refer primarily to the mythic theme of the creation, teeter-tottering, and eventual shattering fall of the cosmic monad. It may even be said that the religious and ethical dilemma for Taoism to a great extent concerns the way of putting Hun-tun or Humpty-Dumpty back together again (Girardot 1983 and Schipper 1982).

The Shapeless Shape of Myth in Chinese Tradition

Despite the presence of mythic units and cosmogonic themes throughout many of the Eastern Chou and Han texts, the deductive and reconstructed existence of originally coherent myths relating to various local cultures, and the necessary assumption of an unrecorded oral tradition of living mythological fabulation at the folk level of tradition, it still must be admitted that the earliest written sources do not preserve integral mythological tales. It is true that the Taoist texts tend to reveal a quantitatively greater deposit of explicit mythological themes and images; for instance, the *Chuang Tzu's* parabolic style testifies to the early literary use of mythologically based story fragments. But these elements in the *Chuang Tzu* represent only the debris of myths, and a text like the *Lao Tzu* is totally devoid of any narrative element or even any proper names.

This general situation is not endemic only to early Chinese literature; the fragmented presence of myths in ancient literature is a problem encountered in many civilizations (Plaks 1976, 16). It may be said, however, that early Chinese literature is rather remarkable in relation to other traditional literatures because of its characteristic failure to emphasize any extended narrative form or detail in its use of mythic lore. In this sense, as Andrew Plaks (1976, 15–16) remarks, the basic aesthetic impulses of early Chinese literature "simply are not geared to the forward thrust of beginning, middle, and end that we naturally associate with patterns of narrative shape in other cultures."

If Chinese literature in its embodiment of mythic elements does not live up to the usual Western literary expectations concerning an overall narrative or epic form of *mythos*, the issue remains as to what kind of "shape" or "structure of sense" Chinese literature sees in human experience. The Chinese, like all other peoples, clearly enjoyed and told individual stories and myths, but the interesting question for the ancient period concerns the general lack of an overarching narrative shape that tied all the small tales together (Prusek 1970 and Egan 1977). Again, in contrast to ordinary Western expectations, the question is really whether the absence of detailed mythic narratives, or a general epic form of dramatic action, in early Chinese literature makes that literature necessarily and essentially nonmythological in nature.

It can be argued that narrative poverty does not indicate that early Chinese literature is radically demythologized or devoid of any mythic structure and intention. Thus it is not mandatory to embrace all of Claude Lévi-Strauss's (1966) grand edifice of theory in order to agree with his idea that mythological thought is primarily a matter of an intellectual strategy of "bricolage" which constantly juggles, rearranges, and transforms assorted mythological signs—bits and pieces—according to a deeper code of relational contrast and dynamic synthesis. At least part of the form and meaning of myth and mythic thought, if not all of its human significance, is to be found at the structural level which perdures beneath the shifting surface dimensions of particular mythic images, themes, and plot developments.

A text can, in other words, be mythic in its meaning and structure without itself partaking of narrative development or without explicitly telling complete mythic narratives. It is instructive to note that nonliterate tribal cultures which would ordinarily be taken as the paradigmatic instance of a living mythological situation do not always, or even characteristically, tell complete mythic narratives (or, for that matter, always have an explicit cosmogonic myth). Rather it can often be the situation that in the cultic life of a tribal culture myths are only sporadically referred to in explicit terms or as whole narratives; when they are told, they may be in the form of fragmentary bits and pieces that gloss different aspects of ritual.

The common conception of "primitive" cultures constantly and necessarily ordering their existence through the actual telling of coherent mythological narratives is in many ways nothing more than another mesmerizing scholarly myth about myth. But because such traditions do not always narratively recite all of their myths, or all of every myth, does not make them fundamentally nonmythic in thought, structure, or shape. These apparently mythologically impoverished traditions can often be shown to harbor an "implicit" mythic structure which acts as an underlying frame for ritual action and the overall religious ethos of the culture (see, for example, Hugh-Jones 1979, 252–60).

This is not the place to enter into a full discussion of the "implicit" structural presence or function of myth in nonliterate and literate cultures. It suffices to say that explicit traditional and oral mythological tales stemming from various local cultures and ancestral clans have become in Chinese literature almost entirely submerged but that this literature is demonstrably no less mythic because of the transformation. Thus, Henri Maspero (1924) in his pioneering study of early Chinese "légendes mythologiques" demonstrated the disguised presence, and continuing mythic functionality, of ancient mythological themes throughout much of the "Classic of History" (*Shu Ching*). It is not, however, just that mythological personages and events have been disguised or transposed by some process of reverse euhemerization, but that in a more radical fashion mythic materials and themes have entered into Chinese literature as a series of extremely abstract, and essentially static, models for organizing and evaluating human life.

Mythic themes in early Chinese literature, in other words, often seem to be reduced to their inner "logical" code or implicit cosmological structure of binary *yin-yang* classification. Moreover, besides the almost constant structural undertow of sublimated mythic themes, the use of explicit mythological fragments in early Chinese literature (and it is the "heterodox" materials like the Taoist texts, the *Ch'u Tz'u*, or *Shan Hai Ching*, that more typically cling to overt mythological elements) also tends to betray an emphasis on the myths' slated, nonnarrative, or classificatory potential for evaluating the present in terms of some exemplary structure from the past. The mythic past in Chinese literature, therefore, tends to inform human behavior by revealing the first, and ongoing, structural principles of transformation, not by remembering the dramatic action or heroic actors of the original tales.

As Sarah Allan (1981, 18) tells us, the ancient Chinese "did not narrate legend but abstracted from it. Aware that the legends were structurally similar, [they] paralleled them to make the repeating themes apparent and continually sought to derive the concepts associated with the signs." But undergirding the gross manipulation and transformation of legends used as signs of the concrete world and of cultural events is a "level at which the signs function as the elements of mythical thought and the legends serve as myth" (Allan 1981, 18). The constantly changing reality of nature and social life only demonstrated to the ancient Chinese that "history," like the Tao, always stays relatively and structurally the same.

I would also like to add that these abbreviated observations on the role of "implicit" myth impinge upon what Mircea Eliade (1968, 21–38) seems to mean when he speaks of the archetypal "prestige" of cosmogonic myth—even when a specific cosmogonic narrative may not be coherently or actually present in the tradition. Since for any cultural tradition the world is always *just* there, it need not be told of, even though it implicitly speaks through the overall symbolic systems and cultic life of the tradition. Just as nonliterate traditions may not tell of "beginnings" by means of coherent mythic narratives, so also ancient literate traditions may not always record clearly delineated stories of world foundation. And when either nonliterate or literate traditions do explicitly refer to mythic "origins," they may often emphasize anthropogonic and cultural beginnings over cosmogony in the strict sense of cosmic origins.

The issue is, then, whether fragmented mythic themes or origin myths of various kinds imply a prior cosmogonic structure of thought even when it is not clearly stated; whether generalized cosmological systems of meaning as revealed in ritual or literature presuppose particular cosmogonic themes. From this perspective, the most intriguing question becomes not the presence or absence of cosmogonic ideas but the question of how and why actual cosmogonic myths come to the fore in some nonliterate and literate traditions. In like manner, a correlative question concerns why and in what way specific cosmogonic themes may be relatively more explicit and important in some forms of thought than in others within the same literate tradition. These considerations, I hasten to add, have a direct bearing on early Chinese tradition and on the difference between Taoist and Confucian understandings of what happened in the "beginning."

Creation and the Conundrum of Hun-tun

To be explicit about what is inferred by the cosmogonic sensibility found in early Chinese literature, I first want to emphasize that the Chinese were certainly interested in the question of origins and in the significance of some original model of individual and social harmony found in the distant past, as the differing golden age conceptions in Taoism and Confucianism disclose (Bauer 1976, 31–33). It can also be shown that the ancient Chinese were concerned with cosmogonic creation in the strict sense, since explicit mythic allusions to the first creation of the prehuman world out of the *hun-tun* condition, as well as an implicit overall cosmogonic frame of reference, are especially found in all of the early Taoist texts (*Lao Tzu, Chuang Tzu,*

Huai Nan Tzu, and *Lieh Tzu*). There are also numerous traces of cosmogonic myths (i.e., the interrelated *hun-tun*, cosmic giant, and primordial couple typologies) to be found in other ancient documents such as the *I Ching*, *Shan Hai Ching*, and *Ch'u Tz'u* (Chang 1976, 157–59, and Girardot 1983). Despite frequent assertions that the "Chinese are the only major civilization without a creation myth" (see Major 1978, 9), the facts of the matter dictate the conclusion that China was never so creatively obtuse as to have avoided the whole issue of creation.

But there is still something different about the Chinese understanding of creation that goes against the grain of what someone like the modern day "creationist," Duane Gish, demands of a good creation story. One way to indicate some of the variance is to say that Chinese cosmogony—especially the cosmogonic tradition and mythic logic associated with the undifferentiated cosmic stuff of *hun-tun*—characteristically refuses to transform the puzzle of creation into an epic, once-and-for all event requiring the causal presence of some "extraordinary individual," narrative persona, or hero of creation who stands outside of the created order. Chinese cosmogony was an utterly mundane affair since the division of the primordial raviolo of *hun-tun*, the passage from one to two, happened spontaneously, by and of itself (*tzu-jan*), and this way of happening naturally continues to resonate (*ying*) throughout the ten thousand things of the phenomenal world. There was no dramatic spectacle of the willful or purposeful activity of a creator at the beginning since, in a sense, nothing happened that is not always happening. This constant creativity of natural life is the graceful "way" of the Tao in the world which can be identified with the cosmogonic activity of the primal wonton of *hun-tun* (e.g., chapter 25 of the *Lao Tzu* which identifies the Tao with the cosmogonic condition of *hun ch'eng* and the creative mode of spontaneous action called *tzu-jan* [Chiang 1970, 166–70, and Chan 1963, 144–45]).

In contradiction to universal folkloric testimony, cultures very often take a decided stand with respect to the question of the chicken and egg. Thus despite periodic protests by such maverick sages as Nicolas Cusa, Isaac Luria, Alfred Whitehead, and Tom Stoppard, it is fair to say that the West has tended to opt for a kind of heroic and divine chicken which bravely and mysteriously produced and brooded over a fallen omelet of a world. As a counterpoint to this entirely defensible position, Chinese thought asserts with equal justification that in the beginning there was no avian spirit hovering willfully and urgently over the chaotic waters; or that at the very least "chickens and eggs have been succeeding each other in one form or another literally forever" (Stoppard 1972:27). Regardless of some sporadic disagreement evidenced primarily in popular currents of Chinese tradition, the Chinese imagination has tended to find cosmic eggs, gourds, and dumplings more metaphysically nourishing than divine chickens (Girardot 1983).

Infinite regression from the Chinese perspective leads back not to a Prime Mover but to a spontaneous Prime Movement (*tzu-jan*) and its continuing Primary Resonance (*ying*) among all things (Robinet 1977, 14–16, 33–34, 231–34, 266–67). Imagining the creation in China gave rise, therefore, to a mythological "nothing" which harbors its own internal promptings toward the somethingness of phenomenal life. And because of the original free gift of life, the first grace of movement, there is an ongoing reverberating interaction and harmonious ordering of all the ten thousand things of the world. Moreover, the Tao as the name for this cosmogonic power and process in and among things continues to create new forms of life by the transformational principle (*fa*) of its periodic regression (*fan, fu, kuei*) back to the original undifferentiated wriggling of germinal life (e.g. chapters 25 and 40 of the *Lao Tzu*). It acts this way by not acting (*wu-wei*) in any purposeful way.

It can be shown that *tzu-jan, ying,* and *fan* represent important technical principles for understanding the Taoist vision of the Great Tao as a primal "action guide." In fact, even more so than for the largely individual, self-generated, spontaneous, or initiating action of *tzu-jan,* it may be said that the harmonious resonance (the *ying*ing or ringing relationship) among all existing things and people comes closest to the Taoist relational understanding of being "good"/*shan*[a] in an ecological and social context—remembering here the Indo-European root associations of "being united, associated, or suitable" and the Chinese associations of "aptness, familiarity, and wholeness" (e.g., chapter 8 of the *Lao Tzu*).

As A. C. Graham remarks, the nearest Chuang Tzu comes "to formulating his implicit [ethical] principle" is found in the next to last parable of chapter 7 (the last story being the ditty of Emperor Huntun) which describes the mirrorlike nature of the sage who knows how "to respond" (*ying*) to any situation (Graham 1981, 14). It is here that the positive proscription of *ying*ing action is first described as the negative method of *wu wei* which avoids being "possessed by name/fame" (Watson 1968, 97; Graham 1981, 98):

> Do not [*wu wei*] be possessed by fame; do not [*wu wei*] be a storehouse of schemes; do not [*wu wei*] be a proprietor of wisdom. . . . The Perfect Man uses his mind like a mirror—going

> after nothing, welcoming nothing, responding [*ying*] but not storing.

In like manner, it can be noted that chapter 7 is given the overall title of "Ying Ti Wang" which, if we follow Julian Pas, might be broadly translated as "responsible/responsive leadership" (Pas 1981, 487). *Ying*ing action, in other words, can be understood as a particular principle of social reciprocity that may be contrasted with other more renowned regimes of social interaction (i.e., the rule of *pao*).

It should also be recognized that in the *Lao Tzu* (e.g., chapters 38, 46, 51, 52, 80) and in the "primitivist" sections of the *Chuang Tzu* (using Graham's scheme, these would comprise chapters 8–10 and parts of chapters 11, 12, 14, 16—see Graham 1981) there seems to be a close parallel between the "early period of the cosmos and the early period of human society" (Munro 1969, 143). This tends to corroborate the notion that the Taoist virtue/*te*[a] of "cosmogonic behavior" clearly has implications for both individual and social ethics. It is in this way, for example, that much of the enigmatic meaning of *wu-wei* as the most celebrated description of Taoist behavior can be shown to allude to the cosmogonic mode of generative action and resonating interaction known as *tzu-jan* and *ying*. These terms have reference to the behavior of the individual sage; but they also frequently allude to a mode of "primitive" social life that can be distinguished from other, more cosmetic and civil forms of reciprocity. Indeed, I will suggest in this essay that there is a distinctive Taoist ideal of egalitarian social reciprocity, an ethic of the "pure gift," to borrow from Malinowski, that may refer to a surprisingly perceptive ethnographic understanding of ancient Chinese social development (on Malinowski and forms of reciprocity, see Sahlins 1968, 139–236).

The moral implications of cosmogony always tend to turn on "the possibility of an absolute standard of judgment in human affairs: the ultimate accountability of man for his action" (Plaks 1976, 18). Unfortunately what is absolute in the Chinese creation myth of *hun-tun* is that there is no permanent, unchanging absolute, which is, perhaps, the reason why some early Confucians, at least, seek substantiation of moral prerogatives in the creation of the fixed rules of human civilization rather than in the ambiguous flux of cosmic and primitive reciprocity. The absolute cosmogonic law, or "constant Tao," is that all things are relative since they spontaneously transform and respond to each other in relation to constantly changing situations.

Both Confucianism and Taoism, it can be said, make use of implicit mythic models; but whereas Confucianism tends to read human nature and society back into the cosmos, Taoism sees the necessity of finding nature hidden in the heart of man. Or to put it another way, both Confucianism and Taoism share much of the same mythological heritage and cosmological tradition, but Confucianism focuses its sweaty ethical attention on the cosmological and civilizational aftermath of the "primitive" creation while Taoism is coolly cosmogonic in its moods and motivations (see Girardot 1983, 40–43). Ethically speaking, the issue for the Taoists is not, then, a matter of moral "accountability" which implies some absolute standard of calculated evaluation, but a question of awareness, of really knowing that all things are rooted in, and in time return to, the undifferentiated impartiality of the beginnings.

Fall and Return: How Far Back Is Enough?

If the way of creation is ever-present naturally as the *creatio continua* of all life, why is it necessary to know of previous creations or even the ultimate creation of this current world? Why and how does one have to "return to the beginning" to discover the constant principles of creation in life? Why and how do both Taoism and Confucianism agree on the need to mend the original "inborn nature" as the basic act of moral cultivation, yet differ in their evaluation of what is finally primal to human nature? Thus is raised the question of why man is no longer "just so," why he is no longer naturally and spontaneously moral.

Sometimes, because of human time, it is not so easy to become aware of, and act on, what should be the self-evident fact that the Tao glides freely within and among all existing things. In other words, by the time of the Eastern Chou period of Confucius and of Lao Tzu, something unnatural had happened, as it had in Douglas Adams' *Hitchhiker's Guide to the Universe:* "lots of the people were mean, and most of them were miserable, even the ones with digital watches." Or, as the *Chuang Tzu* (chap. 12) says, the problem was the civilizational development of a technological consciousness: "where there are machines, there are bound to be machine worries, where there are machine worries there are bound to be machine hearts." The disaster of this turn of affairs is that by following the moral dictates of a "machine heart" men and women tend to lose sight of their more primordial relations to themselves and each other: "with a machine heart in your breast you've spoiled what was pure and simple." They forget, as it says in the *Chuang Tzu,* the cosmogonic "arts of Mr. Huntun" and start to act out of a mechanical and calculated self-interest (Watson 1968, 134–36).

It seems, to paraphrase only slightly the view of Mencius (i.e., part 2 of book 3 where the *Meng Tzu* refers to the deluge period when people, like animals, made nests in trees), that the Taoists, again, were like those in *The Hitchhiker's Guide* whose opinion it was that "they'd all made a big mistake in coming down from the trees in the first place. And some said that even the trees had been a bad move, and that no one should ever have left the oceans." In other words, the mythic issue of the "fall," or the "great confusion" (*ta-luan*), emerges in China as it does wherever men and women abound and rebound (see Girardot 1983, 6–11). In keeping with the early Taoist moral imagination concerning cosmogonic and cosmological development, however, the "fall" was not precipitated by sinful disobedience of some divine imperative, but rather it "just happened" in the course of the appearance of increasingly complex sociocultural forms. It seems, in fact, that for the Taoists a fundamental alteration in the underlying intentionality of individual action coincided with the time in cultural history when disinterested *ying*ing interaction among men and women was replaced by a form of differential and deferential social accountability.

Early Taoism, while at times polemically sounding so, cannot be thought of as a nihilistic tradition that seeks to dissolve man and society permanently back into the seamless unconsciousness of nature. Rather than believing that "no one should ever have left the oceans," it is more realistically a question of the interpretation of what counts as "creation" and "world"—that is, the kind of human intentionality and social interaction that is felt to be most meaningful in relation to the ultimate structures and creative processes of existence. It is an issue of knowing what is the original basis of individual behavior and social life; what way of living and thinking most adequately fosters the well-being and maturation of men and women in relation to the existing worlds of nature and culture.

What may be called the problem of the "fall" for early Taoism is therefore the existential fact of being born into a world already predetermined as "civilizational"—i.e., the aforementioned situation that stresses the calculated advantages of a "machine heart" over the guileless "arts of Mr. Hun-tun." This kind of analysis of the "fall" is particularly emphasized in both the *Lao Tzu* and the "primitivist" sections of the *Chuang Tzu* and continues as a subterranean protest throughout later crystallizations of the Chinese imperial order. It is in this way that the lament of the T'ang dynasty Ch'an Buddhist poet, Han Shan, recalls the ancient Taoist cosmogonic theme of the death of Emperor Hun-tun found in chapter 7 of the *Chuang Tzu* (Watson n.d., 96):

> How pleasant were our bodies in the days of Chaos [*hun-tun*],
> Needing neither to eat or piss.
> Who came along with his drill,

And bored us full of these nine holes?
Morning after morning we must dress and eat,
Year after year, fret over taxes.
A thousand of us scrambling for a penny,
We knock our heads together and yell for dear life.

From the Taoist point of view the boring meddlers "who came along with their drills" to rupture the seamless integrity of Hun-tun were none other than the civilizing Sage Kings, the paragons of moral virtue for Confucianism. The creation of nature, man, and early culture was "good" because it effortlessly manifested the dynamic and relational equality of the Tao; but with the appearance of a world of ranked order, a ritual reciprocity that inculcated and maintained the privilege of the "noble way," or a moral technology premised on a permanent hierarchical system of values, something disastrous happened in the hearts and affairs of men. Most basically for Taoism, it was a matter of forgetting who we were and where we came from—once upon a time.

There is something of a counterpoint between a Taoist nostalgia for the cosmogonical behavior of the "noble savage" that depends on ultimate origins and a Confucian advocacy of a progressivist doctrine of "sacred history" that classically goes back to the first appearance of a civil order—in other words, a fundamental opposition between the "uncarved" and "carved," undifferentiated and discriminatory, cultural orders. For both it is the "creation" of a new "world"—whether primitive or civilizational—that establishes the true principles of order and meaning; and for both the issue is one of the emulation of a paradigmatic model from the hoary past. The important difference is in terms of where that past is located—in myth or history, in an undifferentiated cosmogony or a hierarchical cosmology—and how it is interpreted.

The Taoist and Confucian disagreement over the locative grammar of creation results, therefore, in a crucial difference in how the cosmological models of human behavior are described and how the problems of man are to be ameliorated. From the Taoist cosmogonic perspective the subsequent creation of the civilizational condition by the lordly heroes of the Confucian classics is seen as unnatural, disharmonic, and ultimately destructive of authentic human nature and the organismic balance between social and cosmic life. Because the hierarchical political and ritual order of civilization attempts to suppress and deny once and for all the natural values of spontaneous creativity (*tzu-jan*), and the primitive egalitarian/communalistic relativity of cultural existence (*ying*), there is a definitive break with the sacred transformational constancy of the beginnings.

Once Hun-tun or Humpty-Dumpty is smashed into ten thousand unequal and jagged shards, all the king's horses and all the king's men—all the loyal agents of feudal civilization and all the fussy gentlemen and ministers of Confucian tradition—cannot put them, or broken human nature, back together again. They tend to worry too much about Hun-tun's appearance, respectability, and virtue, and not enough about his protean visage and amorphous kindness (*shan*[a]). Early Taoism, on the other hand, knows the Way to put Hun-tun and Humpty-Dumpty back together so they can be themselves again—at least for a while.

Thus the puzzle of a fragmented Hun-tun and Humpty-Dumpty, not unlike the dilemma of shattered myth, is the fundamental salvational and moral issue posed by Taoism in its refusal to go along with the evaluative moral prescriptions of Confucianism. Similar to the Humpty-Dumpty nursery rhyme in English folk tradition, the theme of *hun-tun*, or the early Taoist cult of the cosmogonic chaos and its primitive freedom, betrays remnants of a political protest based on the imagery and form of a mythic paradigm (Stevens 1968, 67–82). From the Taoist viewpoint, the problem with the Confucians and their moral logic is that they fail to go back far enough in order to discover the real source for cosmic, natural, human, and social order. They fail to see that, like Cook Ting's rule of cutting with a "zip and a zoop" (*Chuang Tzu*, chap. 3; using Watson's translation [1968, 50]), there is no single, permanent, or all-inclusive, action-guide, only a pure action that guides and glides by slipping and sliding between the polarities of life. From the Taoist perspective, they neglect to follow the Tao's spontaneous habit of returning periodically to the undifferentiated condition. Refusing Cook Ting's advice, Confucians fail to "go along with the natural makeup, strike in the big hollows . . . and follow things as they are" (W, 51).

In sum, early Taoist tradition can be thought of as epistemologically mystic, metaphysically cosmogonic, culturally primitive, and ethically relativistic. It remains to describe more clearly some of the nature and content of the relativistic ethos of Taoism, but it can certainly be said that the significance of the Taoist appeal to the cosmogonic logic of *hun-tun* "comes from [their] presumed ability to identify fact with value at the most fundamental level, to give to what is otherwise merely actual a comprehensive normative import" (Geertz 1973, 127). To play with Clifford Geertz's formulation just a bit, the Taoist "ought to become" is felt to be connected with an original and comprehensively factual "is not yet."

Putting Hun-tun Back Together Again

I will not be able to provide all of the textual documentation that is available or pertinent, but by selectively focusing on a few representative passages from the Eastern Chou text known as the *Chuang Tzu* (ca. fourth to second centuries B.C.E.), one can show that early Taoist ideas concerning both individual and social morality, along with the regressive method for putting one's "primitive" nature back together again, are primarily grounded in mythological creation themes. It should be kept in mind that the *Chuang Tzu*, like the other early Taoist texts, is largely an anonymous compilation, sections of which date to different historical periods and situations. The earliest texts of the *Chuang Tzu* and *Lao Tzu*, moreover, do not represent a sociologically distinct or wholly self-conscious "school" of thought in ancient China. Despite such important qualifications, these texts do express a generally coherent philosophical, religious, and moral vision of life that contrasts with the position maintained in Confucian-inspired writings. Indeed, the common intentionality found in these early documents, as well as to some degree the partial continuity between these writings and the later post-Han scriptures of liturgical Taoism, can be said to be rooted in their distinctive appeal to the cosmogonic ideal of *hun-tun* (Girardot 1983).

Kith and Kin

As an illustration of what I have called the Taoist idea of the "fall" I would first of all like to draw attention to one ethnographically interesting "primitivist" passage from chapter 14 of the *Chuang Tzu* (Watson 1968, 164–65, and Graham 1981, 215; I have largely followed Graham's translation) that echoes sentiments found in the *Lao Tzu* and, despite objections from Graham, is not wholly foreign to some of the key ideas seen in the "inner chapters" of the *Chuang Tzu*. This passage is found in the context of a longer and quite famous parable concerning a meeting between Confucius and Lao Tzu. As might be expected, Confucius's initial dissertation on the significance of benevolence and righteousness (*jen, i*) is rudely denounced by Lao Tzu as only so much pretentious "huffing and puffing." Attaching too much importance to having benevolence and righteousness merely serves to "muddle the mind" and misses the point as to what actually motivates the "perfection of virtue" (*chih te*).

Upon reporting this encounter to Tzu-kung, Confucius conceded his complete befuddlement: his "mouth fell open" and "his tongue flew up" so that he "couldn't even stammer." Amazed and intrigued by his master's bewilderment, Tzu-kung decided to go and see Lao Tzu for himself. In the ensuing dialogue, Tzu-kung is concerned not just with Lao Tzu's rebuke to Confucius but with what is even more incredible, his reputed rejection of the "Three August Ones and Five Emperors," who were the foundational figures of the civilizational tradition:

> Lao Tan [=Lao Tzu] said, "Young man [referring to Tzu-kung], come a little closer and I will tell you how the Three August Ones and the Five Emperors ruled the world. In ancient times the Yellow Emperor ruled by making the hearts of the people one. Therefore, if there were those among the people who did not wail at the death of their parents, the people saw nothing wrong in this. Yao ruled the world by differentiating the hearts of the people by the sense of kinship [*ch'in*]. Therefore if there were those among the people who decided to mourn for longer or shorter periods according to the degree of kinship of the deceased, the people saw nothing wrong in this. Shun ruled the world by making the hearts of the people competitive/rivalrous [*ching*]. Therefore the wives of the people became pregnant and gave birth in the tenth month as in the past, but their children were not five months old before they were able to talk, and their baby laughter had hardly rung out before they had begun to istinguish [*shih shui*—literally, had started interrogatively to ask "who" or "whose"] one person from another. It was then that premature death first appeared. Yü ruled the world by making the hearts of the people disputatious [emending *pien*[a]/change to *pien*[b]/disputation, following Graham 1982, 55]. And the more men used their hearts for thinking, the more there was for weapons to do. Killing thieves was not considered killing, they said; every man in the world should look out for his own kind [*chung*, conveying the notion of those springing from the same ancestral stock or "seed"—Graham translates, "each man became a breed of his own"]. As a result, there was great consternation [*ta-hai* or "great terror"] in the world, and the Confucians and Mohists all rose up. When they first started there were rules of ethical behavior [*lun*[a]], but now men are using their own daughters as their wives, unspeakable!" Lao Tan continued, "I will tell you how the Three August Ones and the Five Emperors ruled the world! They called it 'ruling' but in fact they were plunging it into the worst confusion [*luan*]. . . . And yet they considered themselves sages! Was it not shameful [*ch'ih*]—their lack of shame [*wu ch'ih*]."

As the story goes, the problem is one of an interconnected process of cultural devolution that took a decisive turn for the worse when the civilizational heroes of Yao, Shun, and Yü established discriminatory standards for social life. In keeping with the classic Chinese debate over the significance of funeral ritual, this passage starts by saying that while the Yellow Emperor ruled by unifying the hearts of the people and respecting the relativity of mourning practices,[2] Yao came along and ordered the world by the ritual principles of ranked kinship (*ch'in*) so that, for example, the length of mourning had to be adjusted according to the degree of kinship. Yao's method, it should be noted, essentially reflects the Confucian understanding of *li* which stresses one's duty (*i* or righteousness) in relation to the "customary 'appropriateness' . . . of conduct to status" (Graham 1983, 4). With the coming of Shun the customary obligations and affectations of kinship gave rise to "competitive" (*ching*) hearts so that even before its first laughter, an infant had begun to distinguish one person from another. It is when children were precociously inculcated with a sense of status rivalry that the naturally regenerative powers of the life cycle were diminished and premature death appeared in the world.

Finally it is said that "Yü ruled by disputation" and logical discrimination so that in Mohist fashion "killing thieves was not considered killing." Through the constant and heightened use of discriminatory language and behavior, each man (like each word) became a breed unto himself and all relational harmony was lost to the world. Thus the world was in utter panic, and in response the Confucians and Mohists rose up to reestablish the rules of ethical conduct (*lun*[a], which should probably be distinguished from the sometimes cognate term, *lun*[b], suggesting the pure relativity of all linguistic arrangements).[3]

The ironic turn of this passage is indicated by Graham's observation that the "positions credited to Yao and Yü in far antiquity are those of the Confucian and Mohist schools respectively" (Graham 1981, 214). Thus by rising up to establish the rules of ethical conduct based on either the Confucian emphasis on the agnatic custom of *li* or the Mohist standards of utilitarian logic, they were only resowing the seeds that breed discrimination and moral tyranny. What started out on the basis of kinship regulations concerned most fundamentally with the definition and prohibition of incest only ended up, as it says in this passage, with the "unspeakable" practice of "men using their own daughters as their wives." Surely this is "unspeakable," but throughout the passage Lao Tan constantly hints that it was a kind of "speaking"—the use of a discriminating language—that created the problem of social and moral dissonance in the first place. To rise up like the Confucians in the name of a "rectification of language" founded on Yao's rule of customary obligation only helps to reestablish the causes of human alienation.

The allusion here to the relation between language and social reality, the reference to incest, and the specific linkage of mourning customs with the differentiation of kinship degrees suggest some of the anthropological prescience of the Taoist reflections on cultural development. It is in this way that Claude Lévi-Strauss's "elementary structures of kinship" serve as a helpful commentary on this passage. Comparatively speaking, a social institution such as mourning that starts out with a relatively fluid form (i.e., the practice during the time of the Yellow Emperor which may be compared with the parables of the death of Sang-hu in chapter 6 and the death of Chuang Tzu's wife in chapter 18) and with the moral purpose of maintaining a generalized egalitarian social equilibrium, is followed by traditions of ranked kinship that stress rigorous quantitative regulations on mourning structurally disposed to establish and reinforce the ancestral inequality of different descent groups (Lévi-Strauss 1969, 29–145, 311–45; on the developmental relation of egalitarian, ranked, and stratified societies, see Fried 1983, 467–94, and Kirchhoff 1968, 370–81).

At a later stage of cultural development (or structural permutation), potlatching-type contests frequently appear that decisively alter the act of ranked ritual reciprocity into an intensely "competitive" system of status rivalry (Cooper 1982, 103–28, which especially alludes to the theories of Marcel Granet). The "big man" in these situations gives everything up with the ritual intention of storing up and securing the status and honor, name and fame, of his descent group. This, it can be noted, is not unlike other traditional tales of dynastic succession in ancient China that praised Shun and Yü for their initial ceremonial renunciation (*jang*) of the world (Allan 1981). Shun and Yü were paragons of virtue from a Confucian perspective since they acted out of a spirit of ritual propriety and were therefore honored for their humility and merit. From a Taoist point of view, however, this was only a sham of true reciprocity since the language and ritual of renunciation constituted a kind of "potlatch" already structurally predisposed to insure and augment their kingly virtue. This kind of reciprocity, in other words, is marked by a "calculated generosity," since Shun and Yü gave away *only* to be paid back (*pao*).[4]

It is most important to realize that the Taoist analysis of the key factor in the turn toward cultural and personal disaster comes, ironically and latently, at the point when "hearts" were made to feel "a sense of kinship." The term used here is *ch'in*, which connotes and names, like *jen* or benevolence, a quality of shared corporate identity—those feelings, moods, and motivations of intimacy, affection, and familiarity toward one's own ancestral descent group. The root cause of rivalrous competition is, then, found at that stage in cultural development when clan fellowship and the ritual jousting of potlatch emerge as an effective and affective way to differentiate, classify, and control the increasing complexity of human life. When "other" groups are distinguished—even in the rudimentary sense of kin systems of "dual organization," and even though these ritual systems are designed to reciprocate differences and maintain a total continuity of life through marriage transactions—the germ of seeing one's own kind as "other" is already present in the human corporation (Lévi-Strauss 1969, 69–83). This is ritual that gives "face" and "name" to man and, therefore, fosters "face-work" (Goffman 1967) at all levels of human intercourse. The Classics, of course, are designed to ratify the cosmological truth of face-saving behavior based on the principles of customary accountability established by Yao.

The sociology of knowledge proposed in this passage from the *Chuang Tzu* is that the loss of primitive equality, the condition when the hearts were one, comes about in sociocultural development when men start to classify themselves into particular lineage groups, into a cosmological system of order that is formally reflected in an honorific language of respect, in systems of potlatching prestation designed to gain the return of social prestige, and in discriminating modes of thought. When a ritual system of clan-right and face-language appears, therefore, the condition when "each man becomes a breed unto himself" is sure to follow. Thus the face-making intentionality of clannishness cannot help but eventually breed selfishness because both are motivated, intended, or willed "for the sake of" (*wei*) one's own kind—whether defined collectively or individually. "By *wei* is meant," says Hsun Tzu, "the direction of one's sentiments as a result of the mind's reflections" (Munro 1969, 80, 142). The Taoist reaction to this is that it is better to woo the *wei* of action back to a more primitive way of responsive interaction.[5]

The early Taoists do not totally reject the moral significance of benevolence and righteousness. What is condemned is the Confucian effort to classify *jen* and *i* according to the prescriptive code of *lun-li* ethics which rests upon the ranked partiality and ritual formality of a certain kind of patrilinear clan reciprocity (the so-called *wu-lun,* or the five basic relationships: ruler to subject, father to son, brother to brother, husband to wife, fellow/male to fellow/male). Indeed in a passage earlier than the one discussed from chapter 14 of the *Chuang Tzu* (cf. *Lao Tzu,* chap. 38), *jen* and *i* are not completely dismissed as worthless. What is criticized is the discriminatory cast of mind and language that has restricted their meaning to one type of ritual order already based on the habits of face-saving behavior. "Perfect benevolence," like true virtue, says the *Chuang Tzu*, "knows no affection/sense of kinship" (*ch'in*) (Watson 1968, 155). The problem is the sublimated potlatching mentality of *pao*, which "prizes" benevolence as a form of calculated moral investment.

With a good deal of psychosocial justification, the Taoists say that it is wrong to cling to one ritual expression of *jen* and *i* as the potted and permanent action-guide for all human behavior. Because *jen* and *i*, like their functional mechanism of *lun-li* ritual, are but words for human values that defy customary and rational classification, it is wrong to make them into the absolute criteria of virtue. There is no one way to be *jen* and *i*, and laughing or singing at a funeral may sometimes be wholly apt and good—i.e., a proper *ying*ing in relation to that particular event at that particular time and place (cf. *Chuang Tzu*, chaps. 6 and 18). The *Chuang Tzu* (chap. 14) therefore recommends that "benevolence and righteousness" should only be viewed as the temporary "grass huts of the former kings." It is all right to "stop in them for one night but you mustn't tarry there for long" (Watson 1968, 162). Do not, in other words, carry them around like badges of honor and merit. Do not make permanent, and classically sanctified, rules that define moral cultivation; human rules and feelings, like everything else in the universe, are constantly changing. Seek instead to return to that kind of virtuous heart and mind that resonates with the transformational rule of natural life and follows the existentially malleable patterns of social affectation. This represents a primitive form of virtue, the way of *tao-te*, which "discards Yao and Shun and does not act presumptuously [*pu wei*]" (Watson 1968, 155).

One must first turn around if one is to try to retrace the steps that got one culturally and personally into the present fix. In this way, *wu-wei* is a negative behavioral principle with the positive intention of eventually re-creating the circumstances necessary for the rectification of human intentionality. *Wu-wei* is, after all, only the right kind of actively negative response, or *ying*ing rebound, to prevailing conditions. The problem therefore with some scholarly discussions of *wu-wei* is the failure to recognize that the early Taoist texts present *wu-wei* as having a broad spectrum of passive and active, individual and social, meditative and practical, implications (see Ames 1981). Like the impartial "aptness" and "goodness" of the Tao, *wu-wei* "fits" and responds to the particular situation existing at that historical and cultural moment. And when times are bad, one's initial reflex must be essentially negative in character.

Musical Mystery Tour

"Not-acting" according to the ordinary way of civil behavior is only a propaedeutic to the positive activity of *tzu-jan* and the harmonic reactivity of *ying* which, like language, came forth freely from nothing. In keeping with the seesaw mode of cosmic behavior, *tzu-jan* and *ying* describe the authentic nature of virtue (*te*) that sympathetically echoes the original activity of the Tao. Therefore, in relation to the behavior of the Taoist adept who has "discarded Yao and Shun," *tzu-jan* describes the special disinterested quality of his intentionality which is, then, the basis for his free and impartial action in the world. To be "self-so" is to act utterly in concord with one's own inborn nature and without any ulterior motives of self-interest. It is action for others which is its own reward. In the Taoist scheme of things, *tzu-jan* clearly does not refer to a totally individualistic principle of spontaneity which ignores the natural and human worlds. *Tzu-jan*, above all, represents a mode of action qualified by its *ying*ing relation to other things and men. Thus, after the first free sacrificial gift of phenomenal life, the Tao acts in and through the organic interaction of the ten thousand things. The Tao's action is always self-so, but also selfless since it only acts to enable things to be just themselves while being together.

Given the cosmogonic point of view emphasized in the Taoist texts, it is also important to see that the overall strategy of *wu-wei* specifically includes various forms of meditation that are said to accomplish a veritable psychological regression back to the undifferentiated time before Yao and Shun. But the Taoist does not return in order to rest in the dark night of chaos; rather it is a matter of reattaining temporarily the threshold necessary for freshly creative action and undulating interaction in the world and society.

In this regard it is helpful to examine another passage from chapter 14 of the *Chuang Tzu* (Watson 1968, 156–58; Graham 1981, 164–67). This passage lays out the stages of return in meditation, makes use of a number of technically significant terms (*hun, tzu-jan, ying, ho,* etc.), and couches its overall presentation in mythological imagery that refers to a specific kind of cosmogonic logic. Recalling the Yellow Emperor's rule when "minds were one" discussed in relation to the Taoist devolutionary theory, this passage portrays the way to relearn the secrets of the Yellow Emperor's virtuous method. It demonstrates that there are definite stages, a mythic pattern, musical logic, or lyrical structure, involved in the reversal of the "fallen" condition.

After performing the haunting tunes of the "hsien-ch'in music," the Yellow Emperor is questioned by a bemused bystander (who is said to be sequentially "afraid," "weary," and "confused" by the music). In explaining the effect his strangely disconcerting melodies have on his interlocutor, the Yellow Emperor says that in learning to play the music of the cosmos it is necessary to proceed successively through three stages of increasingly profound harmonic chords.[6] "Perfect music," says the Yellow Emperor, must "respond [*ying*] to the needs of man, accord with Heaven's principle, and proceed by the five virtues." But for this kind of harmony to exist in the world of man, there must be a deeper harmony that "resonates" (*ying*) with a primordial "spontaneity" (*ying chih i tzu-jan*). "Only then," it is said, can this music "bring order to the four seasons and bestow a final harmony [*ho*] upon the ten thousand things."[7]

In describing the outer manifestations of cosmic and cultural resonance, the Yellow Emperor implies that the rhythmic reciprocity of human affairs must be guided by individual inborn natures, the original transformative significance of ritual, and the primal qualitative meaning of the "five virtues." But the outer manifestation of harmony in the human world, the kind of music actually heard by men and women, expresses and depends on a second, deeper level of euphony with the inner cosmic pulsation of the two things of *yin* and *yang*, earth and heaven.

Having arrived at the experience of the cosmic vibration of duality, one is prepared to enter a third stage of soundless music which is responsible for the overall symphony of life. The sage in meditation reaches a point where he teeters (in the reverse sense of Humpty-Dumpty's disaster) on the brink of falling into the sonorous void of *hun-tun*. This is the formless stage of utter "confusion" which makes all forms of authentic harmony possible and is described here as the sage's final "intunement with the command of spontaneity" (*t'iao chih i tzu-jan chih ming*). And it is this stage, "complete" even before the creation of the two (cf. Lao Tzu's *hun ch'eng* in chap. 25), that is specifically named with the cosmogonic and embryonic *hun-tun* terminology (it "enwraps" the cosmos). It is a "clouded obscurity" or condition of mind that resembles a "dark tangled mass" (*hun sui*).

The Yellow Emperor concludes his discourse by saying: "I end it all with confusion, and because of the confusion there is stupidity. And because there is stupidity there is the Way, the Way that can be lifted up and carried around wherever you go." In rejecting a morality that rests on the evaluative mind and expresses itself classically, the Yellow Emperor is admittedly "stupid" but, from the Taoist point of view, this is the sacred stupidity of someone foolish enough to question the cosmological motivations and premises of conventional moral wisdom.

The Yellow Emperor's musical metaphor for the Taoist's way of being in the world indicates that true social reciprocity is "harmonic" *(ho)* and, within the flow of biological and social time, there will always be room for qualitatively different sounds and melodies. There is, however, no place for a permanently fixed pentatonic scale of hierarchical order since the creative ambiguity of the cosmogonic principle allows for only a quantum mechanics of social status that is constantly and simultaneously upsetting and blending all polarities (Watson 1968, 156; Graham 1981, 165): "now with clear notes, now with dull ones, the *yin* and the *yang* will blend all in harmony, the sounds flowing forth like light, like hibernating insects that start to wriggle again, like the clash of thunder."

In the light of the Yellow Emperor's muddled method, it may be that the point of Chuang Tzu's celebrated reverie about butterflies (*Chuang Tzu*, chapter 2, entitled "Ch'i Wu Lun[a]" or "The Sorting Which Evens Things Out" [Watson 1968, 49; Graham 1981, 61]) does not really rest on the priority of a cosmogonic anamnesis of cosmic unity or the diurnal dream of larval metamorphosis. Like Kafka, Chuang Tzu is interested in what happens *after* the metamorphosis. The most important sense of Chuang Tzu's dream is what "just" happens at the threshold of creation when, after hibernation, "insects start to wriggle again" and, with a "clash of thunder," the world of things appears. The hinge of Chuang Tzu's parable is when "he woke up" and rediscovered that, after all, "he was solid and unmistakably" Chuang Tzu. After the first undifferentiated chaos there continues a distinguishable "ordered chaos" or vacillating pattern of interrelated differences among all things: "between Chuang Chou and a butterfly there must be some distinction! This is called the Transformation of Things [*wu hua*]."

Eating Humble Pie

I would like to conclude this essay by offering a short ritual meditation on the *Chuang Tzu*'s amazingly pithy parable of Emperor Hun-tun, the mythological ruler of the ambiguous middle ground between all polar distinctions. This is the very last tale from chapter 7 which, to recall the chapter's title of "Ying Ti Wang," has at least something to do with the cosmogonic or *ying*ing behavior of certain legendary emperors—perhaps in this case Ti Hun-tun. Since the passage in question is exceedingly terse, it may be given in full (Watson 1968, 97; Graham 1981, 98):

> The emperor of the South Sea was called Shu, the emperor of the North Sea was called Hu, and the emperor of the central region was called Hun-tun. Shu and Hu from time to time came together for a meeting in the territory of Hun-tun, and Hun-tun

> treated them very generously [*shan*[a]]. Shu and Hu discussed [*mou*] how they could repay [*pao*] his virtue [*te*[a]]. "All men," they said, "have seven openings so they can see, hear, eat, and breathe. But Hun-tun alone doesn't have any. Let's try to bore him some" Every day they bored another hole, and on the seventh day Hun-tun died.

This story is, I believe, only a fragment of some originally more coherent cycle of creation mythology, and I have previously written about how the tale of Emperor Hun-tun informs many of the main themes seen in early Taoist mysticism (Girardot 1983). Here, however, I would like to extend my earlier analysis by indicating how the remaining bits and pieces of the story synoptically rehearse the overall moral dialectic of "primitive" reciprocity suggested in the first part of this essay.

The gist of the tale is that the two emperors of the north and south, Hu and Shu ("Fast" and "Furious" in Graham's translation), from time to time get together in the central kingdom of the strangely amorphous, roly-poly emperor known as Hun-tun. Hun-tun, as a good host, always entertained his guests "very generously." The term used here for Hun-tun's good and very apt behavior toward strangers is *shan*[a], which in this hospitable context may also allude to the cognate term *shan*[b], which has the sense of a ritually shared meal of "savory food." Indeed, the scene that is suggested seems in keeping with the very plausible notion that all forms of social reciprocity ultimately go back to the basic human act of sharing food (e.g. see Lévi-Strauss 1978, 471–95).

Aside from the immediate fellowship of their periodic gatherings, it seems that Hun-tun rather foolishly did not expect any particular return for his good graces. Hun-tun, in this respect, seems similar to the people of the "land of manifest virtue" since, having "few thoughts of self," they "knew how to give, but expected nothing in return [*pao*]" (*Chuang Tzu*, chap. 20 [Watson 1968, 211; Graham 1981, 173]). Because of their ignorance of the principle of *pao*, it is said that they do not know what "conforms to the propriety of ritual [*li*]."[8]

But as the story goes, Hu and Shu were hardly ignorant of the norms of ritual propriety and felt strongly obligated "to pay back" (*pao*) Hun-tun's virtuous actions toward them. At this point it is important to notice that the text suggests the self-interested nature of their deliberations since it uses the term *mou*, which has the sense of the "scheming or plotting" of an appropriate response. Their calculated mode of response, therefore, contrasts both with Hun-tun's innocent graciousness and with the *ying*ing reciprocity described in the immediately preceding parable that characterizes the Taoist sage as someone who "is not a storehouse of schemes" (*wu wei mou fu*).

Following their own rules of clannish reciprocity, Hu and Shu decided to return Hun-tun's humble generosity by precipitously giving him exactly what he did not need—the fixed civility of a human face. Hu and Shu, in other words, saved face by trying to make Hun-tun just like themselves, by initiating him into their ritual ways of propriety. But by worrying about what was the respectable way to respond, they only ended up by boring poor Hun-tun to death (the common fate, it must be said, of most Taoists upon encountering a Confucian gentleman). It is also at this point, of course, that primitive goodness died.

The abrupt conclusion to the parable indicates that the original sin of these well-intentioned and lordly gentlemen was that they willfully plotted their response according to the *lun-li* standards of *pao* (where north is north and south is south without the creative ambiguity of the center). By insisting that respectability was worth more than undifferentiated and spontaneous kindness, Hu and Shu in effect turned the simple human pleasures of a shared meal (when hearts can temporarily be one again), into a veritable potlatching feast concerned fundamentally with status management.

With the death of Emperor Hun-tun, the morality of *pao* as a calculated accounting of social transactions certainly represents, in Chinese tradition, an effectively pragmatic approach to the constant imbalance of civilized human intercourse. The question remains, however, if such an understanding of the evident partiality of human behavior really represents an ultimate basis for morality. In this sense, Taoism's remembrance of Hun-tun's foolish behavior, like modern anthropology's pursuit of the primitive, reminds us that we cannot simply accept traditionally respectable answers. Above all, the search for the beginnings of nature and culture, as well as a concern for the act and significance of creation, represents a primal quest for the essential roots of human morality, an attempt to define a lost human potential.

And even if it is fantastic to imagine that men and women might actually return to the "land of manifest virtue," Taoism suggests that we may at least rediscover the middle kingdom within our own minds and bodies (Schipper 1982). At the very end, Taoism takes us back to the beginning so that we might know how to start living in the world again. Thus, it may finally be said that the early Taoist emphasis on "behaving cosmogonically," modeled on an awareness of the faceless generosity of Hun-tun, expresses what Lévi-Strauss (1978, 508) has called the "hidden ethic" of primitive myth. This is an ethic that affirms that a "sound humanism does not begin with oneself, but puts the world before life, life before man, and respect for others before self-interest."

Notes

1. Some Chinese commentators feel that these lines are misplaced and belong after the first three verses in chapter 63 that refer to the principle of *wu-wei* (see Ch'en 1977, 272).

2. Graham feels that the "primitivist" sections of the *Chuang Tzu* characteristically elevate Shen Nung and condemn the Yellow Emperor along with Yao, Shun, and Yü. While this is most often the case in the "primitivist" sections, my reading of the passage in chapter 14 would identify the rule of the Yellow Emperor with the harmonious regime of Shen Nung before the decline (and it should be noted that in chapter 20 Shen Nung and the Yellow Emperor are linked together as the rulers of the paradise time).

3. *Lun*[b] often has a favorable connotation in the Taoist texts so that, for example, chapter 2 of the *Chuang Tzu* is entitled "Ch'i Wu Lun[b]" or "The Sorting Which Events Things Out" (see Graham 1981, 48). It should also be noted that the collected discourses of Confucius are entitled "Lun[b] Yü."

4. David Nivison (1976) has recently tried to show that *pao*-reciprocity must be seen as the essential functional correlative of *te*[a]-virtue (as well as the interrelated idea that by being virtuous, one "gets"—*te*[b]—an augmented return on one's investment). But the brunt of his argument, based on a comparative, philological analysis of the Shang oracle-bone inscriptions, is that already in the earliest civilizational period of the Shang-Yin, the interrelated ideas of *te*[a] and *pao* referred basically to a kind of religio-political reciprocity that was foundational for the protofeudal system of royal prestige and authoritarian power.

Already during the Shang period (ca. second millennium B.C.E.), and certainly during the Chou, it seems that in aristocratic circles virtue by means of *pao* was specifically a social mechanism for maintaining unequal political status. Along with other interconnected nuances, the *te-pao* linkage in the "world" of civilization signified the "special concrete relation between Lord and vassal." Especially pertinent in this regard, and not developed by Nivison, is the "potlatch" idea that superior prestige and virtue is established and fixed when the magnitude of the gift creates an impossible debt on the part of the recipient. Thus self-interest is furthered in an exchange situation where the gift carries with it the largest possible interest on its return.

In practice the moral idea of yielding, or *noblesse oblige,* tended to be a political fiction for having and keeping both the world and honor, a pretense for maintaining the status quo either as a feudal aristocrat or as a moral paragon. Since the "original" civilizational model of political or moral order was inherently a permanent hierarchical relationship between unequal parties, both the political aristocracy by birth and the Confucian aristocracy by merit were always playing with a stacked deck.

5. The cosmogonic context for this is suggested by Munro (142) when he remarks that: "the events that occur in the universe because of Tao are not consciously or purposively done, as could be said of the natural events decreed by an anthropomorphic deity; they simply unfold in accordance with the laws of change. Similarly, there should be no goal-directed conduct by men. Tao impartially 'produces' all things, or determines all changes; it has no favorites. . . . The man who models himself on Tao should also be disinterested and impartial."

6. This pattern may be compared with the *Lao Tzu*'s reverse cosmogonic pattern of man, earth, and heaven, *tao/tzu-jan* in chapter 25; the creation sequence of 1–2–3–10,000 in chapter 42; and in chapter 2 of the *Chuang Tzu* the "piping of man, earth, and heaven."

7. These lines are often felt to be a commentator's interpolation into the text (see Watson 1968, 156).

8. The relation of *pao* and *li* is brought out in the *Li Chi,* which says that: "In the highest antiquity they prized (simply conferring) good; in the time next to this, giving and repaying [*pao*] was the thing attended to. And what the rules of propriety [*li*] value is that of reciprocity [*pao*]." Quoted by Yang Lien-sheng (1969, 3).

Glossary of Chinese Terms

chih te 至德

ch'ih 恥

ch'in 親

ching 競

chung 種

fa 法

fan 反

fu 復

ho 和

hun ch'eng 混成

hun sui 混逐

hun-tun 混沌

i 義

jang 讓

jen 仁

kuei 歸

li 禮

luan 亂

lun[a] 倫

lun[b] 論

mou 謀

pao 報

pien[a] 變

pien[b] 辯

pu wei 不為

shan[a] 善

shan[b] 饍

shih shui 始誰

ta-hai 大駭

tao-te 道德

te[a] 德

te[b] 得

t'iao chih i tzu-jan chih ming 調之以自然之命

tzu-jan 自然

wu ch'ih 無知

wu hua 物化

wu-wei 無為

wu wei mou fu 無為謀府

ying 應

ying chih i tzu-jan 應之以自然

References

Adams, Douglas
1979 *The Hitchhiker's Guide to the Galaxy.* London: Pan Books.

Allan, Sarah
1981 *The Heir and the Sage, Dynastic Legend in Early China.* San Francisco: Chinese Materials Center.

Ames, Roger T.
1981 "*Wu-wei* in 'The Art of Rulership' Chapter of *Huai Nan Tzu:* Its Sources and Philosophical Orientation." *Philosophy East and West* 31:193–213.

Bauer, Wolfgang
1976 *China and the Search for Happiness.* New York: Seabury Press.

Chan, Wing-tsit
1963 *The Way of Lao Tzu.* Indianpolis: Bobbs-Merrill.

Chang, Kwang-chih
1976 *Early Chinese Civilization: Anthropological Perspectives.* Cambridge: Harvard University Press.

Ch'en Ku-ying
1977 *Lao Tzu: Text, Notes, and Comments.* San Francisco: Chinese Materials Center.

Chiang, Hsi-ch'ang
1970 *Lao Tzu Chiao Ku* (Lao Tzu Collated and Explained). Taipei: Ming Lun.

Cooper, Eugene
1982 "The Potlatch in Ancient China: Parallels to the Sociopolitical Structure of the Ancient Chinese and the American Indians of the Northwest Coast." *History of Religions* 22: 103–28.

Egan, Ronald C.
1977 "Narrative in *Tso Chuan.*" *Harvard Journal of Asiatic Studies* 37:323–54.

Eliade, Mircea
1968 *Myth and Reality.* New York: Harper Torchbook.

Fried, Morton H.

1983 "Tribe to State or State to Tribe in Ancient China." In *The Origins of Chinese Civilization*, ed. David N. Keightley, pp. 467–94. Berkeley: University of California Press.

Geertz, Clifford

1973 "Ethos, World View, and the Analysis of Sacred Symbols." In his *The Interpretation of Cultures*, pp. 126–41. New York: Basic Books.

Girardot, N. J.

1976 "The Problem of Creation Mythology in the Study of Chinese Religion." *History of Religions* 15: 289–318.

1983 *Myth and Meaning in Early Taoism*. Berkeley and Los Angeles: University of California Press.

Goffman, Irving

1967 *Interaction Ritual: Essays on Face-to-Face Behavior*. Garden City: Anchor Books.

Graham, A. C.

1981 *Chuang Tzu, the Inner Chapters*. London: George Allen and Unwin.

1982 *Chuang Tzu, Textual Notes to a Partial Translation*. London: School of Oriental and Africa Studies.

1983 "Taoist Spontaneity and the Dichotomy of 'Is' and 'Ought.' " In *Experimental Essays on Chuang Tzu*, ed. Victor Mair, pp. 3–23. University of Hawaii Press.

Hugh-Jones, Stephen

1979 *The Palm and the Pleiades, Initiation and Cosmology in Northwest Amazonia*. Cambridge: Cambridge University Press.

Kirchhoff, Paul

1968 "The Principles of Clanship in Human Society." In *Readings in Anthropology*, ed. Morton Fried, pp. 370–81. New York: Thomas Y. Cromwell.

Lévi-Strauss, Claude

1969 *The Elementary Structures of Kinship*. Boston: Beacon Press.

1978 *The Origin of Table Manners: Introduction to a Science of Mythology* Volume 3. New York: Harper and Row.

1966 *The Savage Mind.* Chicago: University of Chicago Press.

Major, John S.
1978 "Myth, Cosmology and the Origins of Chinese Science." *Journal of Chinese Philosophy* 5: 1–20.

Maspero, Henri
1924 "Légendes mythologiques dans le Chou king." *Journal Asiatique* 205: 1–100.

Mote, Frederick
1972 "The Cosmological Gulf between China and the West." In David C. Buxbaum and Frederick Mote, eds. *Transition and Permanence: Chinese History and Culture,* pp. 3–21. Hong Kong: Cathay Press.

Munro, Donald J.
1969 *The Concept of Man in Early China.* Stanford: Stanford University Press.

Nivison, David S.
1976 "Can These Bones Live: The Concept of Royal Virtue in Shang China." Paper presented at the Workshop on Classical Chinese Thought, Harvard University.

Pas, Julian
1981 "Chuang Tzu's Essays on 'Free Flight into Transcendence' and 'Responsible Rulership.' " *Journal of Chinese Philosophy* 8:479–96.

Plaks, Andrew H.
1976 *Archetype and Allegory in the Dream of the Red Chamber.* Princeton: Princeton University Press.

Prusek, J.
1970 "History and Epics in China and in the West." In his *Chinese History and Literature,* pp. 17–34. Dordrecht: D. Reidel.

Robinet, Isabelle
1977 *Les commentaires du Tao To King jusqu'au VII siècle.* Paris: Institut des hautes études chinoises.

Sahlins, Marshall D.
1968 "On the Sociology of Primitive Exchange." In Michael Banton, ed., *The Relevance of Models for Social Anthropology,* pp. 139–236. London: Tavistock.

Schipper, Kristofer
1982 *Le corps taoiste.* Paris: Fayard.

Stevens, Albert Muson
1968 *The Nursery Rhyme.* Lawrence: Coronado Press.

Stoppard, Tom
1972 *Jumpers.* New York: Grove Press.

Watson, Burton
n.d. *Cold Mountain, 100 Poems by the T'ang Poet Han-shan.* Taipei: Wen Feng.
1968 *The Complete Works of Chuang Tzu.* New York, Columbia University Press.

Yang, Lien-sheng
1969 "The Concept of *Pao* as a Basis for Social Relations in China." Pp. 3–23 in *Excursions in Sinology.* Cambridge: Harvard University Press.

4 *Above, Below, or Far Away: Andean Cosmogony and Ethical Order*

Lawrence E. Sullivan

Hymn to the Creator

O Viracocha, Lord of the world,
whether you are male or female,
you are for certain the one
   who reigns over heat and creation,
the one who can work charms with his saliva.
Where are you?
I wish you were not concealed from these your sons!
Perhaps you are above, perhaps you are below us,
perhaps you are far away in space.
Where is your mighty tribunal?
Hear me!
Perhaps you dwell in celestial waters
or in the waters beneath the world
   and on their sandy
shores.
Creator of the world,
Creator of mankind,
great among our ancestors,
before you,
my eyes grow faint
although I long to see you—
for if I see you,
get to know you,
listen to you
and understand you,
you will see me
and get to know me.
Sun and moon, day and night,
summer and winter,
they wander not in vain
but in prescribed order
to their determined place,
to their goal.
They arrive
to wherever
you lead

with your royal staff.
O hear us,
listen to us,
let it not be that we tire
and die.
O victorious Viracocha,
ever-present Viracocha!
You are unequalled on earth.
You exist from the beginning of the world
to its end.
You gave life and courage to man when you said
"Let this be a man,"
and you gave the same to woman when you said
"Let this be a woman."
You created us and gave us a soul.
Watch over us that we may live in health and peace.
You who may be in the highest heavens among storm clouds
give us long-lasting life,
and accept our offering,
O Creator.

(From Means 1931, 437ff.)

It is the argument of this essay that the Incas developed a profound ontology of periodicities and spatialities into major principles of ethical order. That ethical order is a prescribed experience of the world available in ritual. Far from a cosmogony which places the center of moral gravity over epistemological categories and the question of proper moral reasoning which adjudicates the principles, premises, and processes of the knowledge of good and evil, Andean cosmogony appears to govern, through ritual, the structure of a community's experience of that which is "true" and "real." The result is an ethical emphasis on the discernment of proper relations of individual to group and group to cosmos in space and time. Divining the origins and meanings of these relations determines the moral quality of one's experience and perception. The consequences of improperly structured experience or misperception of cosmic and social relations are not limited to an individual mistake in judgment. They possess dire cosmic effects: infertility of crops, animals, and humans; famine; epidemic; catastrophe. The moral dilemma of improper experience of space and time and the consequences for failure to structure properly the experience of the cosmos are recounted in Andean cosmogonies. The creation myths also reveal the bases of ritual-ethical structures necessary for proper experience and the meanings attached to that experience in ritual. Andean cosmogonies do not provide, in one-to-one fashion, ethical action-guides or contents of ethical codes.

The peoples of pre-Columbian America, especially of Mesoamerica, invested great energy in the development of calendars which provided criteria for divining and ordering proper individual and group behavior. Recently, scholars have confirmed the complexity of the Inca calendar. Although reliable chroniclers were firm in the conviction that the Inca possessed sophisticated calendric schemes (Poma de Ayala 1966, 883, 884; Acosta 1954), little investigation of them was made until well into the twentieth century (Escobar 1967; Zuidema 1977). This essay illustrates connections between Andean cosmogony and the principles of ethical order which organized the Inca capital city of Cuzco, where sociopolitical hierarchies and individual and group behavior were bound up with the experience and interpretation of spatial dispositions and temporal relations manifest in the creation of the cosmos. Examined are several cosmogonic themes which recent investigators argue are pan-Andean: the four ages and three crises of creation; the concept of the mountain body; the dialectical qualities of the celestial creator-god; the cosmogonic role of Holy Mother Earth; and the notion of *Pacha* as a time/space continuum in the *mesa*, a microcosm of powers at work in the universe. The essay then considers the calendar as a systematic ontology of protean elements manifest in the cosmogony. Finally, it looks at the city of Cuzco as a spatial expression of the principles of the calendar: an ordered embodiment of the cosmogonic themes which govern domestic residence, institutional relations, and proper behavior.

Cosmogony and Ethical Order

It is not known at what time the systematics of creation first fascinated the populations of the Central Andes. The earliest chronicles already evidence both a sophistication and a variety of creative scenarios (Rowe 1960; Valcarcel 1964, 2:366ff.). Recent scholars have shed the light of textual and historical criticism on the chronicles of the conquest period (Duviols 1966, 497–510; Duviols 1971) and the documents recorded by explorers and ethnographers over the subsequent four hundred years. By disclosing patterns of cosmogonic thought they render possible the interpretation of their meanings. When Juan Jacobo Tschudi, Julio C. Tello, and Luis E. Valcarcel began their investigations of the archaeology and ideology of ancient Peru in the late 1910s and 1920s, there was little consensus about the substrata of religious ideas prior to the Inca Empire. The history of the Inca tribe itself and the story of its imperial expansion were poorly understood. The growth and structure of its pantheon and the location of Inca religious ideas vis-à-vis those of their geographic and temporal neighbors—of Chimu, Tiahuanaco, Pachacamac, and Huarochiri—had yet to be explored.

In *El Dios Creador Andino*, Franklin Pease (1973) compiles materials on Andean creation and creator gods. He extracts the most ancient versions of the myths from sixteenth-century chronicles (Betanzos 1924 [1551]; Cieza de León 1967 [1550]; Sarmiento de Gamboa 1947 [1572]; Molina 1943 [1575]) and sets them apart from the seventeenth-century versions recorded during the Inquisitional "Extirpation of Idolatries" by Indians strongly influenced by the new image of the world brought with the conquest (Poma de Ayala 1966). Pease (1973, 12) contends that the earliest group of myths offers access to the most ancient Andean cosmogony, one which antedated the Inca state of the fifteenth century. He presents also a third group of cosmogonic variants kept alive in the oral traditions of Andean peoples and recorded by ethnographers between 1891 and 1972.

For the four earliest chroniclers, Cuzco, the Inca capital, was the central focus. It was here that they gathered their information on Inca cosmogony and on the cosmogony of subject peoples in the highlands and on the coast. The following is a creation narrative recorded by Juan de Betanzos in 1551 and summarized by Luis E. Valcarcel.

> 1. In ancient times they say the earth and the province of Peru were in darkness, and that there was neither light nor day upon it. 2. And that in those times this earth was entirely night, they say that Con Titi Viracocha came forth from a lake. 3. That he went to Tiaguanaco. . . . He made the sun and the moon, and that he sent the sun to go along the course which it follows; and then they say that he made the stars and the moon. 4. Previously, he had already created another people at that time when he made heaven and earth. This people had done a certain disservice to him, and in punishment for this annoyance which they did, he ordained that they be turned into stone on the spot. 5. In Tiaguanaco he made a certain number of peoples from stone and a chief who governed and ruled them and many pregnant women and others who had lately given birth; and that the children have their cradles, as is the custom; all this made of stone. 6. He ordained that they disperse, except for two who stayed with him, and whom he instructed: 7. "These will be called such and such and will come forth from such and such a province and will live in it and there they will grow in number and these others will come forth from such and such a cave and will call themselves such and such, and will inhabit such and such a place, and just as I keep them here painted and made of stone, so must they come forth from the springs and rivers and

> caves and mountains, in the provinces which I thus tell you and which I name to you and you will go straightway to this place (pointing toward where the sun rises) separating them, each one unto itself, and making known to them the law they must observe (literally: 'showing them the direction they must take . . .')." 8. They went forth (the messengers of Viracocha) calling forth and bringing out the peoples of the rivers, caves, and springs and highlands and peopling the earth in the direction of the place where the sun rises. 9. From Tiaguanaco, Con Titi Viracocha sent his two companions forth in this way: a) the one by way of the place and province of Condesuyo, the shoulders where the sun rises, on the left hand, b) and the other by way of the place and province of Andesuyo which is on the other, the right-hand side. 10. And he went straightway to Cuzco, which is in the center of these two provinces, coming along the royal road which runs along the mountain range toward Caxamalca. 11. Upon his arrival at Cacha (eighteen leagues from Cuzco), the Canas Indians came out bearing arms with the intention of killing him, without recognizing him. 12. Viracocha made fire rain down from the sky, and a ridge of mountains was burned. 13. Having seen this, the Canas dropped their weapons and threw themselves to the ground. 14. Seeing them do this, Viracocha "took a staff in his hands and went to where the fire was and struck it two or three times," extinguishing it. 15. Viracocha made himself known as a god, and the Canas paid him homage, raising up a sumptuous *huaca* to him. They carved his image out of a huge stone of almost five staffs in length and one staff in breadth, give or take a little. 16. Continuing along his way, Viracocha arrived at Tambo de Urcos, and ascending the highest mountain, he sat down there to rest. From that place there came forth those who put in that place a bench of gold in memory of the god. 17. In Cuzco he made a lord whom he called Allcaviza and he gave his name to the people. 18. He continued his way as far as Puerto Viejo, where he rejoined his two messengers, and with them he went to the sea "where he and his messengers went by way of the water just as they had gone along the earth." (Valcarcel 1964, 2:367–69).

Pease believes that a standard structure of the creation myth may be extracted from Betanzos's text when taken together carefully with the versions recorded by Cieza de León, Sarmiento de Gamboa, and Santa Cruz Pachacuti. These additional chroniclers make much of an event not mentioned by Betanzos—a primordial flood which destroys the world. The Cuzcan versions share a number of features. They are decidedly not solar. They speak of several successive primordial creations by Viracocha, who disappears after each one. An unnamed offense provokes Viracocha's reappearance for a further creative episode. Viracocha moves from the south (Collasuyu) near Lake Titicaca to the north (Chinchaysuyu) and involves himself directly in the creation of these sections, whereas the creation of the quarters called Antisuyu and Contisuyu is presided over by his two helpers. In this one sees in germ the predominance, from the beginning of creation, of the Andean twofold relative division into Hanan ("upper") and Hurin ("lower") which exists at every level of social organization in the Andes from the ayllu (descent group) to that of Tiwantinsuyu, the Inca empire of the fifteenth century. Further, this dual social organization, from its mythic beginnings, offers place of privilege to the moiety whose residents Viracocha personally calls into existence.

Most striking in these variants is the cyclical nature of creation: the four epochs of the world. "Between these ages there existed periods of *chaos*, which can be characterized by the different destructions which the divinities brought about: conversion into stone [earthquake?], water [deluge, flood], and fire" (Pease 1973, 18). This structure of the four epochs is found even in the *Huarochiri* manuscripts and in the seventeenth-century chronicles of Guaman Poma and Santa Cruz Pachacuti. In the *Huarochiri* texts, each of the four ages is connected with a god who acts as a protagonist in the three successive cosmic struggles and periods of destruction. Yanamca Tutañamca is defeated by Huallallo Carhincho; then Pariacaca destroys the world with fire. Finally there appears Cuniraya Viracocha, who bears "an evident Cuzcan influence" and who is linked to the "master weavers" (Pease 1973, 19). I shall have occasion further on to mention the importance of textile-weaving in the formation of the Inca state and its important place in the expression of Inca cosmology (Murra 1962, 710–25). Guaman Poma provides the names of the four ages: Uariuiracocharuna, Uariruna, Purunruna, and Aucaruna (Poma 1966, 49ff., and Pease 1973, 19). Santa Cruz Pachacuti speaks also of four ages: Purunpacha, Ccallacpacha or Tutayapacha, Purunpacha raccaptin, and the age of Tonapa Viracocha (Santa Cruz Pachacuti 1950, 209–11; Pease 1973, 19). In the seventeenth century, Antonio de la Calancha gathered a collection of creation texts all containing a sequence of creative epochs (Calancha 1938, 1:412–13). Even those texts of the Peruvian coast which do not treat of a celestial god like Viracocha but deal instead with the chthonic dema Pachacamac ("lord of the earth" or "lord of the inner earth") testify clearly to four ages of the world's creation with their three respective catastrophes (Calancha 1938, 1:412–14).

Pease concludes that this structure of creation is a most ancient one in the Central Andes. For the purposes of the investigation of the relationship of cosmogony to the principles of Inca ethical order, several points bear recapitulation. First, the three worlds which came to be destroyed were governed by three different principles marked by a common ontological quality: they were of unbroken duration. Each age was governed by a single principle: by unbroken darkness, or by unbroken light, or by the perpetuity of stone. In contradistinction to the fourth world marked by the dynamism characteristic of human action, the first three primordia were temporarily static. The only change possible was the total change from order to chaos—from unbroken structure to destruction. Any change effected the undoing of the univocal governing principle. Totality and chaos stand in necessary relationship as antipodes to the principles marked by unbroken duration. Second, the first three worlds were not only temporally unbroken but were spatially totalitarian. One by one, each eternal principle (light, stone, darkness) ruled unbounded and unchallenged in space: there was no other space where a different principle of being reigned. Third, only during the fourth creation—that of the human world—do we see the appearance of a space fragmented into Hanan-Hurin, "lower-upper," Antisuyu, Contisuyu, Cinchaysuyu, and Collasuyu with their peoples of different kind. Likewise, only in this last world of human inhabitants do we encounter the existence of time fractured into periodicities governed by the spatial dispositions of sun, moon, and stars and their manifestation as rain, clouds, drought, hail, lightning, night, and day. Fourth, although later I will argue that the principles of ethical behavior manage communal experience, perception, and meaning so that, in some sense, they reflect the order of the cosmos, we must here keep in mind that it is the order only of the *last world*—the one which survives—that proper ethical order reflects. In fact, there is a real conflict—a total antagonism—between the world in which humans behave and the worlds of the first primordia. The relationship between ethical order and the early primordia is not a simple one. They are not mirror reflections or inverses of one another; neither antithetical nor complementary opposites. The order of the last world is an algorithmic combination of protean principles found in the first three worlds—now, however, governed by the very ontological quality the first primordia could not support: dynamism. In other words, the fourth creation is made to accommodate change. The principles of ethical order, therefore, are born of the paradox of reflecting an order in flux. The protean beings of the early primordia still appear but now become subject to temporal periodicity and spatial bounds: light by day alone, darkness by night alone, primordial stone only in the huacas. Fifth, the primordia of "unbroken duration" are, ironically, destroyed. Insofar as they now exist, they are vestiges in the world of finite periodicities of day, night, and the human life-cycle. Of themselves, the primordia were unique and unreplicable. In the present world, their cycle resembles the human pattern of multiple births, deaths, and rebirths. Thus, paradoxically, unchanging beings embrace the change inherent in human creation and avoid total destruction brought on by their eternal nature. If before, totality and chaos were the antipodes marking closure of the primordia, now finiteness (death and boundedness of the body, the fixed course of the sun in its rising and setting, the "direction which humans must take up," etc.) and generation (fertility, regeneration, transmigration of the soul, cosmic renewal in the new year) make new apertures of old closures so that the world made up of finite places and times may continue infinitely. We shall see the enormous ethical burden pitched upon humans who reflect this cosmic order which in turn reflects them. Improper behavior may have disastrous consequences for the cosmos—famine, flood, death, disease, infertility, prolonged eclipse. In a sense, the continued fruitful interplay of cosmogonic elements in the cosmos is dependent upon the history of proper human action, which relies, in turn, on observance of cosmic order. Cosmogonic order and human ethical behavior relate dialectically as two necessary antipodes which serve as passages out of the chaos and finitude inherent in their respective natures.

Feeding the Mountain Body

Although space is fragmented in the fourth creation, it is not without integrity. The creation narrative recorded by Betanzos implies a specific order in this world: that of a body complete with shoulders, left hand, right hand, upper and lower parts, and so on. In fact, the organization of the world into a body form is one with a long history in the Andes (W. H. Isbell 1977). What is at best implied in Betanzos's version is made explicit in the ethnography of Joseph Bastien, who studied an Andean community of the Bolivian highlands in 1968 and 1972 (Bastien 1978). By consulting post-conquest chronicles, especially the *Huarochiri* manuscripts, he studied the history of cosmogonic ideas found in the village of Kaata. Comparing them further with other contemporary ethnographies throughout the Andes, he concludes that there is a pan-Andean cosmogonic theme of great historical depth which conceives of the ayllu (the community which lives in hamlets spread across one mountain) as a unified body.

John V. Murra (1972) has demonstrated that Andean social, political, and economic strategy demands control of as many distinct ecological niches as possible. Since altitude is the most dramatic ecological variable in the Andes, Murra shows how a system of "verticality" unites diverse ethnic groups in a network of kinship, exchange, marriage, and migration. A mountainside provides pastures for herds in the highlands (13,500–16,000 feet), rotative fields for oca, barley, and potatoes in the intermediate zones (11,500–13,500 feet) and agricultural areas for corn, wheat, peas, and beans in the lower zones (10,500–11,500 feet). The mountain community (ayllu) encompasses all three zones.

"Kaatans are aware of the body metaphor for their *ayllu*, and most agree on the location of the head, mouth, tits, heart, bowels, legs, and toenails" (Bastien 1973, 169). The three major ecological zones correspond to the three planes of the cosmos and three sections of the body. Uma pacha ("original time and place") is the head, the highlands, from which all souls and beings have their origin. The human soul emerges from the uma pacha, journeys down the outside face of the mountain during life (like the water courses), then in death transmigrates up the inside of the mountain along subterranean waterways to emerge once again from highland springs (puqyos) or caves (pacarinas) considered to be the "eyes" and "mouth" of the body.

The mid-section (uqhu) houses the important inner organs of heart and bowels. Their products, blood and fat, "are the principles empowering things: blood is the life principle and fat is the power principle" (Bastien 1973, 170). In this community *Wiraqocha* means "sea of fat." During one agricultural rite, a llama is brought from the highlands and chicha (corn-brew) from the lowlands to the midlands where a ritual president sacrifices the llama. Its heart, liver, pancreas, and stomach are read to ascertain the vitality of agricultural life (blood) and political power (fat) in the ayllu (Bastien 1973, 172–73). The chicha is poured as libation and also consumed. Bastien shows how a number of similar rituals maintain the unity of the ayllu and the life of the mountain body by feeding its parts with blood, fat, coca, chicha, tobacco, and other food offerings.

Throughout the year the head, heart, arms, and legs are fed in ritual at shrines (huacas) located all across the mountain-body. In addition to those major ayllu feasts, individuals and domestic units observe similar feeding rites for countless numbers of important earth shrines (huacas) erected near the mouths of springs and near rivers in memory of the dead, or on the occasion of misfortune and sickness, or at marriage, birth, and baptism. The rivers which form the boundaries of the mountain-body and which give it corporal integrity allow for the circulation of life principles deposited in them as offerings. It is the circulation of life principles at shrines and in the waters which perpetuates the passage of time because the waters which descend the mountain from their origins in the uma also ascend the innards of the mountain-body. The river becomes an image for the paradoxical experience of history in the Andean world view: "the consecutive, passing-by-places, sense of the river washes away misfortunes; and by another cyclical experience . . . the river returns what has been washed away, again completing the body" (Bastien 1973, 224).

Understanding themselves and the space they inhabit to be created as one body, the members of an Andean ayllu consider it morally necessary to feed regularly the parts of the mountain-body. By offering blood, fat, and chicha beer to head, arms, heart, bowels, legs, and toes, the circulation of life and power is maintained. The parts are fed on the macro-schedule of the agricultural seasons of plowing, planting, and harvesting and also on the micro-temporal schedules of individual life-cycles. For the good of the whole, each individual, and each hamlet, must feed the huacas in its charge in order to perpetuate the circulation of time along the rivers of death and fertility and regeneration. Ritual feeding of the spatial metaphor—fragmented into body parts—makes possible its integrity and continuity in time.

The Moments of Viracocha

Rudolfo Kusch observes that the creator Viracocha is himself fractured in the process of creation (1962). We have seen above that human ethical principles are ordered by a complex relationship to the cosmogonic order resulting from the three destroyed primordia and the present creation. In a similar way, Viracocha's own structure is reordered as a consequence of his successive creations.

The native chronicler Santa Cruz Pachacuti, an Inca noble, considers this consequential structure of Viracocha under the heading of the five *unanchan* ("standards, or signs"). Kusch argues that the Andean view of the creator and creation is of an order which is dynamic in the extreme: a dialectic of five internal relations—the five standards. Under the first standard, Viracocha is "Artificer of the world" (*Pachayachachic*). In this aspect he is the teacher whose knowledge of the world's structures precedes their actual existence. "This supposes that the world and god were opposites" (Kusch 1962, 28). Secondly, Viracocha is "fundamentally rich" (*ticci capac*). He was rich in potency, containing within his being the "conditions for the richness of the world" (Kusch 1962, 28). The world will be rich because Viracocha is. Without him the world would be a "boiling spring" (*manchay ttemyocpa*). Viracocha's richness, though bountiful, is ordered and stands over against the chaos of the world. Under his third aspect, Viracocha is "Tunupa, who is the world in being" (*Ttonapapa-cha (ca)yaspa*). This is the moment of the "march of the god over the earth," calling into actual existence the world order which had remained formal potency in him in the first two moments. Under the fourth standard, Viracocha is sexually dual (*cay cari cachon cay uarmi cachon*). Just as standards one and three are related as potent inner order of the god to manifest world order, in the same way his "fundamental richness" (standard two) relates to sexual duality as potential richness to manifest order of fruitfulness. Generation is not without order: androgyny. Furthermore, the order of sexual duality is cosmic and total, deriving as it does from the very being of Viracocha. There are references to the auto-copulation of Viracocha necessary to the continued existence of the world in several hymns recorded by Santa Cruz Pachacuti as well as in the lingam-in-yoni theme found in depictions of the *orcorara* ("penis-vulva") on the Coricancha (Temple of the Sun) in Cuzco and the Echenique Plaque worn as a medallion by the Inca high priest. The temporal cycles and spatial alignments of continued existence are suffused with a cross-cutting sexual order which must be observed. Under his fifth sign, Viracocha is the "fundamental creative circle" (*ticci muyu camac*). The dialectic correlations of the first four moments, the ordered play of their tensive pairs (1–3, 2–4), point to this fifth moment of Viracocha's full realization as a cosmic flower (*tega*), oriented toward the four fundamental points of space, the four winds, and the four gods of the horizons. "In the image of the flower, Viracocha fills qualitative space, a space-thing which regulates the cosmos" (Kusch 1962, 36). In this moment of manifest relationship of Viracocha to the world, claims Kusch, there enters the moral concept of necessity (the regulation of needs) and the frustration of impulse whether it be human or divine (1962, 37). Both Viracocha and humanity stand opposed to the chaotic possibilities of the world—the boiling spring. They govern it by the principles derived from their own structures: the five standards and the human life-cycle—thought, fecundity, material expression, sexuality, spatial and temporal disposition. It is Kusch's conviction that the principles of moral order were embodied spatially in the Inca capital of Cuzco as *ceques*, "threads" of shrines (huacas) radiating in straight lines from the Coricancha sun-temple in the center of the city to all four quarters (suyus) of creation. We shall have occasion to see how these spatial expressions of the bases of moral order were interwoven with the temporal bases of moral order which derive from the dialectical "moments" of Viracocha's internal relations and his creation.

Pachamama: Holy Mother Earth

The study of the telluric Earth Mother of the Andes complements and complicates the cosmogonic order derived from uranic creators like

Viracocha. Sexual duality, resulting from Viracocha's androgynous richness, expresses itself at every level: from the cosmic to the quotidian. In the rush to expound artistocratic religious notions of the Inca imperial overlords, emphasis in research has fallen on celestial divine figures of the moiety created by Viracocha himself and controlled by his celestial hypostases and associate deities. In the conceptions of myth, this moiety reckons its descent from a mythical ancestor descended, in a distant past, from the "son of the sun" who had come there from another region to form a moiety identified with the masculine principle. Often ignored are the facts which indicate that *Pachamama* (Mother Earth) "was taken to be the sister of this god and ancestress of those groups which believed themselves descended from a progenitor emerging from the very ground where they live. They constitute another moiety, symbolically identified with the 'lower' half and the feminine principle" (Mariscotti de Görlitz 1978, 228). Exactly which moiety Viracocha is identified with in myth is a complicated matter conditioned by the interests, lineage, and social position of the chronicler and his subject (Zuidema 1964, 114–66, and passim).

Ana Maria Mariscotti de Görlitz argues that Pachamama is set beneath the great celestial gods (Viracocha, Inti) in the hierarchy of the Andean pantheon. However, Pachamama and the other goddesses who appear to be functional hypostases or local personifications of her, were prototypical partners with other male celestial divinities in a *hieros gamos* (Mariscotti de Görlitz 1978, 227). Mariscotti de Görlitz ranks Pachamama as hierarchically equivalent to the lord of meteorological phenomena—a polymorphic fertility god who personifies lightning, thunder, rain, hail, and snow and who appears often with ornithomorphic attributes. Notwithstanding the fact that he descends from the theologically important Viracocha, and the imperially imposed Inti, it is the lord of meteorological phenomena, known under a variety of local names, who is the most important masculine deity in the Andes at every regional level and from the point of view of cultic practice (Mariscotti de Görlitz 1978, 228). It is he who is linked to fertilizing rains, lightning, springs, and rivers. It is Pachamama who is his partner. This fact has crucial importance for understanding the criteria for the bases of moral action in daily life.

Pachamama personifies the divinized earth and, as such, is the "mother of all people everywhere" and the moving force of vegetation. She is often the "mother" of specific crops and a tutelary of pottery, spinning, and weaving (Mariscotti de Görlitz 1978, 223). Her cult locus is "the most conspicuous manifestation of rural Central Andean religion: the veneration of natural rocks, menhirs, cairns and geographical features" (Mariscotti de Cörlitz 1978, 224) (huacas), taken as a *pars pro toto* of the divinized earth. These huacas, manifestations of the Earth Mother, "serve as seats for specific local spirits and *integrate systems in which ancient cosmological conceptions are reflected*" (Mariscotti de Görlitz 1978, 224; emphasis added). The antiquity of such a conception is testified to by the continuity of the paraphernalia of her cult, the structures and materials of traditional offerings made at the earth shrines, the behavior of cult officiants, the structure of ancestral rites associated with Pachamama, the structure of her religious feasts (ritual hunts, dances, and mock combats between moieties), the festal calendar, and the structured relationship with her male-god partners (the constellations, weather gods, lords of springs and lakes, the lords of animal flocks).

If we keep in mind Bastien's image of the cosmic mountain-body, kept alive by ritual feeding of its parts at the huacas by individuals and groups whose duty it is to maintain them, we shall see Pachamama's crucial role in the continuity of cosmic order. The huacas are her body: places where the Earth is open and exposed. Feeding the lord of lightning or hail is an act which must be done where Pachamama joins with him. The sexual dualism of creation pervades every moral human act made on behalf of the cosmic and human fertility. Each feeding is a time of union between Mother Earth and her male partner: a *hieros gamos* effected through the cooperation of human agency. This sacred conjugal union of cosmic male and female being is seen most clearly in the agricultural feasts of plowing, planting, sowing, and animal husbandry. These are the times when the whole earth is "open" to be fertilized. Pachamama's fecundity is ambiguous, however, and strict taboos on certain foods, movements, sounds, and relationships must be observed to avoid illness and epidemic or eventual famine from poor crops.

Billie Jean Isbell describes two such feasts (1978, 137–65). In the branding ritual, fertility of the herds is assured because communities "put the cattle to bed" in a mock marriage and "marital embrace" of a bull and cow on a ritual table (*mesa*). During the Yarqa Aspiy festival (clearing the irrigation canals) in September when the "earth is open," phallic crosses of the Wamanis, the potent mountain spirits, are brought down into the valleys while the irrigation canals are being scraped clear. "The women are purified and ready for conception, Earth Mother, *MamaPacha*, is open, cleansed, and awaiting the final act of union with the masculine energy, the moving force of water from the puna-dwelling *Wamanis*. We might even think of the irrigation water as the semen of the *Wamanis*" (Isbell 1978, 143).

Although these annual feasts illustrate *hieros gamos* most dramatically, it may be kept in mind that the union of an "upper" male with the "lower" huaca of Pachamama at the time of even the least significant individual's misfortune-offering of chicha libation, thread, or fat to spirits of springs effects a union of the same order of cosmic sexual dualism. It is the human responsibility to effect this fertile union by feeding them when they unite where the earth is open. A profound knowledge and observance of this *where* and *when* is a serious moral responsibility. Furthermore, humans must avoid causing disunion by transgressing the proper order of kin relations, marriage prescriptions, and sociopolitical hierarchies set up by the upper-lower arrangement of ayllus, verticality, and moieties. It is worth noting that new shrines are established when misfortune strikes as a result of sins of omission or transgression in this regard; i.e. , the effects of improper behavior may be overcome or avoided by establishing a huaca and making offerings.

The Mesa: Pachas and Suyus

We have seen that the Andean cosmogony envisions this world as a continued arrangement of finite places and times, contrary to the total "unbounded" times and places of the first three primordia. The separate parts of the cosmic body must be fed periodically to preserve the unity of the people who are an integral part of it. Similar to the way in which Viracocha's five dialectical moments "reflect," in some way, the processes of the created world, human nature is regulated by (and regulates) the dynamics of cosmic order. This human regulation is manifest in knowledge of the ontological importance not only of specific places (huacas) and times (seasons, life-crises) but also of the union of a sexual order which qualifies cosmic and human space/time at every level. The regulation of such a conjunction manages the structure and interpretation of the ritual experience which is the prerequisite basis of proper moral acts and attitudes. Ignorance of or betrayal of these bases of moral action bears serious consequences which need the formal redress of divination, confession, purification, and healing. "All [these] formal ceremonies begin with the preparation of the '*mesa*,' i.e. , with the spreading of an altar-cloth over a table, or over a platform of stones, or box, or directly on the ground, and the distribution on it of cult objects required for the occasion and the offering" (Mariscotti de Görlitz 1978, 112).

This same mesa is used to divine the causes of moral or physical disorder, to search for the transgressions which cause misfortune. It is used as well during healing rites, propitiation ceremonies, sacrifices, and agricultural rituals (as above during the branding ceremony). The structure of the mesa is judged to be of great antiquity through the Andes. The meanings attached to it, considered of equal historical depth, are intimately tied to cosmogonic structures (Sharon

1978). Since the mesa is a microcosm of the cosmogonic order I have delineated above, it is significant that its structures reveal and redress moral infractions which cause disorder. The cosmic structures themselves provide the bases for the experience of ethical principles. In 1970 and 1971, Douglas Sharon studied the mesa of a shaman named Eduardo Calderon Palomino in Peru. Sharon argues that the mesa has ancient roots in the Andes and traces the existence of its detailed structures in the earliest chronicles. The mesa is a board, cloth, or blanket divided into three zones which Eduardo calls fields ("campos"). The three fields may correspond to the three cosmic zones ("pachas") which we have already seen in Joseph Bastien's explanation of the metaphysical counterparts of the three ecological zones of the cosmic mountain-body. The outermost fields, on the far left (west) and far right (east), are opposed to one another. "The Middle Field represents the core of Eduardo's conscious philosophy, for the opposing forces of the universe—as manifest in this microcosm known as *mesa*—are not conceived of as irreconcilable. Rather, they are seen as complementary, for it is their interaction that creates and sustains all life" (Sharon 1978, 64). Power-objects (huacas) are placed strategically in straight lines across all three fields. In addition to the tripartite division into three zones, the mesa is divided diagonally into four quarters ("suyus"), creating four triangles which converge on the center. Eduardo calls these quarters "the four winds" and the diagonal lines between them "the four roads" (Sharon 1978, 65). Each quarter is oriented to a cardinal point on the compass and embodies a different quality: east, positive rebirth; south, action; west, negative death; north, power. The diagonals run from a crucifix placed in the center of the mesa. Along the northern edge of the mesa are eleven staffs stuck vertically into the ground and one length of San Pedro cactus placed vertically to make a twelfth vertical "antenna."

Once the mesa is activated by ritual, the microworld "becomes the sacred space" in which the human and the cosmos meet and become one (Sharon 1978, 72). Although it is impossible to go into detail here, it is important to note that temporal structures, in multiple layers, are essential to the mesa's operation. In fact, the mesa reflects the existence of what Sharon terms a well-structured "metaphysical clock" (1978, 106). The twelve staffs orient the three pachas (cosmic zones), four suyus (quarters), and their huacas toward celestial phenomena and vice versa.

Altogether they precipitate a cyclical process through the twenty-four hours of the diviner's seance, the twelve months of the year, the stages of an individual life-cycle, the four seasons of the sun (the sunrise and sunset seen as equinoxes and noon and midnight as solstices), and the "ages" of darkness and light. The overall temporal orchestration portrays the shaman as "a sacred being bringing light into darkness" (Sharon 1978, 108). This orchestration regulating the cyclical progression of cosmic spaces (suyus and pachas) through time is all the more striking when we recall the image of the cosmic flower (*tega*) spoken of under the fifth "standard" of Viracocha: the fundamental creative circle (*ticci muyu capac*) which is his full realization in time and space. Though the sun's movements through the heavens preside over the day, Eduardo and Sharon believe that the "earthly model for this vital half [the night] of the metaphysical clock is provided by the magical plants, especially, *the* major plant, the night-blooming San Pedro" (*Trichocereus pachanoi*) (Sharon 1978, 107).

It is in divination at night that the "earth is open" and the shaman's client may "open like a flower" (Sharon, 1978, 107). "The main object of a seance in *mesa* therapy is to make the accounts [the huacas] and the participants 'bloom' at night, in imitation of the San Pedro" (Sharon 1978, 197). By activating the power of huacas on the mesa, the three cosmic pachas and four quarters (suyus) become a single, ordered cycle of time. It is Eduardo the human who activates the sacred microcosm. It is morally necessary for him to take up this vocation of regulating needs. He does so very carefully, for he must respect the order of the cosmos which he helps set in spin. By managing his clients' experience of cosmic structures he discerns what actions are proper and also helps effect them.

Cuzco: Cosmic and Ethical Order

Cuzco is a unique and particularly refined application of pan-Andean cosmogonic understandings detailed above. Built into its very construction and its behavioral prescriptions were the moral principles of cosmic and dynamic order. On one level, Cuzco was constructed in the form of a cosmic body. Its imperial walls still describe the form of a huge feline whose head is the highland fortress of Sacsahuaman. Like the inhabitants of Kaata, the citizens of various sections of Cuzco took their names from the parts of the feline body in which they resided (W. H. Isbell 1977). The city-body lay between two rivers, the Tullumayo and Huatanay. As Bastien and Billie Jean Isbell make clear, this is of more than practical advantage. The two rivers literally incorporate the land—lend bodily integrity to the space and inhabitants between them and make life circulate within them in the form of offerings, sacrifices, souls, and spirits. Though not explicit in the sources, it would not be unhelpful to look upon the city as a large mesa, divided into two opposing moieties: the upper (*Hanan-Cuzco*) and the lower (*Hurin-Cuzco*). The upper moiety, created by Viracocha himself, carries with it his prestige. It has associations with the male gods of the celestial realm, the invading royal houses of the Inca aristocratic lineages, the sky, and the mountains. The lower moiety members were linked genealogically to autochthonous conquered peoples, and to *Pachamama* and her female equivalents, to the earth and the coast. Moiety members took responsibility for the huacas and feasts associated with their own moiety. Moieties were truly ritual communities. In this way the moieties were not only geographic but relational cross-cutting concepts. Every sector had "upper" and "lower" members: the sexual dualism it implies pervaded the whole city-body as it does the cosmos.

The dividing line between these two geographic moieties may be found near the large square in the center of the city: the Aucaypata-Hanan (the present-day Plaza de Armas). On the "upper" side of the square, forming the forelegs of the feline, were the residences of the institutional priests and their novices. On the "lower" side, the hind legs of the feline, were the houses of the "Virgins of the Sun" (W. H. Isbell 1977). The plaza itself and its central buildings served as a kind of smaller "Middle Field" where a synthesis of the two opposing fields was achieved. Here were the belly and navel of the feline as well as the structures dedicated to Viracocha, the androgynous creator-god.

Like the mesa, the city (and the entire empire around it) was divided into four quarters (suyus): Hanan-Cuzco ("upper Cuzco") included the northern quarters of Chinchaysuyu and Antisuyu; Hurin-Cuzco (lower Cuzco) embraced Collasuyu and Contisuya. Although true sociopolitical units, these were above all ceremonial divisions which reproduced the structure of creation. Like the twelve staffs of Eduardo's mesa, there were twelve sucanca, vertical pillars used for astronomical observation, which oriented the city to celestial movements and vice versa (Herrera, in Zuidema 1977, 241).

Across the body of the city were set hundreds of huacas (like the power-objects of Eduardo's mesa, the earth-shrines fed in Bastien's ayllu, and the huacas of Pachamama's cult). The exact number of huacas reported varies according to the chronicler. It seems now that there may in fact be 406 or 408 with the possibility of reducing the number to as low as 328 in order to accommodate calendric schemes. Whatever the final tallies be, the numbers of huacas do not appear arbitrary, for their varying totals served specific calendric functions in which each huaca represented one day of a time unit.

Unlike the huacas of Bastien's mountain community, and Eduardo's mesa, those of Cuzco appear to be grouped very carefully according to number within each ceremonial quadrant (suyu). A number of straight sight-lines (ceques) of shrines (huacas) radiated out like spokes from the temple of the sun (Coricancha) in the center of the city toward the horizon. More numerous than the "four roads" from the center of Eduardo's mesa, these ceques run in some forty-one directions with an average of eight huacas per ceque. The ceques organized huacas in and around the city into a calendric system for ritual and social purposes. The ceques in the upper moiety are enumerated in clockwise direction while those in the lower are named in counter-clockwise fashion. I suggest here a continuation of the tension between the solar (celestial) "metaphysical clock" and the telluric and nocturnal cycle of river through mountain-body described by Joseph Bastien and evident in Eduardo's seance. A basis for this view is founded in Zuidema's further suggestion that a division exists between "major groups" of ceques (containing twenty-nine or more huacas among the three of them in a group) and "minor groups" of ceques (with less than twenty-nine). The two most significant suyus each contain two major groups of ceques. The two remaining suyus contain only one major group each. The distribution of major and minor groups suggests a regular pattern. The number of huacas in the six major groups of ceques would constitute the basis of a calendar of twelve solar months: those major groups in the upper moiety express a solar half-year in terms of solar months; those major groups in the lower moiety express a solar half-year in terms of synodic months. This is in keeping with the cosmogonic order disclosed in the course of this paper. In addition, the minor groups evidence a regularity of a different kind. When the suyus are divided on a north-south axis (rather than the east-west moiety axis), there are seventy-three huacas belonging to minor groups on each of the two sides of the dividing line.

The seventy-three-day period correlates the solar and sidereal lunar years (228 + 329 = 657 = 9 × 73) with the Venus cycle (8 × 73 = 584). It is the sixteen-year period, a significant one in Cuzco, which integrates the solar and ritual systems (Zuidema 1977, 247). Zuidema postulates the following explanation for the tension and the need for the intercalation of ritual metacalendars in opposition to the major groups of ceques with solar relationships: "the minor groups had cthonic relationships, that is, to the heaven at night" (Zuidema 1977, 247). Little is known of the latter system since "probably it was an exclusively female calendar" (Zuidema 1977, 248). The continuous intercalation of male and female times throughout the overlapping kinds of "years" achieves, in the expressions of time, the *hieros gamos* continually achieved in spatial terms by the uniting of celestial male beings with the female huaca seat.

Along each ceque are a certain number of huacas which must be attended to ("fed"), each in order according to the calendar (solar, lunar, sidereal, lunar sidereal, synodic, Venutian) which presides over its manifestations. However, the calendar is not abstract but incarnate in the kinship and resident patterns of the population. The care and maintenance of specific stones, springs, mountains, and other huacas which are grouped along each ceque fall in turn on particular social groups and specific mixed numbers, astronomically correlated, of those groups which make up the stable population of the city and constitute, as a whole, a complete social structure with an accompanying normative ritual experience governing their proper relations. The order of their relations, the feeding of huacas, and the calendar derive from the myths of creation of the world, of Cuzco, and of dynasties and lineages. The ritual experience of the cosmic powers revealed in the cosmogony, particularly the ordered powers which shape time and space, becomes the foundation of social ethics. With huacas being served daily by shifting "ritual communities" brought together in complex levels of relations, the calendar coordinates the mobilization of fixed numbers of fixed peoples in those activities and relations deemed moral par excellence because they perpetuate the existence of "this world." Such mobilizations of population levels affected the concepts of right order from the highest military and priestly institutions to the most humble domestic unit. Unlike the calendars of Mesoamerica, which were recorded in terms of an elaborate system of gods, animals, and color symbolisms, the Inca calendar was expressed in the system of proper relations of its populations. The continued life (calendar) of "this world," contrary to the first three worlds, depends on the propriety of human action. The political system of laws, education, and hierarchies had first to be analyzed before making sense of the ceque calendric system. Such is their interpenetration (Zuidema 1977, 220). The following paragraphs provide representative examples of the ways in which a cosmogonically based calendar governed those behaviors morally incumbent on individuals and groups.

According to Garcilaso de la Vega, the legal system is entirely dependent on the "four quarters of the world" and their divisions (1960, chap. XI). In each suyu the Incas established decuries (groups of ten leaders in the four quarters to equal the forty ceques). Their two tasks were to minister over the subsistence activities of their populations in regard to grain, wool, roads, bridges, housing, and marriages and to report and punish breaches of the behaviors proper to these activities. Indeed, the very *quipu* textiles which have proven to be calendric keys were used to keep records of laws, ordinances, rites, and ceremonies (Garcilaso de la Vega 1960, chap. IX).

Education was regulated by the same spatial and temporal dispositions. The *Acllas* (Virgins of the Sun) followed a rotation schedule on three (double?) lunar-year cycles. Guaman Poma reports that within the ranks of these women were twelve further divisions conditioned by their age-grade, the huacas they served, and the capacity in which they served them. Proper conduct within and between the groups depended on the calendar (Valcarcel 1964, 2:605–6).

According to the spatial disposition of their families (determined by genealogy) and the temporal cycle, a fixed number of young men were sent for priestly education in the palace of the Inca. They were instructed by four teachers for four years. Others, based on further calendric discernment, were provided a military education (Valcarcel 1964, 617). The structures of marriage were centrally controlled by the Inca through appropriate age-grade and residence patterns (Valcarcel 1964, 2:638ff.). One age-class was equivalent to six lunar cycles, equal to five solar years (Zuidema 1977, 249). There were ten age-grades for men and ten for women (Valcarcel 1964, 2:649).

Present evidence points to a single complex calendric system simultaneously using four different kinds of months: synodical, solar, sidereal, and anomalous. The solar calendar was used more during the dry season. This relates to the predominance of the upper moiety (hanan) and its social, festal, and economic activity during this time. The synodic lunar calendar was more important in the rainy season and was associated with the lower moiety (hurin) (Zuidema 1977, 228).

An eight-day week propelled customs on the local level. The weekly cycle of service still used by women on contemporary haciendas around Ayacucho is called a *suyu*, "a word used in the same way by Gonzalez Holguin in 1608 for groups of people in general" (Zuidema 1977, 229). The Inca emperor changed wives every eight days. At that time the wives and their trains were brought from their place of origin (Cobo 1956, in Zuidema 1977, 229). After a year (not a solar year) the first wife would arrive again. The eight-day week similarly motivated the rotation of the market system and the priests in the Temple of the Sun.

There were four seasons of three months each based on a hierarchy of relationships: the king's feast after the March equinox and the queen's before the September equinox; the solstices dedicated to the cult of the sun and the equinoxes to the cult of water. On this seasonal reckoning were fixed the mass celebrations for Viracocha, the moon,

Thunder God, god of war, cult of the dead, and collective dispelling of illness.

There were also three seasons of four months each: rituals of sowing, rites of initiation, and redistribution of land. Two- and eight-year periods are fixed for bridge-building, three or six years for rotation of the Virgins of the Sun, five years for age-sets, four years for communal hunts. Each set of these periods carried with it repercussions in land use, personal relations, and behaviors proper to collectives.

During the two solstice feasts "the ritual and calendrical value and rank of all the *huacas* in the empire, of every village, town, province, were reassessed" (Zuidema 1977, 231). From each hamlet in every suyu came children with gifts to Cuzco. The children walked along their proper ceque. Once in the capital, they were sacrificed or sent home along the proper ceque. During the feast of Cituay, at the September equinox, four hundred warriors, one for each huaca, expelled diseases from the city by following the ceques in groups of ten (Zuidema 1977, 233).

The feast of first plowing involved the Inca, the four chiefs of the suyus ("chiefs of 100,000 families"), forty "chiefs of 10,000 families" (ten from each suyu to equal the ceques), and forty groups of ambassadors from the provinces. In this case the leaders of each hierarchical level, each in their turn, carried on the first plowing of the ritual field. First the Inca, the four councillors, and the forty chiefs (conceptually related to the forty-one ceques since the suyu chiefs were conceptually one), then the forty groups, continued the work (Zuidema 1977, 236–37).

Every genealogical group took care of a group of ceques for the occasions celebrated in connection with its huacas. The population was divided into twelve groups, each with a name of its month and each assigned an occupation in regard to given huacas on given occasions. Since the ten political divisions of each suyu (to match the forty ceques) do not always correspond to the group of a given month, there appear to be overlapping systems of divisions and assignments. There seems to be a hierarchy of groups based on first the rank of its suyu and its groups of ceques based upon the cosmogony; second, mythical ties to former Inca kings based upon the origin myths of lineages which are sequels to the cosmogony; and third, the religious cult specific to each group and its ancestors (Zuidema 1977, 240).

This small set of examples shows that on all levels of their society the Incas orchestrated a determined ritual experience of the spaces and times revealed in the cosmogony. This ordered experience of being, together with its meanings made known in myth, becomes a sine qua non of moral life.

Proximity to Being

This essay illustrates the links between pan-Andean cosmogonic structures and an ethical system in Cuzco. The link between creation myth and proper behavior is maintained in the ordered experience of ritual. The vitality of Andean cosmogonic thought sustains its distinctive character even through local variety, Inca overlordship, and Spanish domination: an all-pervasive sexual dualism and a deep appreciation of the capacity for ontological expression of spatial and periodic symbols. Furthermore, the cosmogonic order of the Andes is not static but dynamic and changing. The nature of the human being, in a way parallel to that of the creator god, relates to the order of creation as both agent and consequence. Human "history" or action is itself an integral and integrating cosmic element. In feeding celestial spirits at earth shrines (huacas) at properly discerned times and places, individuals properly related in space and time unite the dual cosmos and nourish the circulation of its parts in order to perpetuate its generative cycle of birth and death. The city of Cuzco sets itself the task of accomplishing what the protean beings of the first three primordia could not do: keep being in existence. This is the human destiny for which it harnesses itself to an ethical order.

The foremost element of this ethical order is the experience of ordered change—the indefinite play of finite parts. The calendar is the wisdom acquired in discerning the order. Cuzco, like a macro-mesa, is a sacred field where the basic structures of moral order are made geographically clear on the divinized earth in relation to the proper order of the heavens. When powers are activated by human ritual agency, harmful transgressions of dynamic cosmic structures are divined; these transgressions are "changed" through efficacious redress which render proper the relations which ought to "be" between the cosmos and humanity, and among human beings.

As a historian of religions, in this essay I have chosen, as John Reeder (1978) suggests, to take a "propadeutic" position to philosophical description. Nevertheless, while using history of religions to help ground the discussion of moral bases, I have employed a number of insights current in religious ethical circles. In "Religious Ritual: A Kantian Perspective," Ronald Green (1979) makes clearer the relationship which Joachim Wach once postulated between cult ritual and moral instruction: cult and code (Green 1979, 229). I have found suggestive Green's consideration of ritual as an effort to use "complex symbolic and group activity for the purpose of expressing and vivifying fundamental moral conceptions" (229). Important differences emerge between us when I suggest that moral behavior structured in the calendar points to an underlying religious ontology which serves as ground for both moral principles and religious belief rationally construed in the way Green would have it. Ritual provides—indeed, imposes—an experience of this fundamental order of being. Such an experience is intrinsic to the moral life.

In addition, Green draws out the moral dimensions of ritual, warning: "to reduce the rich panoply of acts and utterances associated with any religious rite to moral instruction and exhortation seems absurd" (Green 1979, 230). Instead, ritual behavior assembles social reality. In the Andean case, the moral commitment the rituals vivify, to use Green's terms, extends beyond the ritual moments (which already fill a large portion of time in Cuzco) to the proper regulation of relations necessary to prepare the rites: weaving textiles for offerings, gathering and carding wool for them, guarding herds for wool, respecting the highland pastures, "feeding" them and revering the wamanis there, rotating the herds through multiyear grazing cycles. Preparing chicha (corn brew) for libations requires marking off fields and distributing land properly, plowing, planting, and harvesting corn with proper reverence of dema, fertility, and meteorological beings. Installation of warrior age-sets and classes of Virgins of the Sun or groups of children gathered for child-sacrifice requires previous observance of fertility and sex acts proper to conception. Successful potato crops, absolutely necessary to survival, depend upon the timely performance of freeze-drying rituals "when the earth is open," storage rituals, observances of taboos regarding the dead present in seasonal winds and in the chullpa buildings. In short, moments which vivify moral principles radiate out from formal rituals at the huacas to the "rituals" of quotidian activities with no clear demarcation between them. Consequently, work, sex, and relations in general become extended parts of the ritual even if its focus be at a huaca. Cuzco becomes, in Frank Reynolds's terms, a ritual community whose ethical "modes of action mesh together to form the basis for a dynamically functioning religio-social system" (1979). This article attempts to take seriously Reynolds's call for "more intensive studies of the structure, dynamics, and social implications of the normative modes of action . . . manifested in their community life" in order to offer an "inner perspective" (1979, 23).

Green has pointed out that Christian baptism reenacts a central mythic event which carries enormous moral significance—the death and resurrection of Christ. In the Andean case, I have tried to show that any of the celebrations carried out at a huaca point as well to the complex central mythic event of the cosmogony: the uniting of male and female valences and the coordination of the finite fragments of space and time left over from the first three ages into the pachas and suyus of the present age of creation. These are the cosmogonic lines of moral force which pull the daily actions of the community into ordered alignments.

This appeal of morally expressive ritual to a deeper and more fundamental religious ontology explicit in the creation accounts and the creation theology of Viracocha, parallels an ethical direction outlined by Franklin I. Gamwell (1978) in "Ethics, Metaphysics, and the Naturalistic Fallacy." In proposing to "open the door" for ethical discussion between analytical philosophers and theological ethicists by delineating Charles Hartshorne's view of metaphysics, Gamwell demonstrates that one view of metaphysics entails that "defensible ethics are metaphysically based" (1978, 53). He points then to the necessity of the good as, he terms, "a cosmic variable" against which moral principles of the local good might be evaluatively compared. Cosmic variables are "forms universal throughout all existential possibility" (49), "thus the breadth of the [cosmic] variables is that of the whole universe of what is and what might be" (Hartshorne 1937, in Gamwell 1978, 49). "Cosmic variable" may be a suitable term for the dynamic ontology of protean being found in the ages of creation and expressed as a processual structure of the fourth age and its Cuzcan calendar. The structures of Andean moral experience are based in an ontology of periodicities and spatialities made manifest in the four ages of creation. The parallel with Gamwell's exposition is all the more striking because the moral order maintains a relationship to the cosmogonic order in a way similar to the relationship which Viracocha enjoys with the world—as a process. The ethical order, ritually expressed, is metaphysically based. In this respect, following the analysis of Henry B. Veatch, the Andean ethical system bears less in common with analytical ethics and existentialist ethics based on what Veatch terms "structural history" than with those "ontologically grounded in nature" (1971, 84).

The article follows Twiss and Little's lead by interesting itself more with what they term the structure of moral "codes" construed *as a pattern* and less with the content of that "code." The mesa-diviner, as one kind of Andean ethicist, is less intent on situational application, validation, and vindication than on a "transcendental turn" of experience, to use Veatch's phrase; for "ethical principles, though neither evident in themselves nor susceptible of any direct rational justification, may nevertheless be justified indirectly on the grounds that they are principles which we simply cannot dispense with in that primordial ordering of our experience" (Veatch 1971, 86–87). The mesa-diviner attempts to reunite himself and his client with the protean ontological elements at play on his mesa (a space/time microcosm). It is this unity—the "awakening" or "blooming" of the flower (tega) that is both cosmic and human order—which defines moral action par excellence.

In sum, these three above-mentioned themes from the Andes share common concern with contemporary ethical discussions: first, rituals are expressive of the moral order. Second, this ritual-ethical order is based upon a metaphysics. Third, this metaphysics is an ontology revealed in the cosmogonic account. A fourth theme should be added: the nature and structure of time revealed in the cosmogony becomes the foundation for ordering the ritual experience of being. What is as striking as the importance of time in Andean experience is the flexibility of its nature, structures, and meanings.

It is unfortunate that space does not permit a presentation of the history and structures of divination and its associations with cosmogony. In cases of moral uncertainty or conflicting moral demands, recourse is made to a diviner. On the level of day-to-day action, divination makes clear that which is morally binding. The diviner uncovers those who furtively, or even unintentionally, disrupt the moral order. Since moral breach carries with it serious social and cosmic consequences, it is often misfortune or fear of misfortune which prompts one to seek the counsel of a diviner who, by reason of supernatural election or innate gift, possesses the ability to discern the "causes" of disorder. He seeks these "causes" not in the epistemological examination of competing premises and deductions but in the relations which appear disordered when measured against those relations which "ought" to obtain as revealed in the cosmogony and myths of origins. A variety of techniques are available to diviners, including reading the inflated lungs of sacrificed animals, the innards of guinea pigs, casts of coca leaves, casts of kernels of maize and colored maize flour, the scintillations of stars, the movements of constellations, and so on. Second opinions are frequently sought, especially in cases of persistent misfortune. Some of these forms of divination are simple and inexpensive. They are consulted frequently and casually (e.g., for coca-leaf casting see Bastien 1978, 14–15, and passim; for star scintillation see Urton 1981, 93).

Divination is an important guide to moral action. The power of the diviner to guide moral action rests with the ability to make clear the relationship between the client, the community, and cosmic powers. This, in turn, depends upon the diviner's own proximity to these beings as well as his understanding of their origin and nature. Disordered relationships between the individual or group and cosmic forces are discerned during ritual, for divination itself is a ritual even if some forms of it are highly informal. Further remedial rituals are often recommended by the diviner. The appropriate ritual experience prescribed by the diviner succeeds in redressing moral breach by symbolically reconstituting right relations. As we have seen all along, the emphasis does not fall on the processes of how one knows (i.e., a mode of moral discourse which guarantees) what is morally certain but on how one experiences (i.e., the form of symbolic action which effects) those relations which are proper to "this world."

Since the spaces and times of this world are constantly in flux, as revealed in the cosmogony, proper relations change over time. Such a dynamism can inspire the moral uncertainty which seeks divination. In Cuzco, during the feasts of the solstices, major reassessments of the value of all major huacas adjusted relations of peoples and places to the shifting realities of the cosmos. Twice each year the populations of the city reoriented their interactions in relation to cosmic powers. All ritual relations were subsequently reordered. They did this in order to experience "realities" as they are and thereby see to the continuance of life and the calendar, the infinite play of the finite parts of this world.

To decide ethical conflicts or dissolve moral uncertainties by divination is to construct an argument of images. The orderliness of one imaged realm of experience, the cosmogonic, is used to edify—to "constructure"—an apparently unclear domain of experience, that of moral action. I use the term "constructure" because the two realms of experience are obviously contiguous; they are not entirely separate analogs of one another. In divination, such a process appears to be an example of what James W. Fernandez has called "edification by puzzlement" (1980; 1982, 512–513, 521, 572).

> The kind of thought going on here, with its emphasis on images, is a devaluation or a rejection of language-based thought and the discursive reason that lies in it and that has been the intellectual power tool of modern technical-rational man. (Fernandez 1982, 569–70)

Divination is an elliptical mode of thought which demands that the diviner interpolate, extend, and recontextualize his client's dilemma. The conflict is not broken down for analysis but is drawn into a larger context. When the diviner maintains the cosmogonic context as a frame of reference, the puzzling techniques of divination return him to an experience of the ordered whole. What Fernandez says of Bwiti seems true of the mode of thought at work in Andean divination:

> The trick seems to lie in taking what is apparently incoherent or plainly contradictory and then discovering some higher principle . . . which demonstrates that the opposition is only apparent and is subsumed in that higher principle. In the suggestion of these higher principles, Bwiti operates not so differently from any science whose discovery procedure is to take things apparently incoherent and inconsistent and show them to be instances of a more general principle. The important difference is that Bwiti only suggests and does not aim to discover and state those principles in so many verifiable and refutable words. Rather than struggling relentlessly to deprive puzzles of their mystery, Bwiti works with puzzles. (Fernandez 1982, 571–72)

The higher principle is that *ordered sense of the integrity of being* intuited in ritual.

Cosmogony and Ethical Order

Cosmogony, the attempt of the religious imagination to portray the coming-into-being of the cosmos, appears to be a singularly powerful insight into the nature of being. It both enables and constrains wholly different ways of estimating the value of thought and the meaning of acts. The different cosmogonies presented in this volume frame distinct modes of conceiving reality and asking questions about it. In general, these cosmogonies do not seem to prejudice the content of particulars in an ethical code. However, they do seem to raise up, as a figure-to-ground, the kinds of constraints which ethical discernment must work with and within which certain kinds of questions make sense and others do not. The myths lay out the presuppositions, the truths about being, of ethical expression. For example, the detailed interactions of the populations of Cuzco are calendrically coordinated. Such a detailed calendar of proper behaviors *makes sense* in the world order manifest in the cosmogony. Indeed, the ritual behaviors of the calendar impose an experience of that cosmogonic order. However, nowhere do the cosmogonic texts dictate the specifics of the ritual behavior of Cuzco. The cosmogony does not impose, in a one-to-one fashion, the particulars of the ethical system. It only reveals those things which *are* and the dynamics which power their relationships. These are the realities which must be reckoned with in the ethical order. In the Andes, it is the return to the ontological level revealed in cosmogony which grounds moral life. This transcendental turn is made through the symbolic forms of action which clarify the order of experience. In this volume, Norman Girardot makes a similar statement in regard to those Taoists who undergo a momentary experience which returns them to an undifferentiated condition of being. This return to a particular cosmogonic state of being (*hun-tun*) transforms one's mind, will, and intention. The experience enables one to act morally without necessarily setting into motion the epistemological exercises of thinking about acting morally. In the contrary case of William Paley, presented by Robin Lovin, the link to the ontological level appears to be made through this epistemological activity—the enunciation of symbolic statements about the nature of reality with the aim of clarifying the knowledge of the order of intelligible being.

It appears that there exist quite different cosmogonies which create and bequeath to civilizations key existential dilemmas as well as the meaningful values at stake in the varied solutions to the dilemma. The assumption that cosmogonies are logical constructions drawn from rational designs has been questioned. In fact, we cannot assume that cosmogonies are the edifices of reflective reason. They appear to share the nature of fundamental insights of the first order—primordial insights. As Bernard Lonergan's study of insight demonstrates, it is, somewhat by definition, not constructed through deduction from premises. Its appearance better approximates disclosure—an instance of radical imagining or imaging. Like Bronowski's "axiom" or Gödel's "theorem," it is capable of founding new systems of thought and action, but it cannot be the deductive fruit of such a system. Understandably, such primordial insights about the coming to be of the cosmos are more rare than frequent. The undetermined manner of their appearance, the process of imaginative insight, may account for their portrayal in religious systems as revelation.

In the Andes one finds a cosmogony with ethical bearing. It does not reveal the premises and processes of knowledge which determine the discursive adjudication of ethical decisions. It reveals a structured process of coherent experience available in ritual. The ritual return to primordial experience of the cosmogonic order becomes the effective foundation of moral action in the more inchoate domain of social experience.

References

Acosta, José de

1954 *Historia Natural y Moral de las Indias.* Biblioteca de autores españoles, vol. 73. Obras del Padre José de Acosta. Madrid. [Written in 1590.]

Aveni, Anthony F., ed.
1979 *Native American Astronomy*. Austin: University of Texas Press.

Bastien, Joseph
1973 "Qollahuaya Rituals: An Ethnographic Account of the Symbolic Relations of Man and Land in an Andean Village." Ph.D. dissertation, Cornell University.
1978 *The Mountain of the Condor: Ritual and Metaphor in an Andean Ayllu*. Minneapolis: West Publishing.

Betanzos, Juan de
1924 *Suma y narración de los Incas*. Colección de libros y documentos referentes a la história del Peru. Second Series, 8. Lima. [First published 1551.]

Calancha, Fray Antonio de la
1938 *Crónica Moralizada del Orden de San Augustín en el Perú*. Excerpt in *Los Cronistas de Convento*, edited by Pedro Benvenutto Murrieta, pp. 15–138. Paris: Biblioteca de Cultura Peruana. [First published 1639.]

Cieza de León, Pedro
1967 *El Señorío de los Incas. Estudio preliminar y notas de Carlos Araníbar*. Lima: Instituto de Estudios Peruanos. [First published 1660.]

Cobo, Bernabé
1956 *Historia del nuevo mundo*. Biblioteca de autores españoles, vols. 91–92. Madrid. [First published 1653.]

Duviols, Pierre
1966 "La Visite des idolâtries de Concepción de Chupas." *Journal de la Société des Americanistes*. N.s., vol. 55, no. 2: 497–510.
1971 *La Lutte contre les religions Autochtones dans le Pérou colonial: "L'Extirpation de l'Idolâtrie" entre 1532 et 1660*. Travaux de l'Institut Français d'Etudes Andines 13. Lima.

Escobar, G.; R. P. Schaedel, and O. Nuñez del Prado
1967 *Organización social y cultural del sur del Peru*. Mexico City: Instituto Indigenista Interamericano.

Fernandez, James W.
1980 "Edification by Puzzlement." In Ivan Karp and Charles

S. Bird, eds. *Explorations in African Systems of Thought*, pp. 44–59. Bloomington: Indiana University Press.

1982 *Bwiti: An Ethnography of the Religious Imagination in Africa*. Princeton: Princeton University Press.

Gamwell, Franklin I.

1978 "Ethics, Metaphysics, and the Naturalistic Fallacy." In W. Widick Schroeder and Gibson Winter, eds., *Belief and Ethics: Essays in Ethics, the Human Sciences, and Ministry in Honor of W. Alvin Pitcher*, pp. 39–57. Chicago: Center for the Scientific Study of Religion.

Garcilaso de la Vega [El Inca]

1960 *Primera Parte de los Comentarios reales de los Incas*. Biblioteca de autores españoles, vol. 83. Madrid. [First published 1609.]

Green, Ronald

1979 "Religious Ritual: A Kantian Perspective." *Journal of Religious Ethics* 7 (Fall): 229–39.

Hartshorne, Charles

1937 *Beyond Humanism*. Chicago: Willett, Clark and Company.

Herrera, P.

1916 *A punte cronológico de las abroas y trabajos del cabildo a municipalidad de Quito desde 1534 hasta 1714 (primera epoca)*. Vol. 1. Quito.

Isbell, Billie Jean

1978 *To Defend Ourselves: Ecology and Ritual in an Andean Village*. Austin: University of Texas Press.

Isbell, William H.

1977 "Cosmological Order Expressed in Prehistoric Ceremonial Center." In *Actes du XLII Congrès International des Americanistes, September 1976*, vol. 4, pp. 269–98. Paris.

Kusch, Rudolfo

1962 *America Profunda*. Buenos Aires: Libreria Hachette.

Latcham, Ricardo E.

1929 *Las creéncias de los antiguos peruanos*. Santiago de Chile.

Lehmann-Nitsche, Robert
1928 "Coricancha, el templo del Sol en el Cuzco y las imagenes de su altar mayor." *Revista del Museo de La Plata* 31: 1–260.

Little, David, and Sumner B. Twiss
1978 *Comparative Religious Ethics: A New Method*. San Francisco: Harper and Row.

Mariscotti de Görlitz, Ana Maria
1978 *Pachamama Santa Tierra: Contribución al estudio de la religión autóctona en los Andes centro-meridionales*. Berlin: Ibero-Amerikanisches Institut.

Means, Philip
1931 *Ancient Civilizations of the Andes*. New York: Charles Scribner and Sons.

Molina, (El Cuzqueño), Cristóbal de
1943 "Fábulas y ritos de los Incas" (1575) in F. A. Loayza, ed., *Las crónicas de las molinas*. Los pequeños grandes libros de historia americana. Series 1, vol. 4. Lima.

Murra, John V.
1962 "Cloth and Its function in the Inca State." *American Anthropologist* 64: 710–25.
1972 "El 'control vertical' de un maximo de pisos ecológicos en la economía de las sociedades Andinas." In John V. Murra, ed., *Visita de la Provincia de León de Huanuco en 1562*. Huanuco, Peru: Universidad Nacional Hermilio Valdizan.

Pease G. Y., Franklin
1973 *El Dios Creador Andino*. Lima: Mosca Azul Editores.

Poma de Ayala, Felipe Guaman
1966 *El Primer Nueva Crónica y Buen Gobierno*. Lima: Imprenta GráFica Industrial. [Written between 1584 and 1614.]

Reeder, John P.
1978 "Religious Ethics as a Field and Discipline." *Journal of Religious Ethics* 6 (Spring): 32–53.

Reynolds, Frank
1979 "Four Modes of Therevāda Action." *Journal of Religious Ethics* 7 (Spring).

Rowe, John
1960 "The Origins of Creator Worship among the Incas." In Stanley Diamond, ed., *Culture in History: Essays in Honor of Paul Radin*, pp. 408–29. New York: Columbia University Press.

Santa Cruz Pachacuti Yamqui Salcamaygua, Joan de
1950 *Relación de antigüedades deste reyno del Pirú*. Asunción del Paraguay.

Sarmiento de Gamboa, Pedro. [First published 1613?]
1947 *Historia de los Incas*. Buenos Aires: Biblioteca Emecé. [First published 1572.]

Sharon, Douglas
1978 *Wizard of the Four Winds: A Shaman's Story*. New York: The Free Press.

Urton, Gary
1981 *At the Crossroads of the Earth and the Sky: An Andean Cosmology*. Austin: University of Texas Press.

Valcarcel, Luis E.
1964 *História del Antiguo Perú*. Lima: Editorial Juan Mejia Baca.

Veatch, Henry B.
1971 *For an Ontology of Morals: A Critique of Contemporary Ethical Theory*. Evanston: Northwestern University Press.

Zuidema, Reiner T.
1964 *The Ceque System of Cuzco: The Social Organization of the Capital of the Inca*. Leiden: E. J. Brill.
1977 "The Inca Calendar." In Anthony F. Aveni, ed., *Native American Astronomy*. Austin: University of Texas Press.

Part II

Cosmogonic Traditions and Ethical Order

5 Cosmogony and Order in the Hebrew Tradition

Douglas A. Knight

The Hebrew Bible opens with a double cosmogony. This can easily tempt one to conclude that cosmogonic thinking plays a foundational role in the biblical literature, in orthodox Yahwistic religion, and in the life of the ancient people. Actually, these are much-debated points, which any study of the cosmogonic tradition in Israel and the ancient Near East must take into consideration.

The issue is whether, and when, cosmogonic myths existed in ancient Israel as important, independent intellectual constructs. Gerhard von Rad (1966) has argued forcefully—and persuasively for many—that creation theology plays mostly a secondary and supportive role in Israel's religion. According to him, from the earliest times onward the people viewed the land as the gift of God, the great blessing which YHWH bestowed on the chosen people. Thus the Israelite religion opposed the Canaanite nature and fertility worship through reference not to YHWH as creator but to YHWH as redeemer in specific historical acts. The doctrine of creation scarcely existed as a theme in its own right, von Rad concludes. It was subordinated to the doctrine of salvation, so much so that at times the words "to create" and "to form" are virtually identical with the biblical concept of "to save." Nowhere is this clearer than in Deutero-Isaiah, but even the stories in Genesis 1–3 must be understood as a part of the massive Pentateuchal account of Israel's chosenness and deliverance. The connection is implicit: "from protology to soteriology" (von Rad 1966, 133).

This thesis has not fared well under closer scrutiny. To be sure, one can agree with von Rad that the mere placement of two classic cosmogonic myths at the head of the biblical canon does not necessarily mean that creation theology was highly important throughout ancient Israel's history. However, the fact is that cosmogonic imagery occurs widely in the Hebrew Bible throughout diverse literary traditions, and this would suggest that its importance and independence are not to be underestimated. Moreover, cosmogonic thought is not simply subsidiary to soteriology. At all points in the cosmogonic traditions, even in places where Israel's election or deliverance from enemies is involved, there is a more fundamental level of meaning: the nature of reality itself. The people experience—or want to experience—a world which is orderly, intentional, and good, and they believe that their God makes it so. This becomes expressed symbolically in myths which may be set chronologically at the birth of the cosmos, at later points in history whenever the order becomes threatened, or at the very end of time when chaos again needs to be conquered and a new world created. The nature of reality addressed here clearly involves, in addition to any themes of election and victory, other fundamental matters with which the people have to deal daily: joys and sufferings, work and pleasure, intimacy and alienation, reward and punishment, fertility and sterility, good times and hard times, order and disorder. Of such things are cosmogonic myths also made.

I. Typology of Hebrew Cosmogonic Myths

Before turning specifically to the connection between cosmogony and ethical order, we would do well to gain an overview of the types of cosmogonic myths which can be found throughout the Hebrew Bible.[1] The schematization which follows should not, however, give the impression that these varied cosmogonies are necessarily in competition with one another or are mutually exclusive of each other. They all are acceptable traditions in Israel, often overlap with each other, and share certain values and worldviews in common, as will be indicated later. They differ from each other primarily in terms of three factors: the symbolic form, the dominant social-historical location, and the intention.

1. God creates the entire cosmos and all life within it, everything structured and ordered carefully according to the divine will. Such is the symbolic picture presented in, for example, Genesis 1:1–2:4a, Psalm 104, and Psalm 74:12–17. Its social location would likely have been

the centuries-long institution of the priests, eventually centered in the urban context of Jerusalem. The cosmogonic picture intends, above all, to call the believers to worship and to praise this deity who has been responsible for creating the whole well-structured world. While several hymns indicate this (e.g., Psalms 148; 33; 8), the point could scarcely be made more poignantly than to have the inauguration of the sabbath fall on the final of the seven days of creative activities (Gen. 2:2–3). Beyond this there are other extensive priestly traditions which describe God's establishing the tabernacle and the temple (Exod. 25–31 and 35–40; Ezek. 40–48), thereby extending the cosmogony to include the creation of the cult as well.

2. God focuses intimate attention on the creation of humanity, and all else—plants, rivers, animals, birds—are formed with human delight in mind. This myth receives its clearest expression in Genesis 2:4b–25, with Genesis 3 as an intrinsic part of it; there is a variant version in Ezekiel 28:12–19. The symbolic elements in this cosmogony seem to suggest an agrarian base among the people, where it would have been most natural to view the earth principally as life source. The story's intention is to probe the nature and meaning of human existence, both the beauty and the brokenness of it.

3. God creates a just system of cause and effect in the world, according to which any given action will necessarily lead to its appropriate consequence (goodness yields blessing, whereas evil brings curse) depending on whether the act maintains or subverts the created harmony of all reality. This system, however, may often seem to humans to be shrouded in mystery, or in fact even to have broken down at times; hence the problem of theodicy. The proverbial literature contains some of the clearest articulations of this cosmogony, but it is very much present also in Job, Qoheleth, Psalms, and the prophetic and narrative collections. This orderly conception of reality, while probably having a very wide popular base in society, was especially promulgated in the wisdom school, an urban institution with roots reaching deep in ancient Near Eastern history. Its intention is both didactic and contemplative: to underscore the justice of God's ways.

4. God creates a people, chosen from among all other nations, for a special divine-human relationship. This soteriological symbolism occurs at several points: in the call of Abraham, in the exodus from Egypt, in the founding of the monarchy (2 Sam. 7), and especially in the new exodus from Babylonian exile (Deutero-Isaiah). The cult would have made use of this election/salvation image, and it was at various points in history promulgated also in monarchic and prophetic circles. It seems to have primarily an ethnic intention: to indicate that

the people of Israel have a distinctive role to play in the world and that they have special cultic and moral requirements to fulfill.

5. God engages in warfare against the opponent—whether this be chaos, other gods, or enemies of the people—and out of this holy war emerges an orderly creation over which the divine warrior then rules as king. There is a correlation here between the mythic cosmic war and the historical wars of Israel, not the least the conflicts associated with the exodus (e.g., Exod. 15) and the conquest. The theophany of the divine warrior is accompanied by disturbances of nature (e.g., Judg. 5:4–5; Ps. 68:7–8 [Hebrew, 68:8–9]; Hab. 3:3–15; Isa. 40:3–5), but he rules in peace after the battle (Ps. 24:7–10). This Israelite image of the creator-warrior-king was closely aligned with the view of other ancient Near Eastern religions, and it was nurtured in ritualistic activities of the cult, as evidenced in numerous textual fragments throughout the Hebrew Bible (for the basic studies see Cross 1973 and P. D. Miller 1973). Its intention is to establish YHWH's sanctuary and kingship, to interpret Israel's deliverance from foreign rule, and thereby also to legitimate the Israelite cult and monarchy.

6. The once-conquered chaos has broken forth again, and God will intervene at some future point to subdue chaos or dissolve the present creation and will then re-create the world and the chosen people. This eschatological vision, not unrelated to the hope for a better future which existed throughout Israel's history (Preuss 1968), acquires special importance in late prophecy and apocalypticism (Preuss 1978; Hanson 1979). YHWH will lead in a final, cataclysmic battle against the evil forces (e.g., the Gog and Magog oracles in Ezekiel 38–39; also Zechariah 9–14; Daniel 7–12) and after the victory will rule as universal king—similar to the divine warrior-king symbolism of early Israel (see the preceding paragraph). But here the lines between good and evil are much more clearly drawn, and the emphasis falls on the new order which will emerge: a new heaven and a new earth (Isa. 65:17; 66:22), a new Jerusalem (Isa. 60; 65:18–19; Zech. 2:1–5 [Hebrew, 2:5–9]), the people restored to the land (Amos 9:13–15), a new heart in the people (Ezek. 36:26; cf. Jer. 31:33), and a tranquility reminiscent of the idyllic garden of Eden (Isa. 11:6–9; 65:19–25). The social location for such visions of the new creation is usually to be found in groups which have their roots in the prophetic or sapiential traditions, and which are living in a time of powerlessness or oppression. This suggests, then, that the intention is to envision an alternate world in which God will vindicate their suffering and establish them in peace.

These six main types form a varied profile for ancient Israel's cosmogonic conceptions. They cannot all be collapsed into a single intellectual idea or ideology without doing them violence. However, they can be compared together in order to ascertain their common and their distinctive elements and the views of reality to which they all give voice. As perplexing as it might at first seem to us to find six very different cosmogonic myths in one culture, it becomes understandable when we consider that ancient Israel's history spanned some one thousand years and that during that period there was a full range of political, social, economic, institutional, and religious variations. The multiple cosmogonies correspond to various aspects of this history. As symbols, the myths and rituals "store up" fundamental cultural meanings (see, among others, Geertz 1973, 127), and the variety of meaning complexes in Israel's history issued in a similar variety of cosmogonic visions. When a given myth originated, it would have been passed on and retained as long as later generations saw in it a proper interpretation of their reality. But this was not a passive preservation, for tradition-historians have discovered that each one of the six cosmogonic myths—sketched above only in their typical forms—experienced change, accretions, and reinterpretations as it was being transmitted through the centuries. They were not all synthesized or reduced into a single controlling symbol. At this point, as in many others, Israel was able to maintain and affirm pluralism as a distinct aspect of her heritage and identity.

In light of such diversity, we should only expect that there will be different ways in which order in the world is perceived—or perhaps, rather, that various myths will highlight different aspects of the order. These need to be ascertained before we can ask about any ethical dimensions of this order. The comments which follow will largely be restricted to the first three types of myths. These are more narrowly tied to the birth of the cosmos and the initial creation of all life, whereas in the biblical tradition the latter three envision essentially a re-creation or a new establishment of order.[2]

II. Cosmogony and Order

Archaeologists, working in the Near East for one and a half centuries now, have brought to light numerous creation texts, representing a wide spectrum of cultures in the ancient world. Together they constitute an intriguing assemblage, each part of which must be understood in light of the distinctive environmental, social, political, and religious experiences which the particular cultures expressed in mythic form. Regrettably, these various cosmogonies have not been as fully exposed to the current methods of the history of religions and the phenomenology of religion as they have been studied in light of the narrower interests of the various historical fields. For example, the Babylonian creation epic from the early second millennium B.C.E., the

Enūma eliš (Pritchard 1969, 60–72), has been frequently examined in its own right and in comparison with the Israelite texts. Yet despite all who have followed in the wake of Gunkel's landmark religionsgeschichtlich work, *Schöpfung und Chaos* (1895; see also Anderson 1967 and Westermann 1972), one does not have the impression that one understands fully how this classic text and its articulated experiences of reality are to be related to the many other cosmogonic expressions in ancient Near Eastern narratives and poetry.

Common Ground in the Ancient Near East

In lieu of such a needed study, one can highlight several points of comparison among the extant texts from Near Eastern antiquity, in the area extending from Egypt to the land of the Two Rivers.[3] It should be emphasized at the outset that the following features are not necessarily to be found in every text, but where not expressed they seem to be implicit in the diverse cosmogonic myths or are at least not inconsistent with the respective cultures' views of the origin of the world. The common ground which the bulk of these conceptions of creation shares would seem to include the following.

1. The cosmos is viewed as a closed, three-storied whole, all parts of which are under the control of divinity, either polytheistically or monotheistically conceived. Among the various cultures there are several specific points of similarity, e.g., the division of the waters between those above and those below, a sharp delineation if not antagonism between the waters and the dry land, the essentially solid dome of the heaven, the importance of the heavenly bodies of light, the idea of a netherworld, the pillars of the sky and the pillars of the earth.

2. The cosmos and humanity did not come into existence by chance or without intention, for some creator god(s) is directly responsible for its existence. The underlying theological pattern, then, is that the divine sphere established the mundane sphere—not merely its physical characteristics but in fact also the fundamental nature of reality, the social institutions, the human features which are lasting. To the extent that order prevails in the world, it is of divine origin. There was virtually no controversy over this point; the "only" problem which one meets is whether and how humans are able to understand the means and the principles by which the deity created.

3. There is virtually no sense of *creatio ex nihilo*. Something exists prior to the creation act—whether it be the elements of chaos in Babylon, or the waters of Nun from which the Egyptian creator-god Atum emerged, or "the face of *tĕhôm*" in Genesis 1, or the dry land

in Genesis 2. Creation out of nothing is not articulated in the Hebrew tradition until ca. 100 B.C.E. in 2 Maccabees 7:28.

> I beseech you, O child, to look at the sky and the earth and, having seen everything that is in them, to recognize that God did not make them out of things that existed [or: God made them out of things that did not exist]. Thus also humanity comes into being.[4]

But before this notion was given currency at this late date, it was common to hold that some type of substance existed prior to the creation act and that it either belligerently opposed the creator or impassively provided the raw material out of which all else was fashioned. To use Aristotle's categories, the material cause preexists the creator, whose role is more that of the efficient than the formal cause, but never the first cause as understood in Christian philosophy. In good mythopoeic fashion, the material cause—be it the waters, the land, the darkness, or the rebellious forces—remains throughout the culture's life as an entity with which the people must contend if their social and individual existence is to continue. As at creation, the real threat always is not "non-being" but "anti-being."

4. For this reason, the creation act is not limited to bringing the physical world into existence. Rather, above all it establishes the proper order of things in this world. This appears to be a fundamental characteristic of ancient Near Eastern cosmogonies. The principle of order achieves its most detailed development in Egyptian thought, as Schmid (1968; see also 1973) has demonstrated, even though it is clearly present elsewhere as well. The Egyptian term *maʿat*, commonly translated as "(world-) order," "correctness," "truth", or "righteousness," embraces all aspects of reality—cosmic, social, and moral. *Maʿat* was established at creation and in Egyptian religion was deified as the daughter of Re. No distant rubric, *maʿat*—hence order—is directly and repeatedly associated with six spheres of life: law, wisdom, nature and fertility, war and victory, cult and sacrifice, and kingship. Royalty, in fact, is the bracket that holds the other five areas together inasmuch as the king is charged to maintain the order in all. Schmid demonstrates in his study that this essential principle of order prevails also in the other ancient Near Eastern cultures, especially in Sumer, Ugarit, Asia Minor, and the Hittite kingdom. In each setting, the principle is associated with cosmogony and underlies the important aspects of cosmic and human existence. It is no less apparent in Israel, where

the parallel term is *ṣedeq/ṣĕdāqâ*, commonly translated "righteousness." In the Hebrew tradition the principle is not essentially supramundane, hypostatized as deity itself. Nor is it commonly associated with YHWH's role as leader of the people, i.e., the image of the ancestral god. Nonetheless, it is basic to the worldview of the people and to the orderly workings of society. Humans are expected to act in harmony with this order, and a system of rewards and punishments awaits their actions. "Righteousness" for humans is thus not fundamentally a stance of piety but a pattern of behavior which supports rather than subverts the cosmic and moral order.

5. The underlying theological pattern of the various creation accounts is the sovereignty of the gods, and usually also the identification of the creator or "high god" with the one who ultimately maintains and judges the world. The interest in the dawn of the universe does not in the first instance stem from a curiosity about the "how" of the process, although we should probably not underestimate the people's fascination with such details. Theologically, references to creation especially underscore the role of the creator as the one to whom the world owes its existence. Yet far from leaving the matter as a datum of past time, the worshipers extend this divine sovereignty over the nations and over rival gods. Since each people perceives a special relationship between their creator god and themselves—the Babylonians and Marduk, the Egyptians and Atum or Re or Ptah or Khnum, the Canaanites and El, the Israelites and YHWH—this is effectively an affirmation of that people's own supreme importance among other nations. It is also typical for the people to believe that, while this creator god can and does intervene directly in human affairs both within that country and in foreign lands as well, the king or some other leader has the divine charge to maintain the order on behalf of the gods.

6. Throughout the ancient Near East, there appears to be no single literary form which alone is considered especially proper for speaking about the creative activity of the gods. Rather, the believers draw on the full arsenal of literary means in order to give it expression: epic myths, hymns, proverbial sayings, contemplative discourses, disputations, motive clauses, and more. The imagery is rich; the creativity boundless. In virtually all instances, the articulations do not intend simply to recount the past, but above all to interpret its meaning for the present. Such reactualization complements also the praise and awe that similarly undergird these statements about the creator.

Distinctive Ground in Ancient Israel

As significant as is this common ground among the various cosmogonies of the ancient Near East, we cannot understand them adequately unless we ascertain also the distinctive ground for each. In this context, I will restrict my observations to the Hebrew tradition. I should, however, note that each of these points will not necessarily be unique in the whole ancient Near Eastern world. What makes them distinctive is their appearance together in the total configuration of ancient Israel's view of the world's origin.

1. Cosmogony and anthropogony are in Israel strictly divorced from any theogony, whereas the creation of the gods often figures quite integrally in the cosmogonies of her neighbors. To be sure, one could simplistically attribute this difference to Hebrew monotheism. However, we have learned not to speak facilely about monotheism in Israelite theology, certainly not in any abstract or philosophical sense. For the Israelites apparently reckoned with the existence of other gods from the beginning of their history, stipulating only that the deities worshiped by other nations were decidedly weaker than YHWH—until the exilic and postexilic periods when these foreign gods could be discredited to the point of being virtually denied existence. Yet throughout their history the Israelites do, in fact, posit other heavenly beings with whom even YHWH associates. These include: the angel/messenger figures (especially frequent in the Elohist tradition); the curious "sons of God" in Genesis 6:1–4; the spirit, which normally in the early periods is identified with the dynamic essence of YHWH yet which in places seems to acquire a nearly independent, personalized form, such as the "evil spirit" of 1 Kings 22:19–23; "the adversary" of Job 1, Zechariah 3, and 1 Chronicles 21; the heavenly cortege described in Isaiah 6 and elsewhere; the image of a heavenly counselor in Deutero-Isaiah and elsewhere; and even the oblique reference to a divine plurality in Genesis 1:26a ("and God said, let *us* make humanity in *our* image, according to *our* likeness"; cf. also 3:5, 22; 11:7). If for the Israelites the supremacy of YHWH over everything was of paramount importance, one might suppose that this could have been most effectively established by picturing YHWH as creating all other heavenly beings just as YHWH also formed the world and humanity. One can only speculate that this is lacking because it could too easily have allowed for an enfranchisement of these other divine beings alongside YHWH. If there ever existed in Israel any theogonic myths, they have carefully been denied access to the Hebrew Bible. Note only the passing references to God's creation of Behemoth (Job 40:15), of Leviathan (Ps. 104:26), and of the serpent in the garden (Gen. 3:1)—all probably mythical remnants (cf. Childs 1960, on the concepts of "broken myths" and "mythical vestiges"). Nor is there any statement about God's own origin that can approximate that of the Egyptian creator-god Atum: "I am the great god who came into being by himself" (Pritchard 1969, 4, from "The Book of the Dead").

For the Israelites creation consisted solely of cosmogony and anthropogony.

2. Furthermore, the Hebrew myths dealing with the birth of the cosmos envision no struggle between the creator and any other beings or substances. In stark contrast is the Babylonian *Enūma eliš* where Marduk must first slay the rebel deities Tiamat and Kingu before the cosmos and humanity can be formed. The Hebrew Bible, as Gunkel discovered in his pioneering work (1895), contains fragments of such conflicts, above all in the poetic literature. "The deep," *tĕhôm* (without the definite article, almost as if it might be a proper name: "Deep"), is tantalizingly suggestive of Tiamat, the salt water, and the primordial mother.[5] Yet this and other mythic figures offer no resistance whatsoever to YHWH, not even as a narrative foil. With the most detail to be found anywhere, the poet in Job 40–41 can picture Behemoth and Leviathan as mere playthings in YHWH's hands. The motif of the splitting of the sea, reminiscent of Marduk and Tiamat, is appropriated by the Israelites primarily as an image of God's saving power in separating the exodus waters. YHWH's holy wars and apocalyptic conflict (types 5 and 6, above) are described as occurring in the course of history. At the dawn of time YHWH's sovereignty and freedom prevail without contest.

3. The Hebrew tradition preserves, in close connection, two different styles of creative activity, each of which separately has an ancient Near Eastern counterpart but which together are not found elsewhere. In Genesis 1 God, with magisterial manner, creates the world and all life in it, calling everything into existence by means of verbal precatives, as might an architect. A somewhat similar creation by word is to be seen in the Egyptian piece known as "The Theology of Memphis" (Pritchard 1969, 4–6). In contrast with the Priestly view in Genesis 1, the Yahwistic narrative which follows in the next chapter describes a drama in which YHWH is much more anthropomorphically involved. The grand scale of the universe is passed over with the slightest comment, and instead we hear of a god who creates and then associates intimately with human, plant, and animal life. Such direct involvement and even dialogue in the creative processes can also be found elsewhere in the ancient Near East. Yet by juxtaposing both texts in Genesis, the Hebrews are suggesting that neither is complete in itself, that both grandeur and intimacy belong to creation, that both can be willed by the same deity.

4. In the Hebrew myths, we meet a view of humanity which at one and the same time presents us with a task, a liberation, and an indictment. Since these will be treated in more detail below, I need only to call attention to them briefly at this point. In Genesis 1, hu-

manity is presented not simply as the epitome of the creative acts, as many have thought, for all parts of the created order have full worth and proper place. However, to humanity falls the task of dominion, of ruling after the manner of YHWH. The liberation is apparent in both Genesis 1 and 2, for humans are created not of evil or rebellious substance (as in the *Enūma eliš*), nor with a tragic flaw (as in some Greek tragedies), nor with a divided being part of which longs for the supramundane world (as in some Greek thought), but with a fundamentally good and cohesive nature far removed from any evil principle (see Ricoeur 1969, for a study of these diverse mythic alternatives). Humans are thus bestowed with a free will, an independence to act as they choose. The indictment, as expressed in Genesis 3, is that humans, despite the divine intentions, exploit their freedom to ill ends. Nonetheless, it is a high view of humanity that is expressed in wonder and praise by the authors of Psalm 8: Given the expanse of creation, it is remarkable that God gives so much attention to humanity. Whereas elsewhere in the ancient Near East it is not uncommon for the universe and humanity to be created especially for the benefit of the gods—in order to provide them with sacrifices, to adore them, to care for their needs—in Israel humankind is set as beneficiaries in responsible partnership with YHWH.

III. Ethical Dimensions

What do these multiple cosmogonies in the Hebrew tradition, with their affirmations of the orderliness of the world, have to do with ethics?[6] None of them is presented, in a formal sense, as an ethical treatise, yet all have aspects which bear on the moral nature of creation. The key to discerning these ethical dimensions lies in the nature of myth itself. A synthesizing symbol, myth brings together a specific group's fundamental views of the world, its concept of self and society, its sense of the quality of its life, and its moral and aesthetic evaluations. Myth presumes to be able to "identify fact and value at the most fundamental level," such that "the powerfully coercive 'ought' is felt to grow out of a comprehensive factual 'is' " (Geertz 1973, 127, 126). The myth does more than present the presuppositions for morality, for in a more general sense it synthesizes the worldview and the ethos of the people. This suggests that the order symbolically depicted in the myth correlates to the people's vision of order in the world, and thus also to the quality of life in this world and to the moral terms for living. Myth is less likely to dictate specific moral directives than it is to inscribe the general—and normative—contours for moral life. It can point to the nature and locus of the good, to the cosmic context of moral behavior, to the nature of moral agency, and to the significance of moral community. At the same time, the myth may indicate the threats to the moral life as well as the consequences of immoral conduct.

Our task at this point, then, is to ascertain the ethical dimensions of cosmogonic order according to the Hebrew tradition. Again the emphasis will fall on the first three types of cosmogonic myths, with occasional reference to the latter three. There are five main areas in which the Israelite myths seem to link value with fact, i.e., morality (also aesthetics) with ontology and cosmology.

The Integrity of the Created World

According to the Israelites, the order and purposefulness that prevail in the world are the immediate result of the divine plan. This is perhaps most apparent in the Priestly creation story in the opening chapter of Genesis. There God, like a cosmic architect, brings all things into existence in an orderly manner, assigning each its place and proper role. Despite a slight telescoping of creative acts, which suggests an earlier rendition with its own but different pattern (Westermann 1967–74, 111ff.), the structuring of the process into the typical folkloristic pattern of seven parts conveys an unmistakable sense of completeness and symmetry. This is reinforced by reporting the events of each day in a strict fivefold scheme: announcement ("And God said"), command ("Let . . ."), report of the creative act, evaluation ("And God saw that it was good"; lacking only in the second day), and temporal framework ("And it was evening and then morning, a [third] day"). The Yahwistic account of creation which follows immediately in the second chapter makes, by comparison, considerably less use of repetition and literary framing devices. Yet it also displays a clear, inner coherence, accomplished through dramatic heightening, foreshadowings, and character interaction. Most important, the story in Genesis 2 is incomplete without Genesis 3, the eruption of disorder into the pristine garden. Perhaps remarkably to us, the ancient Israelites were not disturbed by differences between Genesis 1 and Genesis 2–3: in the final Pentateuchal edition they let the two stand side by side, rather than conflating them into a single story as they did for the two accounts of the flood. In this manner, each narrative interprets the other, adding details and perceptions where the other is silent. The one with massive proportions and stately style is complemented by the other with its deceptive simplicity of form and content. In the one, we see the cosmos in its fullness and breadth; the other portrays human existence in its complexity and depth.

Throughout both, God is in full control—except where humanity chooses otherwise.

In affirming YHWH's overwhelming, indeed absolute responsibility in creation, the Hebrew tradition is able to emphasize that the cosmos and therewith all reality are founded in accordance with the divine will and order. The world acquires thereby a fundamental integrity and is depicted as being the proper and only sphere for human existence. There are no elaborate scenes of divine activities in some heavenly sphere, nor is there an otherworldly home from which humans have been driven or to which they should return. God meets men and women on the earthly plane, and it is here that human life is to be played out and enjoyed, not shunned. The details of such a this-worldly orientation are more evident elsewhere in the biblical literature than in the cosmogonic myths: Landedness is better than disenfranchisement; the owning of goods than poverty (i.e., so long as the former does not cause others to be poor); companionship than solitude; having children than barrenness; sexual fulfillment (in marriage) than celibacy; strength than weakness; enjoyable moderation and prudence than uncontrolled excess; health than sickness; longevity than early death. The cosmogonic myths correlate with these widely attested moral values, and the myths themselves need do little more than symbolically describe the goodness of creation.

Hebrew *ṭôb* does not mean "good" merely in a moral sense, however. Its semantic field in the biblical literature is very wide: quality in the sense of beauty or pleasantness; usefulness and functionality; reliability; suitability; correctness according to a given standard; moral or religious value. One can reasonably expect that many, if not most, of these meanings are present when the created world is judged to be "good." The cosmos and the human sphere are founded not on caprice or evil but on righteousness, justice, covenantal love, and constancy (Ps. 33:4–7; 89:2,14 [Hebrew, 89:3,15])—an orderliness attributed by some to the role of wisdom in the creative processes (Jer. 10:12; Prov. 3:19–20; 8:22–31). The myth of the act-consequence syndrome (type 3) underscores this sense that the "good creation" is functional, reliable, and moral at its most basic level. Consequently, whenever there is an apparent dysfunction of this global entity, the people make their complaint directly to God, the creator, for immediate deliverance and restitution during their lifetime. The Hebrew Bible is filled with such cries to the heavens for help, and this is even implicit in the fourth and sixth types of cosmogonic myths. The belief in order is connected directly with the problem of theodicy.

Caesura in the Created World

Thus while they affirmed the fundamental integrity of the universe, the ancient Israelites experienced disorder as well, and they had to interpret it for themselves. This disorder can be classified in two types—natural or moral—and both are associated in some manner with the creation motif in the literature. What they have in common is that they represent a caesura in the orderly flow of reality, an incursion of violence and alienation into the intended harmony of things.

Natural disorder is as a rule attributed to the action of the deity, both among Israelites and among their neighbors. Earthquakes, volcanoes, storms, devouring fires, plagues, rampaging animals, swarming insects—such things do not "just happen." The same is the case even for disorders in human behavior—e.g., Saul's evil spirit (1 Sam. 16:14–23), the lying spirit in false prophets (1 Kings 22:19–23), the hardened heart (Exod. 4:21 and Deut. 2:30), quarrels (Judg. 9:23), involuntary homicide (Exod. 21:13). Since the divine sphere is held to be the cause of all that is inexplicable in the human sphere, these go to the account of the gods. In polytheistic religions demons or malicious gods are typically held responsible for the bizarre occurrences. Yahwism, however, could not tolerate any divine or semidivine rivals to God, even though it is not unreasonable to suspect that Israelites earlier in their history—and quite probably in the popular religion of later periods as well—believed in demons also. Presumably the demonic element became "absorbed" (on this suggestion, see Volz 1924) into the developing conception of YHWH, associated with his numinous character, and identified frequently with his theophanic appearances, not the least in the image of the divine warrior. As amoral as YHWH's wrath may at times appear, it became increasingly interpreted—especially by the prophets and the Deuteronomists—as punishment for human evil. In this manner, natural disorders lose their anomalous character. They become understandable as expressions of divine power and will, the same sort which marked the creation of the world.

However, the caesura with more ethical immediacy is the experience of moral disorder, thus the problem of evil. How is this to be reconciled with the fundamental affirmation of this world and the high view of humanity that we find in the Hebrew tradition? Humanity is not created out of evil substance, from the blood of the slain rebel-god Kingu, as is the case in the *Enūma eliš*. Nor is this earth, together with humanity, considered in itself to be evil, or even merely undesirable. The remarkable feature of the Hebrew view of creation, as Ricoeur (1969, 232–78, on the "Adamic myth") so persuasively argues, is its effort to separate the origin of evil from the origin of the good. What is primordial is the goodness of this world and of humanity; what is radically intrusive is the evil which humanity does. For Israel, YHWH has no responsibility for moral evil, which has no part in cosmogony, or specifically in anthropogony. In the idyll of this world, humans interject discord. The myths do not speculate about why humans do this in the face of such a benevolent, functional universe.[7] The point is that people *do* disobey and *do* disrupt their world. Mythically, this could scarcely be communicated with more dramatic effect than is done in Genesis 3, a story not of a fall but of a separation, an alienation, a drive to try it alone. The case in point is not a heinous crime, an act easily recognized as abhorrent by all, but rather a peccadillo, nothing more than eating a forbidden fruit. Yet this dramatizes all the more wherein the severity lies: an act of rebellion against the creator of the good. Humanity, not YHWH, is thus held fully and solely accountable for the occurrence of evil in this world.[8] Just as important, evil in the Hebrew tradition is assigned no ontological value. It has no essence, no independent existence. It is something which is *done* by humans, deleterious acts directed against other humans and against God. Even the serpent in Genesis 3 does not represent an evil principle. As Ricoeur has interpreted this narrative figure, it symbolizes—among other things—only the *tradition* of evil, the experience which we all share in finding evil *already there* before we each are even born (Ricoeur 1969, 257 58). It is not until its contact with Zoroastrianism, apocalyptic, and early Christianity that the Hebrew tradition acquired a belief in the Evil One who radically opposes both God and humanity.

Yet the Hebrew tradition clearly considers moral disorder to be a lasting existent, at least until the New Jerusalem is established—thus in the period between the *Urzeit* and the *Endzeit*. Moreover, such disorder has attained a magnitude well beyond the mere peccadillo of Genesis 3. The hamartiological legends which follow in Genesis 4–11 are also myths of the beginning in the sense that they project into the primal time the realities which the Israelites experienced. Here we see an escalation of immorality—from disobedience to jealousy, to murder, to rampant evil intentions, to subservience, to collective megalomania. The human world has thereby become recognizable. Yet it remains against YHWH's best intentions for the world, a point which is most clearly expressed in the punishments of Genesis 3:14–19. The suffering and oppression which YHWH sets for the serpent, the woman, and the man are marks of rebellion, not of creation. As Trible has pointed out, they are not prescriptions but descriptions. Such subordination and alienation, far from being sanctioned, represent human perversions of creation (Trible 1978, 123–39). Morality must set as its goal not the continuation of such divisive structures but the overcoming of them. Even YHWH the creator is invoked to act to this end (e.g., Ps. 74:20–23).

The Results of Moral Action

In Israel morally good acts are not considered simply to be essential ends in themselves. One's action is expected to produce a necessary, consistently predictable result. The just, benevolent, appropriate deed brings reward and harmony; the evil, destructive, foolish act yields punishment or deprivation. This closed circuit of act and consequence[9] was discussed above (the third type of the cosmogonic myths) as being a part of the order which was established with the birth of the cosmos. It was basic to the worldview of the sages, and it achieved parenetic force in the Deuteronomists' theology of the two ways: that obedience leads to blessing, whereas disobedience calls forth curse. If one cannot expect it to be so and if there is no afterlife where the earthly accounts will be settled, then why should one feel compelled to be moral?

Precisely this dilemma is faced squarely in the Hebrew literature. There are too many anomalies in the world for them to be disregarded by any but the most pious and the most reclusive. Already in the early proverbial literature, we meet observations that the wicked do not always suffer or the righteous always succeed (see Gladson 1978). The problem, faced also in Egypt and Mesopotamia, is brought to new heights in Israel with Job's cry of protest to God. His concatenation of virtues in Chapter 31 contrasts starkly with the treatment he has had to endure. With no concept of fate to which retreat can be made, Job suspects that the order of cosmic justice has failed him. Qoheleth is no less gentle with his complaint. In an argument which ranges between realism and cynicism, he concludes that a human has no chance to know the system according to which God responds to human action—if there even is any system. One should not trouble oneself with trying to ascertain the divine scheme of things, for it is enough simply to act as best one can and to enjoy the small pleasures in life.

Actually, as Crenshaw (1976, 26–35) has argued, the sages manage ultimately to defend divine justice only through recourse to creation theology. With resolute faith that God established an order which undergirds cosmic and human existence, they expect the good finally to win out. To be sure, retribution may not necessarily come as one might wish, but the wise are willing to accept it in whatever form it may assume: whether through a reversal of fortune (Ps. 37), through an encounter with God (Job 42:1–6), through comfort in the cult (Ps. 73:17), or through a conviction that the wicked rich must surely have to endure nightmares and anxieties (Sir. 40:1–11; note how these are connected directly with the primal myths, especially in vv. 10–11). Even if such arguments ultimately beg the question of evil and suffering, it is not insignificant that creation theology allows the believers to affirm the orderliness of the world and to launch into praise of God who judges as well as creates.[10]

Coexistence within the Human World

Both creation accounts in Genesis picture humans in community as the ideal state of existence. It is, moreover, a cooperation of partners, devoid of subservience until the mythic act of rebellion. The Hebrew Bible seems clear that human meaning can be experienced to the full only when one is in association with others, and moral values elsewhere in the Hebrew Bible tend to be ordered to give priority to the corporate body. Yet this does not lead to a total absorption of the self into the group, for each individual retains integrity, rights, and just due. It also does not mean that human relatedness in itself is necessarily and always a good. When the community assumes structures of abuse, oppression, or exploitation, it falls subject to massive rebuke, as occurs especially at the hands of the prophets.

Whereas these first two types of cosmogonic myths provide a general picture of constructive and destructive human relationships, type 3 embodies a conception of reality which is basic to much of Israel's social ethics. It can be set schematically as follows: YHWH created the world according to *ṣĕdāqâ*, "righteousness," a principle of moral and cosmic orderliness similar to the Egyptian *maʿat*. When *ṣĕdāqâ* prevails, the world is at harmony, in a state of well-being, in *šālôm*. An act of sin in the religious sphere or injustice in the social sphere can inject discord and shatter *šālôm*. It then takes a decisive act of *mišpāṭ*, "justice," to restore the *šālôm* and reestablish the *ṣĕdāqâ*. This *mišpāṭ* is not, as in our judicial system, an impartial judging between the violator and the injured party. Rather, it is an act of partiality which is not concerned simply to punish the guilty but to restore the victim to full participation in the community. Only when all deserving persons enjoy the fullness of life in community can *ṣĕdāqâ* reign. World order is thus not a static concept, an essence which exists impervious to all else. It is predicated directly on full moral behavior in the social world, and YHWH is perceived to be its protector par excellence. With this, we can better understand YHWH's role as the restorer of the injured and the liberator of the oppressed. YHWH acts even here as the creator, the one with the power to reform the world and reestablish life—a *creatio continua*, in a sense. This is most clearly expressed in Deutero-Isaiah, where creation is virtually identified with deliverance and YHWH is proclaimed, in cosmogonic terms, as the creator of Israel (e.g., Isa. 43:15; 44:24). To the community, the moral mandate, as stated repeatedly by the prophets, is to maintain the order and to restore it when it is disrupted.

Coexistence with the World of Nature

Hebrew cosmogonies prescribe a similar symbiosis between humanity and nature, a relationship with distinct ethical dimensions. Both spheres are created by God, and each retains its distinctive connection with the divine realm. The Israelites are quick to praise the beauty of the natural world (Ps. 8; 19; 29; 104), to see it as a model of order (Prov. 6:6–11; Jer. 8:7), and even to marvel at its secrets (Prov. 30:18–19, 24–31). This goodness, however, is not merely aesthetic, for it is given a distinct theological interpretation. To be sure, Israelite religion tended to resist viewing nature as sacral in character, as some of her neighbors did, not least the Canaanites. Yet the Hebrew tradition pictures God establishing the natural order, forming the idyllic garden, and pronouncing it all "good," an affirmation which denotes both functionality and quality. Nature does not possess divine traits or actually reveal God, but it does witness to or proclaim the divine glory (Ps. 19). YHWH's awesome power is especially evident in the theophanies, which typically cause an upheaval of nature, a shaking of the earth's foundations. In Israelite theology, the purpose is to bring all people to the knowledge of the deity and the divine works.

God can use nature to purpose—to bless the people (e.g., Deut. 11) or to curse and punish them by means of famine, locusts, plagues, and earthquake. Actually, it is not simply a matter of God's "using" nature to reward or to punish. The act/consequence cosmogony (type 3) envisions such pervasive order in the closed circuit of creation that whatever humans do, whether for good or for ill, will necessarily have repercussions in nature as easily as among people. The story of the flood illustrates this well. Furthermore, with the cursing of the ground itself in Genesis 3:17–19, YHWH is not declaring a "fall of nature" as if it will henceforth have some fundamental defect. Rather, this symbolizes an alienation between humanity and the natural world, and it is attributed to human error. In the scheme of cosmic order and retribution, the earth must also reflect some of the disarray caused by evil act. How much more dramatic, then, is Jeremiah's pathetic picture of the wasteland after YHWH's anger has struck at Israel's sin (4:23–28). Compare this return to chaos, however, with the return to the idyllic garden in Isaiah's eschatological vision (11:6–9).

Nature does not only provide the context for human life, but it also is due a specific style of human moral behavior. In a sense, this is because it has its own integrity and independence from the world of people. According to Ricoeur (1969, 258), the serpent in Genesis 3 symbolizes, among other things, the fundamental indifference of the cosmos to humanity's ethical demands—not antagonism necessarily, but simply an apathy or disinclination to be of ready service. Yet in another sense, our respect is also due nature because the earth is our mother, the womb from which we emerge and the source of our ongoing sustenance (see especially Mowinckel 1927).

Consequently, the Hebrew tradition sets unambiguous moral demands on humanity's relation to nature. The image used in Genesis 1–2 is that of dominion, a concept which for us can easily be twisted into a privilege of exploitation and manipulation. Yet this runs directly counter to the royal ideal of the ancient Near East (see especially J. M. Miller 1972). Kings, although they often enough rule despotically, are charged by the gods to show compassion and concern for those under their authority. Similarly, when humans are given dominion over animals, fish, birds, and plant life, there are limits set to their rights over the natural sphere. The world must be respected, and its viability must be preserved and enhanced. Otherwise, the consequences for the people will be grave. In the final part of his asseveration of virtues, Job—with a not-so-veiled allusion to the curse in Genesis 3:17–18—proclaims to God (31:38–40):

> If my land cry out against me
> and its furrows weep together,
> if I have eaten its produce without payment
> and caused its owners to sigh,
> then let thistles grow instead of wheat,
> and foul weeds instead of barley.

IV. Conclusion

The expanse of ancient Israel's literary heritage reveals multiple cosmogonic traditions. The six types delineated above all coexist in what is conventionally called Yahwism, but each stems from socially distinct groups or from different sociohistorical periods. They thus give voice to the viewpoints and values prevalent in diverse settings: priestly, agrarian, sapiential, prophetic, cultic, apocalyptic. Considered individually, each one has counterparts in the ancient Near East—usually in its basic structure, but at least in its motifs and elements. However, viewed as a group, they constitute a distinctive set of conceptions not to be found elsewhere. This corresponds to the distinctive historical, social, political, environmental life of the Israelite people.

Did any of these cosmogonic myths *begin* an ethical tradition in Israel? One can find no basis for maintaining that this was the case. Each myth emerged as its group's symbolic and fundamental articulation of the nature of the world, the qualitative relation of the world to human life, and the general mode of behavior expected of humankind. As such, each envisions a comprehensive order of reality, and each attributes this order to the cosmogonic activity of God, who variously displays benevolence and power. There is a frank depiction of humanity's tendency to disrupt this order, yet also an unswerving charge to humans to avoid doing this. Furthermore, the tradition is given full rein in pressing the hard questions of theodicy. Throughout, the myths affirm that the good world is the proper arena for human life. The ethical order of created reality thereby receives its most basic affirmation.

Notes

The author gratefully acknowledges receipt of a Vanderbilt University Fellowship and a Fulbright-Hays Grant, which supported research on this project in Israel during the year 1981–82.

1. Not all biblical scholars would agree that there are any myths in the Hebrew Bible, or specifically that these cosmogonic depictions are mythical. The question turns in part on the definition of myth and in part on one's view of the nature of the Bible. I will resist the temptation to engage these issues in this context and will only indicate cryptically that the understanding of myth as used here is most informed by anthropological studies and the theory of symbolism.

2. The three images of deliverance, holy war and kingship, and apocalyptic cataclysm are all cosmogonies in the sense that they envision the creation of a new reality. However, in each case the myths obviously presuppose an already existing cosmic and human reality that will then be *re*-created. The ethical implications of this, too extensive to receive due attention here, would deal with the nature and origin of the first world and the reason for the eruption of new disorder; the means, especially power and violence, used to overcome it; the preferential treatment of one group of people over others, often at the expense of others; and the dualistic structure of reality, including a sharp distinction between good and evil.

3. The list of studies of these cultures and their literatures is, of course, seemingly endless. I call attention here only to the following general works, which contain references to other secondary as well as primary literature: Westermann 1967–74; Harrelson 1970; Ringgren 1973; Smith 1952; and, on methodological issues of comparison, Frankfort 1951.

4. One can well question, however, whether this text in fact intends the same meaning as does the patristic formula of *creatio ex nihilo*.

5. The scholarly consensus now, however, is that Hebrew *tĕhôm* is not philologically derived from Babylonian *tiāmat;* consequently, a mythic connection cannot be substantiated on etymological grounds. The two words seem each to have stemmed from a common proto-Semitic root.

6. Since the early part of this century there has been little pursuit of the discipline of ancient Israelite ethics, at least not in the sense of ethics as a philosophical or theological system of inquiry. Methodological problems are far from being resolved. For preliminary discussions of tasks facing the discipline, see Knight 1982 and Barton 1978.

7. This seems to be another instance of what Auerbach (1953, 11–12) identifies as a basic characteristic of Hebrew narrative: that the stories are "fraught with background" (*hintergründig*), externalizing "only so much of the phenomena as is necessary for the purpose of the narrative, all else left in obscurity."

8. For a more detailed study of moral agency in another part of the Hebrew tradition, see Knight 1980.

9. The programmatic essay on this *Tun-Ergehen-Zusammenhang* or the *schicksalwirkende Tat* in Israel is Koch 1972. The principle can—with caution but also with some legitimacy—be viewed in terms of a "natural theology"; see especially Collins 1977.

10. Some of the best examples of this are the "doxologies of judgment" in Amos; see Crenshaw 1975.

References

Anderson, Bernhard W.

1967 *Creation versus Chaos: The Reinterpretation of Mythical Symbolism in the Bible.* New York: Association Press.

Auerbach, Erich

1953 *Mimesis: The Representation of Reality in Western Literature.* Princeton: Princeton University Press.

Barton, John
1978 "Understanding Old Testament Ethics." *Journal for the Study of the Old Testament* 9:44–64.

Childs, Brevard S.
1960 *Myth and Reality in the Old Testament*. Studies in Biblical Theology, 27. London: SCM.

Collins, John J.
1977 "The Biblical Precedent for Natural Theology." *Journal of the American Academy of Religion* 45/1 Supplement (March), B:35–67.

Crenshaw, James L.
1975 *Hymnic Affirmation of Divine Justice: The Doxologies of Amos and Related Texts in the Old Testament*. Society of Biblical Literature Dissertation Series, 24. Missoula: Scholars Press.
1976 "Prolegomenon." In James L. Crenshaw, ed., *Studies in Ancient Israelite Wisdom*, pp. 1–60. New York: Ktav.

Cross, Frank Moore
1973 *Canaanite Myth and Hebrew Epic: Essays in the History of the Religion of Israel*. Cambridge: Harvard University Press.

Frankfort, Henri
1951 *The Problem of Similarity in Ancient Near Eastern Religions*. Oxford: Clarendon.

Geertz, Clifford
1973 *The Interpretation of Cultures: Selected Essays*. New York: Basic Books.

Gladson, Jerry
1978 "Retributive Paradoxes in Proverbs 10–25." Ph.D. dissertation, Vanderbilt University.

Gunkel, Hermann
1895 *Schöpfung und Chaos in Urzeit und Endzeit: Eine religionsgeschichtliche Untersuchung über Gen 1 und Ap Joh 12*. Göttingen: Vandenhoeck and Ruprecht. English translation of excerpts, in Bernhard W. Anderson, ed., *Creation in the Old Testament*, pp. 25–52. Issues in Religion and Theology, 6. Philadelphia: Fortress Press; London: SPCK, 1984.

Hanson, Paul

1979 *The Dawn of Apocalyptic: The Historical and Sociological Roots of Jewish Apocalyptic Eschatology*. Rev. ed. Philadelphia: Fortress Press.

Harrelson, Walter

1970 "The Significance of Cosmology in the Ancient Near East." In H. T. Frank and W. L. Reed, eds., *Translating and Understanding the Old Testament: Essays in Honor of Herbert Gordon May*, pp. 237–52. Nashville-New York: Abingdon Press.

Knight, Douglas A.

1980 "Jeremiah and the Dimensions of the Moral Life." In James L. Crenshaw and Samuel Sandmel, eds., *The Divine Helmsman: Studies on God's Control of Human Events*, pp. 87–103. New York: Ktav.

1982 "Old Testament Ethics." *The Christian Century* 99/2 (20 January): 55–59.

Koch, Klaus

1972 "Gibt es ein Vergeltungsdogma im Alten Testament?" Reprinted, in Klaus Koch, ed., *Um das Prinzip der Vergeltung in Religion und Recht des Alten Testaments*, pp. 130–80. Wege der Forschung, 125. Darmstadt: Wissenschaftliche Buchgesellschaft. English translation, in James L. Crenshaw, ed., *Theodicy in the Old Testament*, pp. 57–87. Issues in Religion and Theology, 4. Philadelphia: Fortress Press; London: SPCK, 1983.

Miller, J. Maxwell

1972 "In the 'Image' and 'Likeness' of God." *Journal of Biblical Literature* 91:289–304.

Miller, Patrick D., Jr.

1973 *The Divine Warrior in Early Israel*. Harvard Semitic Monographs, 5. Cambridge: Harvard University Press.

Mowinckel, Sigmund

1927 " 'Moder jord' i det Gamle Testament." In *Religionshistoriska studier tillägnade Edvard Lehmann den 19 augusti 1927*, pp. 131–41. Lund: C. W. K. Gleerups Forlag.

Preuss, Horst Dietrich
1968 *Jahweglaube und Zukunftserwartung*. Beiträge zur Wissenschaft vom Alten und Neuen Testament, V/7. Stuttgart: W. Kohlhammer Verlag.
1978 Editor, *Eschatologie im Alten Testament*. Wege der Forschung, 480. Darmstadt: Wissenschaftliche Buchgesellschaft.

Pritchard, James B.
1969 *Ancient Near Eastern Texts Relating to the Old Testament*. 3d ed. Princeton: Princeton University Press.

Rad, Gerhard von
1966 "The Theological Problem of the Old Testament Doctrine of Creation." In G. von Rad, *The Problem of the Hexateuch, and Other Essays*, pp. 131–43. Edinburgh-London: Oliver and Boyd. German original in 1936.

Ricoeur, Paul
1969 *The Symbolism of Evil*. Boston: Beacon Press.

Ringgren, Helmer
1973 *Religions of the Ancient Near East*. Philadelphia: Westminster Press.

Schmid, Hans Heinrich
1968 *Gerechtigkeit als Weltordnung: Hintergrund und Geschichte des alttestamentlichen Gerechtigkeitsbegriffes*. Beiträge zur Historischen Theologie, 40. Tübingen: J. C. B. Mohr [Paul Siebeck].
1973 "Schöpfung, Gerechtigkeit und Heil: 'Schöpfungstheologie' als Gesamthorizont biblischer Theologie." *Zeitschrift für Theologie und Kirche* 70:1–19. English translation, in Bernhard W. Anderson, ed., *Creation in the Old Testament*, pp. 102–17. Issues in Religion and Theology, 6. Philadelphia: Fortress Press; London: SPCK, 1984.

Smith, Morton
1952 "The Common Theology of the Ancient Near East." *Journal of Biblical Literature* 71:135–47.

Trible, Phyllis
1978 *God and the Rhetoric of Sexuality*. Philadelphia: Fortress Press.

Volz, Paul
1924 *Das Dämonische in Jahwe*. Tübingen: J. C. B. Mohr [Paul Siebeck].

Westermann, Claus
1967–74 *Genesis*, vol. 1. Biblischer Kommentar: Altes Testament. Neukirchen-Vluyn: Neukirchener Verlag.
1972 *Beginning and End in the Bible*. Philadelphia: Fortress Press.

6 Cosmogony and Ethics in the Sermon on the Mount

Hans Dieter Betz

That the New Testament should have anything at all to say on the topic of "Cosmogony and Ethical Order" is not to be taken for granted. Rather, the first thing to be said is that characteristically cosmogony plays only a marginal role in the New Testament and that its ethics is not based on creation narratives as those in the Book of Genesis. Obviously, early Christian theology has broken at this point with a long Jewish tradition, when it bases Christian ethics on christology and soteriology rather than on a creation narrative. This fact is confirmed, not denied, by another and no doubt later development in New Testament theology, when Christ was made the preexistent mediator of creation. In this concept, christology and soteriology were revised again, in order to integrate them with creation theology.

As far as the earlier stages of New Testament thought are concerned, there is only one exception to the general rule of disinterest in creation theology. This exception is the so-called Sermon on the Mount (Matt. 5:3–7:27), in my view one of the oldest textual units in the New Testament. In the Sermon on the Mount (in the following, abbreviated SM), a Jewish-Christian ethics is based directly (and without a christology) on a cosmogony in the strict sense of the term. Before this peculiar concept of cosmogony and ethics can be discussed, it is necessary to describe briefly the nature of the Sermon on the Mount.

I. Introductory Remarks

1. As I have tried to demonstrate in a number of articles, the SM as we find it in Matthew 5:3–7:27 is a *pre*synoptic compositional unit. Its literary genre is that of an epitome of the teaching of Jesus, made for the purpose of instructing the members of a church in those subjects which were regarded as essential by that church (see Betz 1979a). The work was not composed by Jesus himself, but in all probability goes back to a Jewish Christian group or church, perhaps in Jerusalem, where it may have been composed in about the year 50 A.D.

The material is taken from what the author or authors of the SM believed to be the sayings of Jesus. These sayings were selected, reworked, and rearranged so as to conform to the author's idea of an epitome.

2. The author or authors of the Sermon were very close in time and religious and cultural environment to the historical Jesus. If they determined that this teaching was the essence of Jesus' teaching, they had a greater chance with it than with any other text in the New Testament of being right. This does not mean, however, that the words of the SM simply go back to the historical Jesus. When it comes down to individual sentences, a decision is very difficult. No doubt many of the sentences do go back to Jesus. But there are others which certainly do not come from Jesus himself but from early interpreters of his teachings. Other material is not different from Jewish wisdom material or from Jewish cultic traditions (see Betz 1975), so, if Jesus used it, he was simply a transmitter and not the author of new revelations. What we can assume is that the theological positions attributed by the author or authors to Jesus were close to those he actually held.

3. All of the teachings of the SM can be accounted for in the terms of Palestinian Judaism of the early first century A.D. As a result, there are no traces of a christology and soteriology as we know it from the letters of Paul and from the gospels. Jesus figures in the SM as the authoritative teacher and interpreter of Jewish Torah (see Betz 1982). There is no reason to assume that he stood higher than Moses or was regarded as a "new Moses." His authority is based on nothing but the assumption that he interprets the Jewish Torah in an orthodox way. His interpretation of the Torah is set in opposition to the conventional (i.e., literal) and the Pharisaic interpretation on the one hand, and the gentile Christian abrogation of the Torah on the other. Thus Jesus' authority was the authority of Torah as he interpreted it.

4. By contrast, the kerygma of the death and resurrection of Jesus plays no role in the SM: it is never mentioned, although it seems to have been known to the author(s). In Matthew 7:21–23, a group of gentile Christians are shown when they appear before the throne of

God in the Last Judgment (see Betz 1981). They complain of not having been accepted, and they turn to Jesus as their advocate in the belief that he will come to their rescue. But Jesus rejects them because they have not kept the Torah faithfully. They are told that, because they did not adhere to his way of interpreting the Torah, they have no right to appeal to him as their advocate. It is not said but implied that these (no doubt gentile) Christians relied on Jesus as a savior figure—that is, on a gentile Christian soteriology. Therefore, the rejection of the claims of these gentile Christians implies that their christologies and soteriologies, which we know were based on the kerygma of crucifixion and resurrection, were also rejected.

5. Because of the role that Jesus the Messiah plays in public perceptions today, it should also be said that the SM does not contain the slightest interest in messianic ideas. Jesus' authority, therefore, does not rest on a special messianic appointment by God, a special call, or a special charisma. Jesus is viewed in the SM as a Jewish teacher, certainly one of great authority, and it is believed that God will appoint him as the advocate of his faithful in the Last Judgment. These roles were in no way incompatible with the Judaism of his time.

6. The consequence of the foregoing consideration is that the ethical order of which the SM has a great deal to say cannot be based upon Christian ideas of salvation as we find them elsewhere in the New Testament. What is the source from which this ethical order is derived? What is its authoritative basis? To say that the SM bases its ethical order on the Jewish Torah is correct. But what is the decisive point *within* the Jewish Torah? Most remarkably, it is a theology of creation, and, for that matter, a peculiar one. This is remarkable because no other text in the New Testament has its ethical order based directly on a theology of creation. The peculiar nature of this creation theology is that rather than the creation mythology of the Book of Genesis (Eliade's *in illo tempore*), the SM bases its ethical order on God's *creatio continua*, his creational activities in the present.

7. The SM is also a document of polemic, carried on against several adversaries at once. One argument is the inner-Jewish polemic against what is described as conventional Judaism and against Pharisaism. A second is the polemic against gentile Christianity. Finally, there is polemic against the general hellenistic culture. This last polemic, however, is more than simply a rejection of the external manifestations of the hellenistic culture. The SM also includes very succinct discussions of topics which were discussed by hellenistic philosophy. Thus, for example, Matthew 6:22–23 contains an argument against Greek theories of vision (see Betz 1979b).

8. One must also realize that whatever the SM has to say reflects the contemporary political situation (see the statements on paganism

in Matthew 5:47; 6:7, 32). The community represented in the SM is aware of its status in the world as an embattled minority. Thus the outside political world of the Roman Empire is viewed in mostly negative terms. There are, however, no open or direct statements, and there is conspicuous silence on points of obvious concern—certainly a strategy of self-protection. Comments on the political world occur only in oblique form or through metaphors. At the same time, the SM shows no sign of defeatism or inferiority feelings. On the contrary, small and insignificant as this community of the SM may have been politically, its members were convinced that, despite their situation as a suppressed and even a persecuted minority, they had a grip on history (cf. Matthew 5:13–16). In fact, the SM itself is a manual for the survival of the community in both history and the hereafter (cf. Matthew 7:24–27).

II. Analysis of Matthew 6:25–34

The SM as a whole having been described, I now return to the topic of cosmology and ethical order in the SM. Because a full investigation would be too large a task for one paper, my focus must be narrowed to one key passage. This passage will be Matthew 6:25–34, a section usually entitled "On Anxiety" (Aland 1976, no. 67). The title "On Anxiety" is misleading, however. It takes up the first term of the passage as the *Leitwort* with which the passage is primarily concerned, but there are two problems:

1. The term *merimnaō* should not be translated as "being anxious" but as "worry." The reason is that "being anxious" focuses on an inner, psychological disposition, while the Greek term primarily designates external attitudes (consider the so-called "worry beads"). Admittedly, the psychology and the attitudes of worrying cannot be separated, but the failure to distinguish between them has often led interpreters of Matthew 6:25–34 in the wrong direction. The passage, therefore, does not deal primarily with feelings but with practices and attitudes.

2. Although, as the *Leitwort* indicates, the section speaks about worrying, the real topic of the discussion is divine providence (so, rightly, the *Jerusalem Bible*). The terminology of providence does not occur in the SM, but the subject matter was important to Jewish wisdom, from which most of the material of the passage comes. More important, the subject matter of providence was widely discussed by contemporary Greek philosophy, and in these discussions similar arguments and some of the same conceptional, proverbial, and metaphorical material occurs. This strategy of oblique theological and philosophical argumentation is typical of the SM.

Conspectus of the Analysis

25a	I. Introduction
	A. Connection: "therefore"
	B. Doctrinal formula: "I say to you"
25b	II. Exhortation
	A. Imperative (negative)
	1. Observation (presupposition)
	a. Worrying as general human behavior
	b. Object in question: life or soul *(psychē)*?
	2. Prohibition: "Do not worry " (cf. vv. 31a, 34a)
	B. Caricature description of behavior to be rejected
	1. Confusion of *care* for one's life (soul) and *procurement* of the means of life
	a. Eating
	b. Drinking
	2. Confusion of *care* for one's person (body) and *procurement* of clothing
25c–30	III. First argument
25c	A. Theses (rhetorical questions)
	1. Life (soul) is more than means of life (nourishment)
	2. Person (body) is more than clothing
	B. Conclusions (implied)
	1. Foolish to confuse the *procurement* of food and clothing with the *care* for life (soul)
	2. Prudent to care for one's life (soul) and person (body)
26–30	C. Proofs
26–27	1. The means of life: a comparison of animals and humans
26	a. The example of the birds
26a	(1) Appeal to observe their behavior: "Observe " (cf. v. 28b)
	(2) The paradox to be observed with regard to the efforts made
	(a) They do not sow
	(b) They do not reap
	(c) They do not gather into storehouses
	(d) Yet, they eat
26b	(3) Conclusion: God feeds them
26c	(4) Consideration (rhetorical question addressed directly)

(a) Presupposition: the traditional distinction between animals and humans, giving a higher rank to the humans
(b) Conclusion *(a minori ad maius):* If worrying about food is redundant among the animals, so all the more among humans

27 b. The example of the humans (rhetorical question addressed directly)
(1) The paradox to be observed with regard to the future
(a) No one can add a span to one's lifetime
(b) Yet, each day adds to one's lifetime
(2) Conclusion (implied): God measures one's lifetime
(3) Consideration
(a) Presupposition: The future is measured by God, not by humans
(b) Conclusion *(a maiori ad minus):* If the future as a whole is under the control of God, it is improper for humans to worry about a portion of that future, i.e., future nourishment, as if it were under human control

28–30 2. Clothing: a comparison of plants and humans
28a a. Presentation of the problem (rhetorical question addressed directly)
(1) Reference to v. 25c
(2) Reference to worrying as general human behavior
28b b. The example of the lilies
(1) Appeal to observe their behavior: "Learn "
(2) The paradox to be observed with regard to the efforts made
(a) They do not work
(b) They do not spin
(c) Yet, they grow
(3) Conclusion (implied): God lets them grow
29 (4) Consideration
29a (a) Doctrinal formula: "but I say to you"
29b (b) Presupposition: the splendor of the royal garment of Solomon is believed to be unsurpassable

(c) Conclusion *(a maiori ad minus):* The clothing of the lilies, because a work of God, surpasses even the splendor of Solomon, which is human

30 c. The human example (rhetorical question addressed directly)

30a (1) The paradox to be observed with regard to the future

(a) The destiny of the lilies

i. Today they stand

ii. Tomorrow they are thrown into the oven as fuel

(b) Yet, God fashioned them so splendidly

(2) Conclusion (implied): God wastes the future on his creatures

30b (3) Consideration

(a) Presupposition: God measures the future without considering the perishability of his creatures

30c (b) Conclusion *(a minori ad maius):* If God treats his lower creatures in this way, how much more will he treat in the same way his highest creature, humanity

30d (4) Address (simultaneously the transition to the next argument): "men of little faith"

31–33 IV. Second argument

31a A. Connection: "therefore"

B. Repetition of the exhortation of v. 25b

1. Imperative (negative)

a. Observation (presupposition)

(1) Worrying as general human attitude

(2) Object: the future generally

b. Prohibition: "You shall not worry "

31b 2. Caricature description of behavior to be rejected (dramatization)

a. Eating

b. Drinking

c. Clothing

32–33 C. Proofs: a comparison between gentiles and Jews (Jewish Christians)

32a 1. The example of the gentiles

a. Observation of their improper *striving for* the goods of life

b. Comparison and identification with the behavior rejected in v. 31

32b 2. The traditional doctrine

a. Of God's omniscience

b. Of his provisions with regard to the elementary human needs

33 D. Conclusions

33a 1. Exhortation

a. Imperative (positive)

(1) Preferred and proper behavior: *seeking* instead of *worrying* (vv. 25b, 31a, 24a) or *striving for* (v. 32a)

(2) Priority: "first"

(3) Preferred and proper object

(a) Overarching: the Kingdom of God

(b) Specific: righteousness

33b 2. Promise

a. Condition: "and [only] then"

b. Traditional doctrine of divine reward

(1) Eschatological (presupposed)

(2) This-worldly (derived): "everything"

34 V. Third Argument

34a A. Connection: "therefore"

B. Repetition of the exhortation of vv. 25b, 31a

1. Imperative (negative)

a. Observation (presupposition)

(1) Worrying as general human behavior

(2) Object: the future; specifically, tomorrow

b. Prohibition: "You shall not worry"

[2. Shortening of the argument through omission of a description of the behavior to be rejected]

34b–c C. Proofs

34b 1. Gnome with reference to tomorrow

34c 2. Gnome with reference to today

D. Conclusions (implied)

1. *A minori ad maius:* If the future as a whole is under the control of God, it is improper for humans to worry about a portion of that future, i.e., tomorrow, as if it were under human control.

2. *E contrario:* If tomorrow is not under human control, today is.

3. *E. contrario:* If it is therefore improper for humans to *worry,* it is indeed proper for them to *seek* the Kingdom

> of God and the righteousness corresponding to it. This righteousness is identical with the ethics of the Sermon on the Mount.

As the analysis shows, the passage contains a tightly constructed argument. It starts with the main exhortation governing the whole argumentation (v. 25a-b). Then three subsidiary arguments are presented to substantiate the truth of the main exhortation (vv. 25c–30; 31–33; 34). Each time, the initial exhortation is restated with small but important differences (vv. 25a; 31a; 34a). In order to understand the arguments, one must distinguish between the surface of the text, which states only the rhetoric and the course of development, and the presuppositions underlying the points made. Both the presuppositions and the method of argumentation are sometimes stated, but they often are only implied, a method of argumentation which is in keeping with Jewish forms common at the time.

The beginning exhortation (v. 25b) is probably intentionally ambiguous: "Do not worry for your *psychē*. . . ." This prohibition first recognizes a human attitude prevailing at the time. To an immeasurable degree, hellenistic man was a worrier There was, of course, plenty of reasons for such worrying. There were severe political and economic circumstances, which have often been described. It is less frequently realized that people had been taught to worry for several centuries. Hellenistic literature, religion, and philosophy—Jewish as well as non-Jewish—are full of encouragement to worry. What were people worrying about? Worrying went in all directions and confused things to a nonsensical degree. Most important, people were worrying about their *psychai*. This term can be interpreted in several ways, and one should consider the various meanings instead of deciding on one.

The exemplifications make things clearer; that is, they point out the confusions more clearly. People habitually worried about eating, drinking, and clothing. Traditionally, these three items were regarded as the essentials for human life. But here also lies the first problem: the confusion of livelihood, of the means for life, with the essentials of life. If one is concerned about the means for life—food and clothing—this is one thing; if one worries about life itself, one ought to worry about things other than the means for life. Then there is another problem: people have been taught to worry first of all for their souls. Socrates was among the first who taught that the concern and care for one's soul ought to take precedence over every other concern. At the time of Jesus, Judaism was also largely preoccupied with the concern for the soul. Was this a legitimate concern within the Jewish religion?

The second of the exemplifications involves clothing. People were habitually worried about their clothing because clothing was regarded as essential for the protection of the body. So far, so good! The term for body *(sōma)* is again ambiguous, referring either to the physical body or to the human person as a whole. One should not overlook the contrast between *psychē* and *sōma* as fundamental anthropological terms. On the other hand, the confusion of clothing with the whole person was often criticized by moralists as typical of the stupidity of the general culture. Worrying about clothing is one thing; worrying about the person is quite another thing.

The initial exhortation, therefore, takes up general human attitudes which were regarded as preeminent in the hellenistic culture as a whole, both Jewish and non-Jewish. On the one hand, these attitudes were based on fundamental anthropological assumptions and, in that sense, they were considered realistic. On the other hand, there was a profound confusion about what things to worry about. As a result, hellenistic people were given to worry all the time and in all sorts of ways. But according to the SM, the confusion about proper and improper worrying renders their whole attitude ineffectual and even foolish. This confusion must be corrected, and the wrong attitudes must be changed.

The first of the three arguments made in Matthew 6:25–34 includes vv. 25–30. It consists of a general argument, a kind of *theologia naturalis,* pointing out the confusion between the means for life and life itself, and between clothing and the body/person. First a thesis is stated in form of a rhetorical question, with a positive answer implied: "Is not life more than nourishment, and the body more than clothing?" (v. 25c). Certainly they are. Consequently, the implied answer is, while it makes sense to worry about life and body, it makes no sense to do so by worrying about nourishment and clothing.

Why does it make sense to worry about life and body? This question is postponed here and taken up in the second argument (vv. 30–33). The question which instead is taken up at this point is one regarding nourishment and clothing. Two proofs are presented to show why worrying about nourishment and clothing makes no sense.

The issue of nourishment is taken up first (vv. 26–27), then the issue of clothing (vv. 28–30). The argument is made first by an example from nature, specifically the birds (v. 26). In order to understand this argument, two presuppositions must be realized.

1. The culture at large, and the Greek philosophers in particular, made the distinction that human needs involve both the natural and the artificial, both the elementary and the luxurious. The traditional

formula "eating and drinking" designates the natural and elementary needs.

2. There existed a long tradition of both Jewish and non-Jewish thought concerning the origins of human civilization and concerning the differences between the human world and the animals. In accordance with this tradition, it was assumed that, regarding the natural needs, mankind is not different from the animals, since the animals have only natural needs. At this point, therefore, observation of nature is recommended as illuminating because it reveals some paradoxes. The birds do not do what human farmers do: sowing, harvesting, and gathering into barns. Yet birds are able to satisfy their natural needs. How do they do it? The conclusion to be drawn is that God feeds them through nature. Therefore, if mankind would do what those who worry about eating and drinking say they do—that is, if they would confine their needs to the natural ones—God's provisions through nature would prove sufficient to take care of these needs. Worrying about natural needs makes even less sense if it is granted that, in the hierarchical order in nature, human beings are far superior to the animals. Since animals do not need to worry about eating and drinking, why should humans?

The fact is that those who profess that they are worried about eating and drinking are in reality striving after the better goods of civilization, if not the luxuries. Nature does not satisfy those needs, of course, nor does God provide for them. Such needs arise from artificial expectation and must be supplied by human toil against the odds and vicissitudes of both the natural and the civilized worlds. To worry about them would indeed make sense, but this is not what in reality is done by those who worry about eating and drinking. In fact, by talking about eating and drinking, they avoid the fact that the real object of their concern is not what is necessary, but what is desirable.

A second example is then taken from the human sphere (v. 27): "But which of you who worry is able to add one cubit to his span of life?" Again, this example serves to expose human pretentiousness. The example referred to shows paradoxes similar to those in the animal kingdom: life is extended day after day, but human beings are prevented by their limitations from adding even the shortest span to their lives. The fact that day after day life is extended is not due to human efforts, therefore, but to God's provision of time.

Those who worry about life also imply that they worry about the lifetime, the future. But, as the example shows, the lifetime is not under human control. Therefore, it is pretentious to worry about lifetime. And yet, this is what the worriers do. Why do they do it? The real reason is that they want to avoid that which they should worry about: God. They ought to worry about God, the provider of time, and about themselves who, day after day, receive his time. But human beings are clever. They do not want to worry about themselves and God, so they worry about something over which they have no control.

One ought to keep in mind that these incisive criticisms are not directed at the superficial masses, the pagans, the conventional Jews, but at the people of the church of the SM themselves, who are constantly addressed in the second person plural. In other words, those addressed are the concerned, the morally and theologically sensitive, the disciples of Jesus. They are led to self-criticism by the exposure of the clever escape routes of their thoughts. Their worrying about life, about eating and drinking, and about the future has turned out to be nothing but a clever avoidance strategy.

What, then, about the issue of clothing (vv. 28–30)? "Why do you worry about it?" An argument analogous to that in v. 26 is made, now taking as the example the lilies of the field. They do not toil or spin, but they do grow and are in fact more beautiful than even Solomon in all his royal splendor. The paradox is easily explained: God clothes them through his providence in nature. An additional consideration is adduced in v. 30, showing the destiny of the beautiful lilies. Accordingly, God not only makes the lilies beautiful but he also allows them to end up as fuel for stoves. Human practicality classifies them as "brush," takes them for granted today, and uses them for fuel tomorrow. This treatment of the flowers by humans not only stands in stark contrast to their treatment by God but also makes a travesty of the human concern for the future.

The conclusions from this observation are then drawn in the end of v. 30. The first conclusion is that if, through nature, God provides for the clothing of plants, it makes no sense to worry about the clothing of humans. At this point, however, a problem arises because of implicit presuppositions. The reason clothing is treated differently from nourishment is that all antiquity was aware of nakedness as the natural state of human beings. Nature can be expected to feed humans, but not to clothe them. How can one expect God to provide clothes to cover human nakedness and to protect the human body from cold and heat? The answer "through nature" would make no sense in terms of ancient anthropology. The answer given by the SM is contained in the address at the end of v. 30: "You men of little trust in God." In other words, God's providence is not simply observable as occuring through the phenomena of nature, but the issue of clothing shows that it is a matter of having trust in God.

Such trust in God is not entirely unwarranted. In the Book of Genesis, the same problem is treated in a similar way following the fall of Adam and Eve. First there is the invention of primitive civilization (Gen 3:7):

> Then the eyes of both were opened, and they knew that they were naked, and they sewed fig leaves together and made themselves aprons.

Where did the fig leaves come from? Adam and Eve took them from the trees, which nature had fortunately provided beforehand. But this primitive outfit was not considered adequate. So God himself gave Adam and Eve a crash course in tailoring (Gen 3:20):

> And the Lord God made for Adam and his wife garments of skin and clothed them.

Now this is the beginning of civilization, and God started it. Hence there is reason to trust that, if God does not provide through nature, he will do so through civilization.

But of course this is not what people who worry about clothing really worry about. Neither fig leaves nor animal skins would satisfy them. According to v. 25, they worry about their *sōma*. It is recognized that clothing does more to the *sōma* than to cover its nakedness. There is an intricate relationship between clothing and the human personality. This was nothing new to say in antiquity but simply part of the presupposition of ancient thought.

Thus, those who worry about clothing really worry about the human personality. What one wears determines who one is taken to be, and this determines who one is. Hence worrying about what to wear seems prudent.

The exposure of this rationale as a self-delusion was no news in antiquity. The wisdom of the proverb "Clothes make the man" was known to antiquity in a multitude of forms. This is indeed what people tend to believe and practice, but it is nothing but foolishness, and everyone knows it. The SM exposes this side of the human foolishness but, at the same time, it values positively the concern for the human *sōma*.

What does it mean to worry about the human *sōma?* This typically Greek noun, for which no Hebrew equivalent exists, can refer to the external body, or to personhood, or to both. In the present context, the consideration of both is typical. How does one worry about personhood legitimately? This question is treated in the next argument

(vv. 31–33), but one should observe already at this point that concern for the *sōma* is approved in principle.

By way of exclusion, therefore, the SM does not endorse a concept of concern for the soul. This is obvious, especially because of the overarching imperative: "Do not worry about your *psychē* " (v. 25). Indeed, this is the view of the SM: "Worry about the *sōma,* not about the *psychē* " In view of the fact that the characteristically Greek notion of care for one's soul was so widespread even in Judaism, the position taken by the SM implies a rejection of the Greek notion. This conclusion is confirmed by the affirmation of the traditional doctrine of the resurrection of the dead in Matthew 7:21–23, where people appear before the throne of God at the Last Judgment as persons *(sōmata),* not as souls *(psychai).*

The second argument (vv. 31–33) addresses the question in a positive way: How should one care for one's *sōma?* This argument presupposes the first (vv. 25c–30) by taking up its results and moving beyond them. First, however, the initial imperative is repeated. The connection is made by the conjunction "therefore" (v. 31a). The imperative "You shall not worry" is now stated in the future tense, and the exemplifications are now dramatized in self-mockery: " 'What shall we eat?' or 'What shall we drink?' or 'What shall we wear?' "

As in the first argument (vv. 25c–30), the exhortation of the second (vv. 31–33) is also justified by proofs, but here the proofs are of a quite different nature: they are theological. The first proof (v. 32a) is syncritical, comparing the attitudes already discussed and rejected with the attitudes of "pagans." The term "the nations" is to be interpreted from the Jewish point of view, as elsewhere in the SM. The conclusion is thereby reached that the behavior prohibited in the previous section conforms to that of pagans, thus constituting assimilation forbidden to faithful Jews.

The second proof (v. 32b) states the traditional Jewish doctrine of divine providence. In this statement, the reality of human needs is acknowledged (cf. vv. 25 and 31), and God's omniscience is said also to include the knowledge of these needs. The question then is, How does God provide for these needs? The answer to this question is given in v. 33, which sums up the theology of the SM as a whole. This answer has two parts. First is a positive exhortation, in which the terminology of worrying is replaced by that of "seeking": "First seek the Kingdom and its righteousness. . . ." Man the worrier is replaced by man the seeker, a change which has far-reaching implications.

The object of the seeking is then named as God's kingdom and its righteousness, objects that do not seem to have anything to do with real human needs. But this impression is deceptive. Seeking the Kingdom and its righteousness is, of course, the main theme and purpose of the SM as a whole. According to the SM, it was also the sole purpose and goal of the teaching of the historical Jesus, a judgment which can be confirmed by many of the older traditions about Jesus. In other words, by taking the precepts of the SM seriously, and by implementing its principles in daily life, the disciples of Jesus are in the proper way seeking the Kingdom and its righteousness. Thus, at this point, the argument made in the section 6:25–34 coincides with the overall purpose and intention of the SM.

The second part of the answer (v. 33b) is a divine promise stated in the form of the passive future: "Then these things shall be provided to you in addition." The promise states the doctrine of merit held throughout the SM. Accordingly, there are two kinds of merit which the faithful disciples can expect: the eschatological reward at the Last Judgment, and a this-worldly reward. The most important reward is, of course, that sought at the Last Judgment, and therefore the Kingdom and its righteousness is to be sought first (v. 33). While that is being done, God's providing for human needs as defined in vv. 25 and 31 is promised "in addition." This "in addition" must not be overlooked, for otherwise the entire argument is falsified. In other words, How does one worry properly about life, the needs of and means for life, and the future? By first seeking the otherworldly Kingdom of God through the implementation of the precepts of the SM. Then God will not only grant the eschatological reward at the Last Judgment, but he will, in addition, satisfy the real needs of human life. Of course, it is understood that the this-worldly rewards will not be granted without the eschatological reward. Therefore, it would be counterproductive to focus one's concern for life directly upon the needs of life, the means for life, and the future. For the SM, there can be no concern for this life except through the concern for eternal life, and there can be no concern for tomorrow except through the concern for the eschatological future. On the other hand, the concern for the eternal life will be rewarded "in addition" by meeting the needs of the real life (eating, drinking, and clothing). And the concern for the *sōma* as personhood will be rewarded "in addition" by meeting the needs of the human body. The expectation of the resurrection of bodies thus corresponds to the recognition of the earthly bodies and their needs. That this correspondence is implicitly directed against the Greek doctrine of the soul should be obvious from the fact that in Greek tradition the preference for the divine soul goes hand-in-hand with the devaluation of the human body.

The third argument (v. 34) is extremely short and concerns itself with the topic of time, a matter mentioned in vv. 27 and 31. Concern for the future means, of course, concern for time. What conclusions about this concern should be drawn from the preceding? How does one properly care for time? In what way does "seeking first the Kingdom and its righteousness" lead to an adequate relationship to time? This question was not answered in v. 33b. First, the imperative prohibition of v. 31 is repeated: "You shall not worry." Then the time is divided into today and tomorrow: "You shall not worry about tomorrow." Not only is worrying the wrong attitude regarding tomorrow, but worrying also makes one take today for granted. This neglect of the present was already criticized in v. 30 as taking the beauty of the flowers today for granted, while using them tomorrow as fuel for the oven. The conclusion drawn is that, in daily life, people who worry about the future really worry about the next day and thereby overlook the present day. Since no human being has control over the next day, it makes no sense to worry about it. At this point, the conclusion reached in v. 27 is repeated. How, then, should one relate properly to time?

The answer is given by the citation of what must have been popular maxims or proverbs. By reference to life-experience, these citations substantiate the judgment. The first sentence predicts: "Tomorrow will worry about itself." The second concludes: "Sufficient for the day is its own trouble." In interpreting these *sententiae,* one should avoid seeing in them a pessimistic mood. There is no intention to be pessimistic but every intention to be practical and realistic.

In practice, we are told, human beings deal with a day at a time, and that is realistic. Every day has enough trouble to face up to, and it is best to come to grips with it. What the trouble will be tomorrow we hardly know, but when we get to it, we will know, and then we will deal with it. And so on. In other words, the proper concern for the future Kingdom of God does also enable a person to deal adequately with time. By ceasing to worry about tomorrow, one is enabled to appreciate the present day and to deal with its problems. When tomorrow becomes that present day, one will do the same. Instead of worrying about the uncontrollable tomorrow, one is freed to divide the time into sections and to deal with the nearest one effectively. Instead of living for a perpetual tomorrow, that tomorrow is allowed to become today. Its imagined threats are allowed to become concrete troubles, which can be handled by concrete actions.

III. Cosmogony and Ethics

What does Matthew 6:25–34 reveal about cosmogony and ethics in the SM? It shows first of all that what the SM has to say on the subject of cosmogony and ethics is said in response to a profound crisis in the belief in the divine providence. This crisis is nothing peculiar to the early Christian movement, but was a common phenomenon of the time. The fact that people at that time were habitual worriers indicates a deep-seated erosion of confidence in the divine providence. All theological and philosophical movements, Jewish as well as non-Jewish, took positions with regard to this crisis and made recommendations for overcoming it. The position taken by the SM is peculiar in several ways. There can be no doubt that it was developed against the background of alternative concepts within Judaism, and also most likely against the background of Greek thought. Within Judaism, the SM draws heavily on the wisdom tradition, but v. 33 goes beyond, and indeed against, Jewish wisdom theology. "Seeking the Kingdom and its righteousness" is not a concept found in Jewish wisdom. In fact, the very special way in which this concept is interpreted in the SM makes it unlikely that, although entirely Jewish, it has any exact parallels in other known theologies of Judaism at the time of early Christianity. It is also surprising in its distance from apocalypticism, a movement that was extremely concerned about the apparent failures of God's providence.

How, then, does the theology of the SM propose to overcome the crisis of confidence in God's providence? The answer to this question is that some very old Jewish ideas have been rethought and appear in a new light, mixed with some new ideas. The prominent concept is that of the Kingdom of Heaven, in which God rules not like a hellenistic monarch but as a Father. There can be no doubt that this concept comes from the historical Jesus. For the SM, God's fatherhood becomes manifest through his *creatio continua* and through his relationship to the disciples of Jesus.

In stating the doctrine of *creatio continua,* the SM does not shy away from some archaic mythology. According to Matthew 5:45, God "makes his sun rise over the bad and the good, and he rains upon the righteous and the unrighteous." This mythical idea is traditional, to be sure, but its repetition in the SM is nevertheless astonishing. Likewise, in Matthew 7:11, human fatherhood at its best is almost provocatively pronounced to be identical with the fatherhood of God: both fathers give good things to those who ask for them. Thus, quite contrary to the hellenistic tendency of depersonalizing the deity, the SM, no doubt intentionally, prefers to speak of the Father in strong anthropormorphic terms. The confidence in the goodness of nature is also unusual in many ways. This goodness of nature is not justified by referring to the creation account of the Book of Genesis, however. Apocalypticism, with its cosmic visions, as well as Philo and Josephus, with their apologetic interpretations, demonstrate that the crisis of skepticism has made simple affirmations of the Genesis accounts impossible.

Nor does the SM imitate the speculative cosmologies of the Greek philosophers, as Philo does. Rather, the goodness of nature is seen manifested in its daily occurrences: sunrise and sunset, the rain, the daily bread, and so on. If such goodness of nature suffices as manifestation of divine creation, the SM has turned away from speculation to what every eye can see every day. If skepticism professes to count only on what can be seen, heard, touched, and tasted—so be it. Nevertheless, the world as viewed by the SM is a dangerous and hostile place. Both the beginning of the SM and its conclusion make this clear. At the beginning, the ten beatitudes (Matt. 5:3–12) state the evils man must confront: poverty, death, brutality, injustice, lack of compassion, rottenness of heart, war, persecution of the righteous, harassment of the church, and martrydom. At the end of the SM, the double parable of the houses built on rock and on sand (Matt. 7:24–27) invokes the old images of the rain, the floods, and the storms. These images represent the unpredictable and turbulent forces and vicissitudes of history. What makes the world such a dangerous place is human sinfulness, not the fatherhood of God. It is God's mercy that keeps the cosmos going, despite the evil and unrighteous who abuse his benefits. It is because of his mercy that there continues to be the chance for repentance. All this is, of course, good traditional Jewish faith.

As far as the church is concerned, the heavenly Father relates to it as to his sons. The disciples of Jesus are regarded by the SM as God's sons. Like all sons of all fathers, they have to grow up, and to help them do this is the purpose of the SM. These sons are taught to imitate their heavenly Father and thus to live as sons of their heavenly Father ought to live. This is the central point of the reconstituted doctrine of providence as presented in the SM. Faith in God's providence can be affirmed no longer in a general way but only in a special way, that is, embedded in a relationship in which a group of Jewish Christians define themselves as sons of their heavenly Father. As they treat him in their lives as Father, so will he treat them as his sons.

The ethics of the SM corresponds to this peculiar form of cosmogony. On the one hand, this ethics is like all Jewish ethics, obedience to the Torah. On the other hand, the emphasis on God's fatherhood, on his providence, on seeking the Kingdom, on the imitation of God and on sonship, all this is a peculiar form of Jewish theology. And so is the concentration on the daily experiences and the turning away from speculation and visions.

Cosmologies alone do not seem to suffice any longer as a general basis for ethics. The cosmic order is no longer credible as a generally acknowledged truth. For the SM, ethics is no longer based on laws and rituals, but it is reinterpreted as imitation of God's fatherhood. Ethics is learning to love the world in the way God loves his creation, in consonance with the daily rising of the sun over the bad and the good, and with the rain sustaining the righteous and the unrighteous.

References

Aland, Kurt, ed.

1976 *Synopsis Quatttuor Evangeliorum.* 9th ed. Stuttgart: Deutsche Bibelstiftung.

Betz, Hans Dieter

1975 "Eine judenchristliche Kult-Didache in Matthäus 6, 1–18." In Georg Strecker, ed., *Jesus Christus in Historie und Theologie, Festschrift für Hans Conzelmann,* pp. 445–57. Tübingen: Mohr, Siebeck.

1979a "The Sermon on the Mount: Its Literary Genre and Function." *The Journal of Religion* 59:285–97.

1979b "Matthew vi. 22f. and Ancient Greek Theories of Vision." In Ernest Best and Robert McL. Wilson, eds., *Text and Interpretation, Studies in the New Testament presented to Matthew Black,* pp. 43–56. Cambridge: Cambridge University Press.

1981 "Eine Episode in Jüngsten Gericht (Mt 7, 21–23)." *Zeitschrift für Theologie und Kirche* 78:1–30.

1982 "Die hermeneutischen Prinzipien in der Bergpredigt (Mt 5, 17–20)." In Eberhard Jüngel et al., eds., *Verifikationen, Festschrift für Gerhard Ebeling,* pp. 27–41. Tübingen: Mohr, Siebeck.

7 Ethical and Nonethical Implications of the Separation of Heaven and Earth in Indian Mythology

Wendy Doniger O'Flaherty

Our tacit acknowledgment of the validity of the subject on which we are writing in this volume—cosmogony and ethical order—implies that we all believe that there is a relationship between the two. To take the two terms in their most basic sense, I would regard a cosmogony as a theory (or, better, a narrative, a myth) about the origin of the universe, and ethical order as a structure relating to the rules of human conduct. (The inadequacy of such definitions is the subject of several of the essays in this volume, but that is a problem which I will not address.) Is there necessarily a relationship between cosmogony and ethical order? If so, does one necessarily give rise to the other? If not, is there usually a relationship between the two? Certainly there are many cosmogonies devoid of any explicit ethical formulations; but do these presuppose an implicit ethical order? Ethical deliberations can make use of all sorts of given cultural structures, including myths and rituals; but this does not mean that myths and rituals must be intrinsically infused with ethical meanings. The Indian materials indicate, I think, that some cosmogonies do correlate, in various ways, with ethical issues, but that others do not. Indeed, there are instances where we can see a clear development from a nonethical cosmogony into an ethical cosmogony, and other instances in which changing ethical needs lead to a major reformulation of the underlying cosmogony out of which they originally arose. Moreover, ethical order may be totally irrelevant to cosmogony, or a cosmogony may support a pattern of human conduct that actively conflicts with the ethics of its society. Thus, whether or not cosmogony and ethical order *must* correlate, in India they often *do* correlate, though this correlation may be one of conflict rather than of mutual support.

Let us first consider cosmogonies to which ethical order is apparently irrelevant. It is, I think, true that all cosmogonies are formulations of order; they explain how a certain sort of order came into the world, and what that order means. There is a limerick to this effect:

> The bases of every cosmology
> Are symmetry, law, and homology.
> For micro- and macro-
> Are front-row and back-row
> Within the one church of Ontology.

But is all order necessarily ethical order? Only in a sense so broad as to redefine *all* philosophical and scientific concerns as ethical, which is, I think, a pointless move. To say that order has meaning is not to say that it has ethical meaning, that it implies certain rules of human conduct. This assertion can, I think, be supported by evidence that one particular cosmogony may have two very different ethical glosses, may be interpreted by different people within the culture as laying down two very different sets of rules for human conduct; and, contrariwise, two very different cosmogonies may be used to support the same ethical formulations. This essay will present several examples of such circumstances.

Some cosmogonies are nonethical. But cosmogonies may also be unethical; that is, a cosmogony may paint a picture of the world in which certain laws are at work which flagrantly violate the rules of conduct held sacred by the people who use that particular cosmogony. Such a conflict may result from historical circumstances: a cosmogony may remain static and therefore become out of joint with a changing view of ethical order. We will see instances where this tension is met by an adjustment in the cosmology, an adjustment intended precisely to change the shape of the cosmos in order to make possible a new ethical development that the older cosmos could not accommodate. But an unethical cosmogony may not necessarily be the result of a historical conflation or lapse; it may simply express a view that the world is itself driven by unethical forces and that therefore certain unethical actions are justified, if we are to fit ourselves to the shape of the universe.

Most cosmogonies represent the evolution of order out of chaos, and in most cosmogonies chaos is evil (antithetical to survival) and even unethical (antithetical to the laws of human conduct); the cosmogonic act is therefore necessarily ethical. But in some cosmogonies, chaos is good and order is evil (O'Flaherty 1975, 11–14; 1980, 293–94); the cosmogonic act is therefore unethical, destructive of the good, and the distinctions made by such an act are invidious, producing inequalities and conflicts that this religion is designed to overcome. To the extent that, as we will see, the cosmogonic act often involves the separation not only of chaos into order but, more specifically, of undifferentiated space into heaven and earth, it may easily be assimilated to the moment of the separation of god and man—a process regarded as evil. Another limerick expresses this thought:

> When the sky and the earth were still close,
> And the planets were wedged tail to nose,
> All would have been well,
> Without heaven or hell,
> Had the gods not become otiose.

Thus the cosmogonic act may be productive of ethical good or of ethical evil.

Finally, ethics may be more or less relevant to a cosmogony depending upon whether evil is viewed as a substance or as a process. If evil is a process, the cosmogony may tell us how that process began, but the actual shape of the cosmos (the cosmology) that emerges from the primeval act of ordering is not necessarily relevant to the problems of human conduct. The way that you think the world *is* may or may not determine the way you think you ought to act; a cosmogony is a structural tool, a way of thinking about anything that concerns you, but it does not necessarily determine the content of that thought. If evil is a substance, however, then the cosmogony is more likely to have direct ethical implications: the shape of the world determines how much room there is for evil, and where it can be distributed. We will see myths of both these types in India; the first (evil as a process) renders the early myths of the separation of heaven and earth nonethical; the second (evil as a substance) renders the later myths of the infinite extension of heaven highly ethical.

Frits Staal (1979) has argued that ritual has no meaning; in a similar sense, one could argue that cosmogony has no ethical meaning. But ritual inevitably attracts meaning to itself, often in the form of the myths that accompany many rituals. Ritual thus provides a blank check on which people cannot help writing meaning, a mold into which people are irresistibly driven to pour meanings; ritual is a black hole which sucks meaning into it, a vacuum into which meaning keeps falling, which meaninglessness abhors. So, too, cosmogonies provide a structure for ethical thinking, a stage on which ethical dramas may be enacted, though the cosmogonies themselves may be ethically transparent. Thus, incest in creation myths does not primarily reflect anxiety about some real or imagined human situation—seduction of children by their parents; it is, rather, an inevitable result of the human tendency to attribute human qualities to the originally abstract sources (a.k.a parents) of the world. Incest, in this sense, is a poetic fallacy.

There are two basic cosmogonies so widespread that scholars tend to regard them as universal, though they are not. They are certainly Indo-European, however, and are attested in many non-Indo-European sources as well. One is the myth of the separation of heaven and earth; the other is the myth of the marriage of heaven and earth. The separation is usually more abstract than the marriage; in the separation, heaven and earth may not be personified at all, while in the marriage they usually are. Because of the greater anthropomorphism of the marriage model, one would expect to find it more susceptible to ethical and psychological glosses than the separation, and this is generally the case. But when the separation is anthropomorphized, and particularly when it is then combined with the marriage, peculiar ethical conflicts arise. For, if the combination of heaven and earth is regarded as a marriage, a fruitful hierogamy, then what can the separation of heaven and earth be but a divorce? And how can this be held as a good thing by the same society that, by sanctioning the marriage cosmogony, affirms marriage as a good thing? These are some of the problems raised by the transformations in the myths of separation and marriage of heaven and earth in the *Ṛg Veda* (c. 1200 B.C.) and the *Jaiminīya Brāhmaṇa* (c. 900 B.C.).

We can, I think, see a definite ethical development in the treatment of cosmogonic themes during this period. The earliest myths, in the *Ṛg Veda,* regard the separation of heaven and earth as a good thing (for the cosmos), and do not draw any conclusions at all as to its implications for human ethical behavior. But other *Ṛg Veda* hymns speak of another kind of separation of heaven and earth, the separation of man from god, which is directly connected with the first, more mechanistic cosmogonic split. This second split is a bad thing, which has clear implications for human ethical behavior: man must act in such a way (offering sacrifice, keeping the laws of the gods, speaking the truth, and so forth) that the gods will not go away. Only the first of these actions—the sacrifice—remains significant as a way to prevent the cosmogonic separation in the next set of texts, from the *Brāhamaṇas,* and sacrifice, by this time, is largely devoid of its ethical components. Instead, it is a means to power, and the "reunion" of man and god is regarded as a way in which power may flow from man to god (through the offering) and from god to man (through the rain and all that rain symbolizes). Finally, however, the *Brāhmaṇas* begin to consider a more subtle problem arising from the cosmogonic divorce and to recast it in terms that foreshadow the genuine ethical deliberations of later Hindu theodicies.

The myth of separation is told repeatedly in the *Ṛg Veda,* where kathenotheism gives credit for the action to various gods on various occasions. A hymn to Varuṇa refers to "the famous Varuṇa who struck apart the earth and spread it beneath the sun. . . . He stretched out the middle realm of space in the trees. . . . Varuṇa placed the sun in the sky. . . . Over the two world-halves and the realm of space between them Varuṇa has poured out the cask. . . . I will proclaim the great magic of Varuṇa the famous Asura, who stood up in the middle realm of space and measured apart the earth with the sun as with a measuring-stick." (RV 5.85.1–5; O'Flaherty 1981, 211). But Prajāpati, too, is the one "by whom the awesome sky and the earth were made firm, by whom the dome of the sky was propped up, and the sun, who measured out the middle realm of space" (RV 10.121.5; O'Flaherty 1981, 28). And when the poet turns to Viṣṇu he says much the same thing: "Viṣṇu, who has measured apart the realms of earth, who propped up the upper dwelling-place, striding far as he stepped forth three times . . . " (RV 1.154.1; O'Flaherty 1981, 226). Indra, too, is the one "who made fast the tottering earth, who made still the quaking mountains, who measured out and extended the expanse of the air, who propped up the sky" (RV 2.12.2; O'Flaherty 1981, 160), and the Soma plant actually *is* what separated earth and sky, "He who is the pillar of the sky, the well-adorned support, . . . the one who by tradition sacrifices to these two great world-halves" (RV 9. 74.2; O'Flaherty 1981, 122; Wasson 1968, 47–48).

The starkness of most of these verses, coupled with the fact that it does not seem to matter who gets credit for the cosmogonic split, lends an air of abstraction and impersonality to the Ṛg Vedic myth. Yet there are verses where the separation of heaven and earth is given vivid anthropomorphic (and theriomorphic) overtones, where the sun is regarded as the child of the sky-father-bull and the earth-mother-cow: "The son of these parents milks the dappled milk-cow and the bull with good seed; every day he milks the milk that is his seed. Most artful of the artful gods, he gave birth to the two world halves that are good for everyone. He measured apart the two realms of space with his power of inspiration and fixed them in place with undecaying pillars" (RV 1.160.3–4; O'Flaherty 1981, 203). The last sentence lands us back in the formulaic, straightforward, abstract version of the myth, but before we get to it we must fight our way through some fairly obscure and riddling verses that bristle with sexual ambiguities: the cow has seed, but one also milks the bull; "he" (almost certainly the sun, but also perhaps the Creator) is the son of the two parents, but he gives birth to them. Such sexual paradoxes are a favorite stock-in-trade of the Vedic poets, but their use in this hymn may be a result of the poet's awkwardness in grafting a set of symbols onto a myth that was not originally a sexual myth. These images of children and seed at first seem more appropriate to the myth of the marriage than to the myth of the divorce of heaven and earth, a marriage that is not directly narrated at all in the *Ṛg Veda.* Yet the sky-father and earth-mother are regarded as parents, and as parents who give birth, even in verses that refer to the usual myth of separation: "The two full of butter, . . . sky and earth have been propped apart, by Varuṇa's law; unageing, they are rich in seed. . . . You two world-halves, rulers over this universe, pour out on us the seed that was the base for mankind. . . . Sky and earth, the all-knowing father and mother who achieve wondrous works—let them swell up with food to nourish us." (RV 6.70; O'Flaherty 1981, 206–7). Evidently, heaven and earth give birth to the sun by separating, not by uniting; later texts express this through the metaphor of the egg that "gives birth to" or releases the yolk (the sun) when the upper shell draws away from the lower. Moreover, in this last hymn, as in the earlier hymn about the cow and the bull, it would appear that the "seed" is not primarily the substance from which children are made but, rather, the substance with which children are fed; the seed is milk. By contrast, in the Upaniṣads the seed is unequivocally procreative seed; a man who is about to impregnate his wife should say, "I am heaven, you are earth; let us embrace and place together the seed to get a male child, a son" (*Bṛhadāraṇyaka Upaniṣad* 6.4.20). The Vedic hymns cited above refer to earth and sky as separate, and to the children that they have engendered and that they continue to feed; they do not describe the union of the two parents, though they refer obliquely to the begetting of the child. Yet what can "the seed that was the base for mankind" mean except the seed shed by sky and earth to produce the human race? Presumably, this seed, like the sun, was creatively shed at the moment when the two worlds separated. These ambiguities suggest that the mixing of the metaphors of separation and marriage had begun even at the time of the *Ṛg Veda.*

These problems are compounded by the existence of two other, secondary myths of cosmogony in the *Ṛg Veda,* myths that can be seen as forming a kind of prologue and epilogue, a temporal frame, to the two that we have begun to examine (see fig. 1). The first is the

Figure 1. Stages of Indian Cosmogony

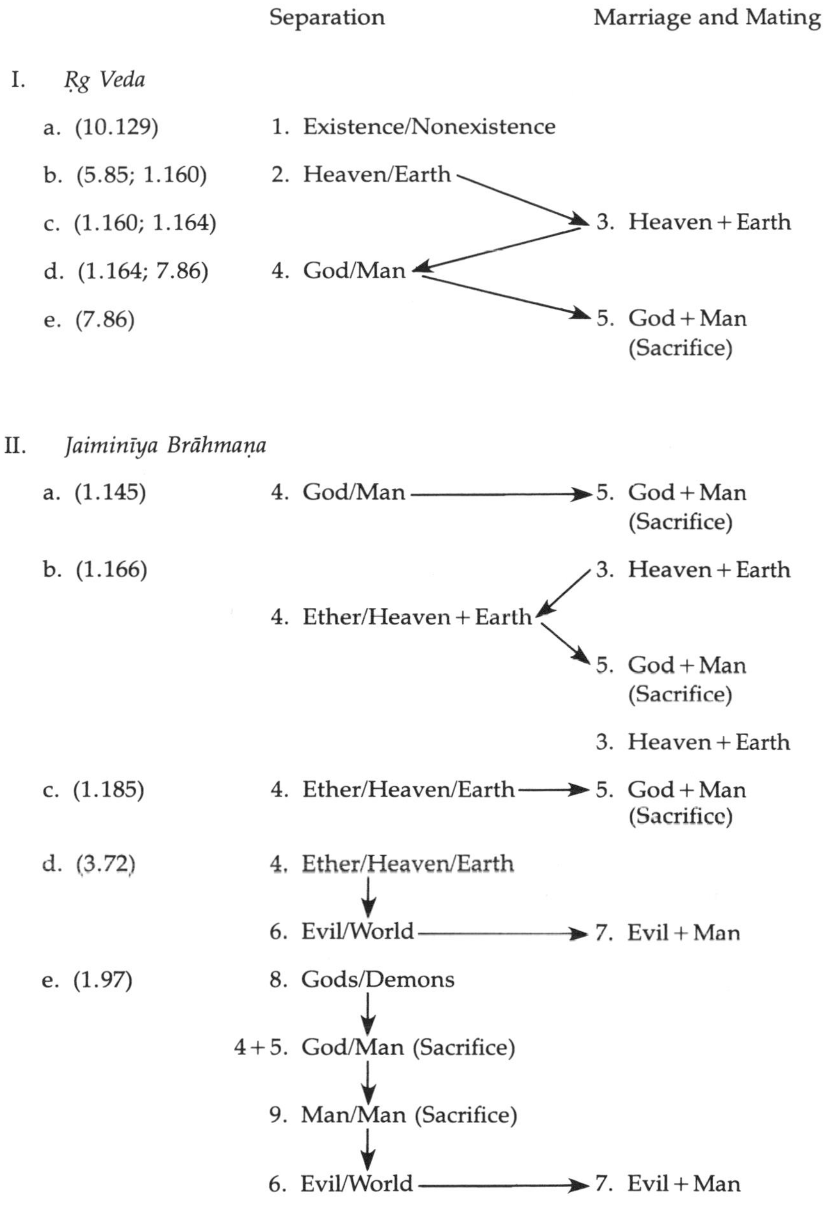

	Separation		Marriage and Mating
I. *Ṛg Veda*			
a. (10.129)	1. Existence/Nonexistence		
b. (5.85; 1.160)	2. Heaven/Earth	↘	
c. (1.160; 1.164)			3. Heaven + Earth
d. (1.164; 7.86)	4. God/Man	↙ ↘	
e. (7.86)			5. God + Man (Sacrifice)
II. *Jaiminīya Brāhmaṇa*			
a. (1.145)	4. God/Man	⟶	5. God + Man (Sacrifice)
b. (1.166)		↙	3. Heaven + Earth
	4. Ether/Heaven + Earth	↘	
			5. God + Man (Sacrifice)
			3. Heaven + Earth
c. (1.185)	4. Ether/Heaven/Earth	⟶	5. God + Man (Sacrifice)
d. (3.72)	4. Ether/Heaven/Earth ↓		
	6. Evil/World	⟶	7. Evil + Man
e. (1.97)	8. Gods/Demons ↓		
	4 + 5. God/Man (Sacrifice) ↓		
	9. Man/Man (Sacrifice) ↓		
	6. Evil/World	⟶	7. Evil + Man

KEY

/ = "separates from" ; + = "joins with"

Letters designate texts; arabic numerals designate themes.

⟶ designates transition within a single myth.

myth that distinguishes between cosmogony and theogony; the second is the myth of the *deus absconditus*.

The *Ṛg Veda* tends to separate the creation of the universe from the creation of the gods: "There was neither non-existence nor existence then; there was neither the realm of space nor the sky which is beyond. . . . Darkness was hidden by darkness in the beginning; with no distinguishing sign, all this was water. . . . Who really knows? Who will here proclaim it? Whence was it produced? Whence is this creation? The gods came afterwards, with the creation of this universe. Who then knows whence it has arisen?" (RV 10.129.1, 3, 6; O'Flaherty 1981, 25). Yet, elsewhere the cosmogony and the theogony seem to have taken place at the same time: "Let us now speak with wonder of the birth of the gods. . . . In the earliest age of the gods, existence was born from non-existence." (RV 10.72.1–2; O'Flaherty 1981, 38). Indeed, this same hymn goes on to describe the procreation of and by the earth-mother—*not*, significantly, in conjunction with the sky-father—in terms both anthropomorphic and self-referential (like the terms describing the sun as both the child and parents of the earth-sky): "The quarters of the sky were born from her who crouched with legs spread. The earth was born from her who crouched with legs spread, and from the earth the quarters of the sky were born. From Aditi, Dakṣa was born, and from Dakṣa Aditi was born" (RV 10.72.3–4). Instead of the male sky, we have four female quarters of the sky. Moreover, these sky deities are the sisters, not the husband(s), of the earth, for all of them are born from the same mother. This mother is simultaneously highly abstract—she has no name and corresponds to no physical part of the universe—and highly anthropomorphic: all we know of her is her posture, which is that of a woman in the throes of childbirth. Finally, the whole group acts out the Vedic riddle of origins, the oldest chicken-and-egg conundrum I know: for earth is then said to give birth to the sky quarters (her sisters, already born from her mother), and then Aditi and Dakṣa gave birth to one another. When the Vedic abstractions are fleshed out, they become embroiled in complex human entanglements. In order for heaven and earth to mate, they must first become created as separate entities; only then can they produce their children. Some cosmogonies emphasize one aspect, some another; so, too, some cosmogonies describe androgynes that must "split" in order to procreate, while others describe androgynes that must "fuse" to procreate (O'Flaherty 1980, 283–334).

We thus have a primeval cosmogony (the making of "distinguishing signs" in chaos), followed by the separation of heaven and earth, followed by the procreative action of heaven and earth, resulting in the creation of the more anthropomorphic gods. The fourth step, the retreat of the creator god or the *deus absconditus*, is also foreshadowed in the *Ṛg Veda.* One Vedic hymn, infuriatingly addicted to sacred riddles, combines this myth with the theme of the sky-father and earth-mother: "The sky is my father; here is the navel that gave me birth. This great earth is my mother, my close kin. The womb for me was between the two bowls stretched apart; here the father placed the embryo in the daughter. He who made him knows nothing of him. He who saw him—he vanishes from him." (RV 1.164.33 and 32; O'Flaherty 1981, 79). Apparently the first "he" in this last sentence is the father, the sky, who vanishes after he has made "him," the sun, the child born of the mating of heaven and earth. The sun is, however, as we have seen, also said to have been born as the result of the *separation* of heaven and earth. Clearly, there is a deep resonance between the two acts; both of them are procreative in the *Ṛg Veda.*

Here the sky-father actually mates with the earth, who is apparently his daughter; this accounts (in the same Vedic hymn) for the fact that earth and sky separated *after their marriage:* the daughter was ashamed. "The mother gave the father a share in accordance with the Order, for at the beginning she embraced him with mind and heart. Recoiling, she was pierced and flowed with the seed of the embryo" (RV 1.164.8; O'Flaherty 1981, 76). Though elsewhere the *Ṛg Veda* seems to identify the daughter as dawn, here she is evidently earth, the daughter of the sky-father but also his consort (as in RV 10.72, where earth is both the mother and the sister of the quarters of the sky). She recoils from him, thus apparently separating heaven and earth forever after. Here, for the first time, we may see a glimmer of moralizing in the theme of separation, though it is certainly a faint glimmer.

On the most elementary level, we can see that the unacceptable results of an incestuous mating, even on the most abstract cosmogonic level, would underscore the prohibition against incest in the human world. On the other hand, we can see that the myth regards incest—again, on the abstract cosmogonic level—as basically necessary. The problem of incest in cosmogonies originally arises from a purely abstract, nonethical problem. If creation begins with some sort of unity (if only the unity of chaos), and then proceeds to a duality that splits and mates, one is stuck with incest, in human terms. That is, the cosmogony is driven by its own, nonethical logic into a position that has unethical implications. We are thus faced with a double paradox, one that applies not only to the incestuous mating but to the original separation: such cosmogonic acts are sinful but necessary, and *both* of them are necessary. The worlds must remain separate, but they also must remain together. The paradox of the creative act that is sinful but necessary appears far more explicitly in India in the myths of Indra, whose slaughter of Vṛtra makes creation possible but also threatens the survival of Indra and hence of the entire cosmos (O'Flaherty 1976, 139–53). More broadly still, it is the paradox of the sacrifice: in early texts, the sacrifice is dangerous to the sacrificer but essential to his survival, and in later texts the slaughter involved in sacrifice is regarded as blatantly evil, yet still necessary. This is not the place to investigate these problems, about which much as been written; but it is appropriate, I think, to point out their relevance to the question on which this essay focuses, the question of the ethical implications of cosmogony.

The double paradox (sinful but necessary, separate but together) reappears in human (albeit theoretical) terms in the dilemma of the *varṇas*. The Kṣatriya must sin, and must be absolved; this problem is expressed in the myths of Indra to which I have referred. The Brahmin, too, has his paradox: he must remain outside the world in order to retain his sacred powers, but inside the world in order that those powers may bear fruit for the sacrificers for whom he acts. Thus, the two paradoxes merge in the Brahmin: his sin *is* his inability to remain both separate and nonseparate. Moreover, together the Brahmin and the Kṣatriya constitute yet another paradox. The two of them must remain separate, by definition and in order to retain their individual identities and powers; yet the Brahmin and the Kṣatriya must intermarry and exchange in order to make it possible for society to function at all (Heesterman 1962, 1–37). Similarly, as Louis Dumont has demonstrated, the "renouncer" must remain separate from society (the fertile cosmogonic divorce), but at the same time he must participate as an outsider (the essential mating) (Dumont 1960, 33–62). More recently still, McKim Marriott has described a complex set of values placed on remaining separate and remaining engaged in Hindu society (Marriott 1976, 109–10). All of these ambivalences of human ethical activity are implicit, *in nuce,* in the Vedic ambivalence toward the necessary, sinful cosmogonic marriage.

A less mysterious and more straightforward form of the myth of the *deus absconditus* appears elsewhere in the *Ṛg Veda*. Here the separation of god and man is depicted in moral rather than physical terms: Varuṇa ignores his transgressing worshiper, though he is not said literally to go away to heaven. Yet, significantly, the withdrawal of Varuṇa is directly linked with his own role in separating heaven and earth:

> The generations have become wise by the power of him who has propped apart the two world-halves even though they are

> so vast. He has pushed away the dome of the sky to make it high and wide; he has set the sun on its double journey and spread out the earth. And I ask my own heart, "When shall I be close to Varuṇa? Will he enjoy my offering and not be provoked to anger? When shall I see his mercy and rejoice?" (RV 7.86.1–2; O'Flaherty 1981, 212–13).

This hymn begins with the primeval situation, unspecified chaos; it describes the beneficial separation of heaven and earth and leaves them amicably separated forever, one assumes. But then it moves on to another, subsequent separation, the separation of Varuṇa and the worshiper, which could not have taken place at the beginning of time but is necessarily secondary. The separation of Varuṇa from the worshiper is clearly regarded as an occasion for great sorrow and regret; it is, moreover, a schism that the worshiper fervently hopes to mend, by giving Varuṇa an offering—the present hymn, as well as some sort of sacrificial offering—that will quell the god's anger and restore the original close friendship. In a closely related hymn, the worshiper recalls the days in which the gods and men lived together happily, before he offended Varuṇa: "Where have those friendships of us two gone, when in the old times we could live together without becoming enemies?" (RV 7.88.5; O'Flaherty 1981, 215). These are among the most heart-searching of the Vedic hymns, highly charged with moral and ethical concerns; and they are linked to the originally a-moral theme of the separation of heaven and earth. The question might then be asked as to whether these two ends of the spectrum (the abstract separation and the *deus absconditus*) are simply logical oppositions of a structural kind or represent a chronological development. A clue to this question might be sought in the further development of the cosmogony, in the Brāhmaṇas.

The *Jaiminīya Brāhmaṇa* is a particularly fertile place to search for such a development, since it is noteworthy among the *Brahmaṇas* for its taste for sex and violence, its tendency to clothe abstract rituals in lurid folktales, and its flair for racy imagery (O'Flaherty 1985). By examining five different *Jaiminīya Brāhmaṇa* versions of the myth of separation and marriage, we can begin to see the ways in which the tradition reinterprets itself in moral and ethical terms that are only dimly adumbrated in the *Ṛg Veda,* even in those hymns that do reflect the procreative metaphor of the myth. Taken in a certain sequence, the *Jaiminīya* texts can be said to describe the marriage, mating, raising of the children, divorce, and rejection of the children (the Vedic custody battle). Moreover, each of these stages is further interpreted by the text itself to point a moral about some other, nonsexual, aspect of human life, usually a ritual aspect.

We begin with the marriage:

> These two worlds were together, but then they separated. Nothing from either of the two reached the other. The gods and men were hungry, for the gods live on what is given from here [offerings], and men live on what is given from there [rain, and from rain, food]. Then the Bṛhat chant and the Rathantara chant said, "Let the two of us get married with our very own bodies." Now, the Syaita chant of fire was the very own body of the Rathantara, and the Naudhasa chant of the Bṛhat. By means of these they got married. The world there gave dawn to this world as a marriage gift, and the world here gave fog to that world. The world there gave rain and the world here gave the divine sacrifice. (JB 1.145. Cf. PB 7.10.1–3; AB 4.27)

The myth begins with a union and a separation. This initial union is not the stage of primeval chaos, however, but a vaguely anthropomorphic union roughly akin to the mating in the third stage of our Vedic scenario. The separation, therefore, corresponds not to the primeval separation (stage 2), but rather to the subsequent separation, stage 4—the *deus absconditus*. The first sort of separation (stage 2) is regarded as a good thing and as *an enduring situation* in much of the *Ṛg Veda* and elsewhere—for even to this day we can see that earth and sky are, in fact, separate. But when the two worlds become anthropomorphic they long for one another as a separated couple would—an early example of the theme of *viraha,* or longing for the beloved: and when they become symbolic of the worlds of gods and men, that *viraha* takes on all the overtones of theological longing, as it will again in medieval *bhakti* literature. Thus the fourth-stage separation becomes a temporary problem that can be solved only by some sort of reunion. In the *Ṛg Veda,* this reunion was achieved through a sacrifice; in the *Jaiminīya* it is a sacrifice conceived as a marriage; more precisely, it is a *re*-marriage. It is not the literal re-union of the two worlds, the complete merging of one with the other as in the situation before stage 1, for primeval chaos would recur (a situation that is, in fact, welcomed in much later *bhakti* texts, but that is anathema to the order-obsessed ancient Brahmins). The reunion is therefore carefully kept in the realm of the anthropomorphic, rather than the abstract, so that mere contact, rather than complete reintegration, is the goal, even as, in *Ṛg Veda* 1.160, the earth and sky procreate by moving apart.

How is this goal to be achieved? The "marriage" is immediately glossed in terms that render inevitable a ritual solution to the problem.

As in the *Ṛg Veda,* the separation of heaven and earth represents a schism between men and gods, and the solution to this is the invention (or, rather, the re-invention) of the sacrifice. For the text implies that once, before the separation took place, men and gods lived in happy symbiosis, like Varuṇa and his worshiper in the *Ṛg Veda*. Elsewhere, the *Jaiminīya* tells of a time when gods and men drank Soma together for the last time, before a great battle took place between them (JB 3.159; cf. JB1.97). Here, marriage gifts are exchanged, and bodies mingle; but the gifts are natural elements (dawn, fog, and rain) plus the sacrifice, and the bodies are twice removed from anything anthropomorphic: the chants stand in for the hymns that represent the worlds of heaven and earth. Thus the *Brāhmaṇa* makes explicit what lies behind much of the *Ṛg Veda* but is never directly linked to the myths of separation there: the sacrifice is the "reunion" that repairs the damage done by the separation from the *deus absconditus.* The sacrifice is the marriage.

A second passage in the *Jaiminīya* expands upon another theme from the *Ṛg Veda:* the depiction of the conjunction of the worlds in terms of human mating. What distinguishes this text from the *Ṛg Veda,* however, is the fact that now there are three worlds, so that the one who is "abandoned" is not the earth or mankind, deserted by heaven or the gods, but the child of heaven and earth, deserted by both parents:

> The light [n.], the cow [f.], and the lifespan [n.] are the three chants. The cow shed her seed in the lifespan; from that was born the light. This world is the lifespan; that world is the cow; and the intermediate ether is the light. They [the lifespan and the cow] made him go in front, as one would a son. This world went up to him and that world went down, and the light went in front of them piled up all around so that we can see. The two of them, wishing to unite, drew the light out to the end [from the middle], just as two parents follow after the son when he goes in front. They became mated chants when they repelled him. The son lies in the middle between the wife and husband. And just as a couple who are about to mate push the son from one side or another, so they did to him. The chants with their plus and minus [penis and vagina] are mated for progenerating. So when anyone desires to have many progeny he should sacrifice with these [chants], saying, "May I have progeny and cattle," and he gets many progeny and cattle. (JB 1.166)

The three worlds of the Brāhmaṇa period—heaven, earth, and the air between—have replaced the two worlds of the early *Ṛg Veda.* (In their turn, they will later be replaced by the three worlds of Hinduism: heaven, earth, and hell.) When the hierogamy is extended to this new Brāhmaṇa model, the roles do not fit easily, and the Brāhmaṇa compounds the confusion by choosing from its ritual storehouse three items that do not correspond to the genders at all; the metaphor is forced. The cow is awkwardly linked with the masculine, both in terms of the world with which she is identified (heaven, usually male) and in terms of her role in the copulation: to impregnate the lifespan. This identification is particularly bizarre in light of the fact that, elsewhere in the Brāhmaṇas, the cow is said to be the earth, an identity that persists throughout later Hinduism, as in the tale of Pṛthu milking the earth-cow (AV 8.10.22–29). But what is most striking about this passage is the way that the Ṛg Vedic myth of the begetting of the sun by sky and earth, through the intermediary link of the begetting of (sun-) light by the cow and the lifespan, becomes the begetting of a third world, the ether, by heaven and earth. This myth, of the productive hierogamy, is then half-transformed into the myth of separation, with a difference: in order that the (original) two worlds can unite (to make more worlds), they must first separate, not from one another but from their first child—they must separate from the (new) world. The myth of the child abandoned by his father is one we have seen in the verse about the sky vanishing from the sun; the myth of the child abandoned by his mother is an important Ṛg Vedic theme that occurs in conjunction with the myths of cosmogony (RV 10.72.8; 1.164.9). This Vedic theme is then introduced in place of the theme of the separation of the two worlds; the marriage becomes a separation. Finally, the majestic Vedic cosmogony plummets bathetically in two ways: first, the world systems are analogized to particular chants, which is standard operating procedure for the Brāhmaṇas (i.e., they revert to abstraction from anthropomorphism). But then the worlds become re-anthropomorphized in a most trivial way: they are plagued by awkward sleeping arrangements (an awkwardness which persists to this day in India). The chants are given sexual organs (the "extra" syllable of one and the "missing" syllable of the other corresponding to what a man has and a woman lacks), and their "mating" is the sacrificial analogue of the marriage of heaven and earth. The greater moral problems of the separation and hierogamy—incest, rejection, the vanishing sky-god—are reduced to the dry linguistic intricacies of the liturgy and the perennial preoccupations of the marriage manuals: waiting till the children fall asleep. Ethics has fallen by the wayside.

The more usual problem posed by these children, however, is the problem of feeding them once they are born; this was already evident in the *Ṛg Veda,* where the "seed" that makes the child is analogized to the "seed-milk" with which he is fed. When the three worlds separate, food separates too:

> Indra gave the ascetics to the wolves. From them when they were being eaten, three boys were left over: Rāyovājas, Pṛthuraśmi, and Bṛhadgiri. They praised Indra and he said to them, "What do you boys wish for, that you praise me?" "Support us, generous one," they said. He tossed them between his two shoulders, and they hung on to his three humps.
>
> Now, these [pl.] worlds, which had been together, separated into three parts, and the eating of food separated after them. They hung on to the three humps of these three worlds, that was the eating of food. Then [Indra] realized, "If they gain control and keep back the eating of food that is the three humps of these three worlds, I will not amass for myself these three humps." He saw this chant and praised with it, and by this means he gained control and kept the food-eating that was the three humps. And by that means he amassed for himself those three humps. . . . (JB 1.185)

The food that was the key to the relationship between the two worlds in our first *Jaiminīya* passage (where it was the basis of exchange between gods and men) is here transformed from a theological to a physiological image: Indra becomes a foster-mother to three orphans and supports them with the food from the "humps" on his back, like the breasts of a mother. Indra is the appropriate one to do this, since he is often likened to a hump-backed bull and since, in later Hindu mythology, he nourishes with milk from his thumb a child who has no human mother (*Mahābharata* 3.126.1–26). But after he has nourished the three boys, in keeping with the reciprocity of the sacrifice, he then fears that they will usurp his powers to give or to hold back food (through his function as the god of rain), and so he undertakes another sort of sacrifice, jealously hoarding the powers for himself. This is a major shift in the ethical stance of the Brāhmaṇas.

For the antagonism between gods and men that lurks behind the scenes of the first *Jaiminīya* myth (bolstered by other myths in which men willfully withhold food from gods, or gods withhold rain from men) is at first displaced in this variant; here, Indra is waging some sort of vendetta against the ascetics, more precisely against a group of ascetics called Yatis. This is a story much repeated, but never fully explained, in the Brāhmaṇas. Indra is said to have committed a number of sins, of which this is the first and the favorite (cf. JB 2.134). The boys are, therefore, orphaned as a direct result of Indra's action in killing their fathers; their "separation from the parents" may be seen as the fallout from one of the earliest skirmishes in the perennial battle between kings and priests, or Vedic gods and Vedantic ascetics, in India. Thus the separation of the three worlds is used to express a schism between Vedic religion and post-Vedic asceticism; and the healing of that schism is expressed through a Vedic image of the parent feeding the child.

What are the ethical implications of this myth? Here, I think, for the first time, one of the gods has become sufficiently anthropomorphic to enable the myth to stand as a model for human ethical behavior. Indra in this tale does not, as in the *Ṛg Veda,* literally and definitively prop apart heaven and earth (though even this, as we have seen, does have ultimately ethical implications, when we regard separation and participation as primary categories for human interaction); here he merely murders a group of ascetics and feeds their orphaned sons. Though one can hardly take this as a positive role model, it does, in fact, epitomize and therefore begin to make possible an ethical approach to one of the enduring problems of ancient Indian law: the problem of the evil king. For, just as the primeval parents *had* to commit incest (when the once abstract pattern of the One who mated with his own creation became anthropomorphized) and yet ought not to have committed incest, so too the king (symbolized by Indra, king of the gods) *had* to kill, and to oppose asceticism (O'Flaherty 1973, 87–89) though it was a sin for him to kill, or to oppose asceticism (Dumézil 1970 and 1973). The solution at this point in the development of Indian moral law was a simple one—feed the gods, feed the priests, feed the survivors of the inevitable debacle. Crude as this was, it was at least an advance on the earlier Indo-European attitude ("Kill as many as you can, take the food, and run like hell") and did provide a basis for later reexaminations of the ethical quandary of the king.

A fourth *Jaminīya* variant on the motif of separation is the most explicitly ethical in its transformations:

> These [pl.] worlds, which had been together, separated into three parts. They sorrowed as one who is split into three would sorrow. The gods said, "Let us strike away the three sorrows of these three worlds." They saw this chant and praised with it, and with it they struck away the three sorrows of these three worlds. And because they struck away the three sorrows of these three worlds, therefore the chant is called the "Three Sorrows" chant. Whoever knows it strikes away sorrow. They [the gods]

> caused them [the sorrows] to enter the impotent man, the criminal, and the whore. And therefore these [three] are sorrowful, for they have been pierced by sorrow. But this does not happen to the man who knows this. (JB 3.72)

Whereas in the *Ṛg Veda* the separation of the worlds was assumed to produce nothing but good, and in other *Jaiminīya* myths the incidental (though inevitable) disadvantages of the separation were eventually explained, here the very first thing that we are told is that separation is sorrow. The anthropomorphism implicit in this attitude is patent. Moreover, whereas in earlier variants the gods were directly affected by this sorrow—for they themselves suffered when the souplines between heaven and earth were cut—here they are apparently benevolent bystanders, acting not out of self-interest but out of pity and sympathy. Or so the text says. Yet a more selfish reason for the gods' concern is apparent when we view the myth in the context of the wider corpus of Brāhmaṇa tales of which it is a part. For the theme of transferring sorrow (or evil) from one person or thing to another is a major one in the Brāhmaṇas, where the person who most often benefits from this transfer is none other than Indra. When he sins, his sin is passed on to volunteers on earth: water, women, trees, and the earth itself (O'Flaherty 1976, 153–60). Thus we may assume that the sorrowing of the three worlds is a situation that harms the gods, that weakens the power of the universe in which their own power must be set. They cannot undo the cause of the sorrow, for they cannot reunite the worlds. Neither could Indra undo his defilement, for the Brahmin dragon had to be slain. What the gods can do is mitigate the ill effects of these necessary evils.

The time-honored way to do this is to palm the evil off on a scapegoat who deserves it anyway: a sinner. Who are these sinners? Significantly, one is excessively neuter, one excessively female, and the third excessively male. The word I have translated as "impotent man" *(klība)* covers a multitude of sexual inadequacies from the Indian viewpoint, including temporary impotence, permanent impotence, castration, and homosexuality. The word for "whore" *(puṃścalī)*, literally "man-chaser," is clear enough. The third word *(kitava)* means a gambler, hence a cheat, hence a drunkard, hence a madman (a thumbnail sketch of the road to ruin in ancient India). We can, I think, assume that the female earth became the whore, the male sky became the criminal, and the neuter ether became the impotent man—the rejected child. These analogies fit far more neatly than did the neuter "light" and "lifespan" and female "cow" in the other *Jaiminīya* myth.

But what is their meaning? Clearly, in this variant the author is taking a stance that will become the prevalent one in later Hindu theodicies: that evil recoils upon the evil; that evil, once created, cannot be destroyed but can be distributed; that the structure of the cosmos is moral, so that evil naturally gravitates to those who are doomed to be evil anyway. In this early text, "evil" is not yet explicated as specifically as it will be in the later lawbooks, where the punishment must fit the crime. Here, evil is still more or less identified with sorrow, the villain of the piece in most of the Brāhmaṇas. The first part of this myth fits a pattern that is dear to the heart of the authors of the Brāhmaṇas: a situation is imagined in which the protagonist is in danger or sorrow or need; he sees the chant that gets him out of that situation; and so the chant is called a danger- or sorrow- or need-dispelling chant. The present text fulfills the requirements of this pattern by the phrase "Whoever knows it strikes away sorrow," a formula which concludes many Brāhmaṇa episodes. But in this case it goes on, to tell us *what happened to the evil*. This extension destroys the usual pat ending of the myth and starts a chain reaction that leaves many loose ends dangling. What did the whore and the eunuch and the criminal do, in the past, to make them deserve to have these sorrows placed on them? Or, if you accept the aptness of these distributions, what did they do to become a whore, etc., in the first place? What could they then do to get rid of the evil that had been placed in them? If one had to devise an ethical system equipped to deal with questions such as these, questions that arise here for the first time, one would probably end up inventing the theory of karma. Although the *Jaiminīya Brāhmaṇa* does not invent such a theory, or even provide the elements from which it was ultimately invented, it does at least paint itself into an ethical corner, so to speak, from which it could only be rescued by the kind of basic rethinking that led, in the Upaniṣads a few hundred years later, to the new theory (O'Flaherty 1980a). The loose ends dangling from the Brāhmaṇas provided fuel for centuries of later Hindu theodicy.

The predominant Hindu theodicy actually begins right in the *Jaiminīya Brāhmaṇa*, in a myth that takes the structural motifs of the cosmogonic divorce entirely out of the context of the ancient myth and transforms it into a tale of the *deus absconditus* type. Here, as in the previous example, the text shows concern for the fate of evil itself, but now that evil is not taken from the worlds, but put into them:

> The gods and demons were striving against one another. The gods emitted [from themselves] a thunderbolt, sharp as a razor, that was man [*puruṣa*]. They hurled this at the demons, and it

> scattered the demons, but then it turned back to the gods. The gods were afraid of it, and so they took it and broke it into three pieces. When it had been shattered into three pieces, it stood right up. Then they took hold of it and examined it, and they saw that the divinities had entered into that man in the form of hymns. They said, "The divinities have entered into this man in the form of hymns. When he has lived in this world with merit, he will follow us by means of sacrifices and good deeds and asceticism. Let us therefore act so that he will not follow us. Let us put evil [*pāpman*] in him." They put evil in him: sleep, laziness, anger, hunger, love of dice, desire for women. These are the very evils that attach themselves to a man in this world.
>
> Then they enjoined Agni [fire] in this world: "Agni, if anyone escapes evil in this world and wants to do good things, trick him and harm him utterly." And they enjoined Vāyu [wind] in the ether in the same way, and Aditya [the sun] in the sky. But Ugradeva Rājani said, "I will not harm men whom I have heard that these three highest gods harm. For the man whom these divinities harm is harmed and harmed indeed." But the divinities do not harm the man who knows this, and they do trick and utterly harm the one who tricks and harms the man who knows this. (JB 1.97–98)

The world split in three survives in this text only in the reference to the appointment of three divinities as watchdogs in each of the three worlds; indeed, a reader of this myth might well see no cosmogony in it at all, were that reader unaware of the supporting corpus of myths that we have seen. The cosmogonic theme is, however, further enhanced by the reference to the splitting apart of the man *(puruṣa)*, a variant of yet another Ṛg Vedic theme of creative separation—the dismemberment of the cosmic Man (Puruṣa) (RV 10.90; O'Flaherty 1981, 29–31). Structurally, this text falls into line with those other texts, but there the resemblance ends. Ethically, they are worlds apart. For where the dismemberment of Puruṣa was done with his apparent consent to the self-sacrifice, and for the benefit of the universe, the dismemberment of the thunderbolt-man is done in order to destroy him and to protect the gods. By this time the concept of sacrifice as a force that joins together the powers of gods and men, through the exchange (of rain and offerings) that reunites the severed worlds, has been transformed into the concept of sacrifice as a force that has the potential to drive gods and men apart, by making men so good, and hence so powerful, that they rouse the jealousy of the gods (O'Flaherty 1976, 83–93).

This jealousy appears first in the gods' fear of the weapon they have created, that rebounds against them as such weapons so often do in mythology (as in life); they separate the weapon from itself, into three parts like the dismembered cosmos, in order to destroy it. When this fails, they divide up something else—not the thunderbolt-man, but evil itself. The same evil that was said to result from the separation of the three worlds in our previous text, and to be taken away from them to be put into people, is here said to be created (apparently *ex nihilo,* or perhaps left over from the previous cosmic divorce) and put into people in order to cause sorrow to mankind. The dynamic is still technically the same: the sorrow, or evil, is taken out of the triple world (or away from the frightened gods) and put into mankind. To the three sorrows-evils of the previous text (impotence, gambling, lust) three more are added: sleep-laziness (a rough equivalent of impotence), anger-love of dice (the gambler's problems), and hunger-desire for women (lust). If there be any doubt in our minds that the gods are behaving unethically, and presenting men with the seeds of all their future ethical problems, the text tells us of a man, a human being (identified unmistakably as such by the formulaic patronymic given only to human sages in the Brāhmaṇas and Upaniṣads) who expressly refuses to behave as badly as the gods behave. Thus we are given the reason for both the generally unethical behavior of mankind (the gods made man prey to the vices that destroy him) and the possibility of ethical behavior among men, for it is possible for people like Ugradeva Rājani to resist, if not to counteract, the efforts that the gods make to ruin us when we try to do good things.

These few examples from two ancient Indian texts merely begin to indicate the rich variety of ways in which the basic structures of cosmogony, constantly reinterpreted, served as an armature on which sages of subsequent generations continued to sculpt their musings on the basic issues of human life. A final set of examples of some of the ways in which ethical order interacts with cosmogony is provided by the Purāṇas, in two sets of myths that I can but briefly summarize here. In the first, we are presented with the assumption that evil is a substance (as it was in the myths of the distribution of sin) and that, therefore, space is a problem. That is, the Hindu universe or "egg of Brahmā" is conceived of as a sealed, perfectly enclosed space with a given amount of good and evil, and a given number of souls. Into this universe is born a good demon, who produces an excess of goodness (virtue, piety, ascetic heat) in the world of the demons; or else a shrine is established, one so good, so powerful, that it sends all who visit it directly to heaven. In both of these instances, there is a problem: there is too much goodness collected in one place (the wrong place), and so, in the first instance, there is no night, nor any job for the god of Hell, or, in the second instance, it becomes so crowded in heaven that people there have to stand with their arms above their heads. Given the original shape of the cosmos—closed—there are only a few ways in which the situation can be remedied: the demon can be corrupted or entirely destroyed; or the demon can be turned into a god, so that he can be transported to the world of the gods, and the rest of the demons then lapse back into their properly evil ways.

This is what happens in the early myths. But under the influence of a major shift in Hindu concepts of ethical order—to be precise, under the influence of *bhakti,* or devotional religion—these problems begin to be met in an entirely new way. The God to whom the demon is devoted comes to him and announces that he, and *all the demons,* are to be taken forever to the heaven of the God, which is infinite and can accommodate all reformed sinners. Similarly, where the shrine in the early myths would be destroyed, or limited so that only a small group of the elect could use it to get to heaven (the equivalent of taking only the one good demon and making him into a god), in the later myths the shrine lets everyone go to heaven—which has now become infinitely expandable. Examples of such a development may be seen in the myths of the good demon (O'Flaherty 1976, 94–138), in the stories of crowds in heaven (O'Flaherty 1977, 248–71), and in a pair of texts in which a virtuous king first is not, and then is, allowed to release all the sinners from Hell (O'Flaherty 1980a, 32–33; *Mahābhārata* 18.2 and *Mārkaṇḍeya Purāṇa* 15.57–80). In all of these cases, it appears at first that certain ethical developments are forestalled by the given set of cosmogonic circumstances, the shape of the arena in which the ethical battle must take place. But as the centuries go on, these developments do take place after all (major developments that involve a challenge to the caste system and a major reformulation of the doctrine of eschatology and salvation), and in order to accommodate them the shape of the arena itself is changed. In this way, cosmogony and ethical order can remain divorced, as it were, for long periods; ethical order develops in its own sphere, in complete isolation from a cosmogony which thus becomes successively more and more outmoded or irrelevant or even inconsistent with that order, until, one day, the two join again in a marriage that changes them both.

A breakthrough in ethics may lead to a breakdown in cosmogony, and to a regeneration; both of these stages may result in new myths. Certain cosmogonies are designed to introduce order, and therefore certainty, into our perception of the cosmos, and when they break down they must be replaced by equally reassuring cosmogonies. But other cosmogonies (the chaos-affirming ones) are designed to assert or even to inspire uncertainty on the level at which traditional cosmogonies inspire certainty; this is true not only of several Buddhist and Hindu cosmogonies but also of the cosmogonies of Heisenberg and Gödel. Yet even these cognitive upheavals offer to those who embrace them some sort of shared conceptual system, an assumed principle of what is most important on some deeper level. Thus even the denial of order becomes in itself an assertion of a different kind of order.

Cosmogonic myths may indicate ambiguity rather than certainty—many kinds of ambiguity, including ethical ambiguity. Even when a cosmogonic myth provides us with an orderly way of viewing the world (and all cosmogonic myths do this, to one extent or another), that world may not in itself be orderly—or ethical. Yet the very fact that ambiguity or disorder is *expressed* by the cosmogonic myth lends a kind of centripetal movement to these conceptions, which otherwise would remain in the centrifugal realm of private dreams and lonely fragmentations. Thus, at the very least, even the myths of chaos offer the structure of their formality as *myths*—linguistic, cognitive, shared, sacred views of the world—and thereby constitute a framework on which ethical understanding may be built.

References

AB *Aitareya Brāhmaṇa*
AV *Atharva Veda*
JB *Jaiminīya Brāhmaṇa*
PB *Pañcaviṃśa (TāṇḍyaMahā) Brāhmaṇa*
RV *Ṛg Veda*

Dumézil, Georges
1970 *The Destiny of the Warrior* (Chicago)
1973 *The Destiny of a King* (Chicago)

Dumont, Louis
1960 "World Renunciation in Indian Religion." *Contributions to Indian Sociology* 4: 33–62.

Heesterman, Jan
1962 "Vrātya and Sacrifice." *Indo-Iranian Journal* 6.1: 1–37.

Marriott, McKim
1976 "Hindu Transactions: Diversity without Dualism." In Bruce Kapferer, ed., *Transaction and Meaning*, pp. 109–10. Philadelphia.

O'Flaherty, Wendy Doniger
1973 *Asceticism and Eroticism in the Mythology of Siva*. Oxford.
1975 *Hindu Myths: A Sourcebook*. Harmondsworth: Penguin.
1976 *The Origins of Evil in Hindu Mythology*. Berkeley.
1980 *Women, Androgynes, and Other Mythical Beasts*. Chicago.
1980a Ed. *Karma and Rebirth in Classical Indian Traditions*. Berkeley.
1981 *The Rig Veda: An Anthology*. Harmondsworth: Penguin.
1985 *Tales of Sex and Violence: Folklore, Sacrifice, and Danger in the Jaiminīya Brāhmaṇa*. Chicago.

Staal, Frits
1979 "The Meaninglessness of Ritual." *Numen* 26:1.

Wasson, R. Gordon, and Wendy Doniger O'Flaherty
1968 *Soma: Divine Mushroom of Immortality*. New York.

Part III

Multiple Cosmogonies and Ethical Order

8 Multiple Cosmogonies and Ethics: The Case of Theravada Buddhism

Frank E. Reynolds

Does cosmogony play a significant role in the Buddhist tradition?[1] The question arises since some very knowledgeable scholars have given, either explicitly or implicitly, a negative answer. But other scholars, equally knowledgeable in Buddhist studies, have discussed several Buddhist cosmogonies. The basic question, at a preliminary level at least, is not so much a matter of interpretation as definition. The term "cosmogony" obviously refers to a conception or accounts of the origins of a cosmos. But what kinds of origins, and what kinds of cosmos?

The notion of origins that informs the term "cosmogony" may be rather narrow or it may be much more inclusive. If the narrower understanding is adopted, "cosmogony" refers only to accounts or descriptions that locate the occurrences they describe "in the beginning" when the cosmos or cosmic process was first created or ordered. But if a more inclusive understanding of origins is accepted, the term also encompasses accounts or conceptions of the way in which the cosmos or cosmic process is continually being generated and structured. For scholars concerned with many religious traditions, this distinction is of very little practical significance. Within the religious orientations they study, the time "in the beginning" when the cosmos was created or ordered is a mythic time, an *illud tempus* that involves not only the past but the present as well. Within other religions, however, a definite distinction is made between a primordial creation or ordering of the cosmos and a continuing process of cosmic generation. For those who study such traditions—among which Buddhism is certainly one of the most prominent—a choice between the narrower and the more inclusive conception of origins is required.

The notion of cosmos that is operative when scholars use the term "cosmogony" also exhibits a considerable degree of ambiguity. Is the cosmos that is implicated when the term is used the universe in its broadest extent? Or is it the particular natural/social world system within which we are presently living? For scholars involved in the study of many religions such questions do not arise because the distinction between the universe in its broadest extent and a particular world or world system within it is not made by the traditions they study. But for those who are interpreting a tradition such as Buddhism in which a distinction between the universe and particular world systems is fundamental, a decision regarding the proper referent or referents of the term must be made.

A further problematic inherent in the notion of cosmos concerns its relevance to the new, salvation-oriented worlds generated by particular deities or religious founders. Clearly the new worlds generated by figures such as Jesus and the Buddha are quite different from the kinds of cosmos that have, according to their respective traditions, preceded them and provided the arenas for their activities. At the same time, the terminology and symbolism used to describe these new worlds strongly suggest that such founders have, in fact, established a new kind of world in which the destinies of human beings (and, in some versions, of nature and society as well) are or will be fulfilled. Once again, the interpreter is confronted with a decision; he must decide whether or not such soteriologically oriented "new worlds" are properly encompassed within the category of cosmos.

For the most part, those scholars who deny that cosmogony has played a significant role in the Buddhist tradition have done so by virtue of a particular set of decisions regarding the three specific issues I have raised. Without necessarily saying it explicitly, they have decided that an authentic cosmogony must recount originating occurrences that are set in a time that is mythic or past. They have decided that a cosmogony must concern the creation or ordering of a cosmos that is identified with the universe in its entirety. And they have decided that the category of cosmos cannot appropriately be applied to new, soteriologically oriented worlds brought into being by a deity or religious founder. Scholars who have affirmed that cosmogony does play a significant role within the Buddhist context have resolved one or more of the three definitional issues by inclusion rather than exclusion. Those who have decided that the notion of origins implied in "cosmogony" may include not only occurrences in a mythic or past time but also the continuous generation of the cosmic process, have recognized that the central Buddhist teaching concerning *paṭiccasamuppāda* (the dependent coorigination of all phenomenal entities) is itself a cosmogonic doctrine.[2] Those who have decided that the notion of cosmos implied by "cosmogony" may include not only the universe, but also particular world systems within that universe, have identified the important Buddhist myth of the devolution of the present world system as a cosmogonic myth. Those who have decided that the notion of cosmos implied by "cosmogony" may also apply to new, soteriologically oriented worlds created by deities or religious founders, have recognized that the accounts of the creative and ordering activities of the Buddha constitute still another kind of cosmogony.

For purposes of the present discussion I have adopted the most inclusive understanding of origins and of cosmos, and therefore of cosmogony. Thus, I will begin by considering the Buddhist cosmogony that is expressed in and through the doctrine of dependent coorigination. I will then consider the Buddhist myth of the devolution of the present world system in which we live. I will turn next to the cosmogony which recounts the Gotama Buddha's establishment of a world that is properly ordered in accordance with the *dhamma* (saving truth). Finally, I will consider the cosmogony that will be consummated by the future Buddha Metteya. In the course of the discussion I will focus attention on the ways in which each cosmogony implicates, either positively or negatively, norms for ethical motivation and action.[3]

The Saṃsaric Cosmogony

The religious and intellectual environment within which the Buddha lived and early Buddhism took form (northeastern India, sixth to fifth centuries B.C.) was characterized by social and intellectual flux, and by a plethora of conflicting teachings. Brahmanic traditions concerning cosmogony, salvation, and ethics were represented, but they were by no means dominant. Long-established, non-Vedic traditions dealing with the same kinds of issues were also very much in evidence. Moreover, the interaction between these various inherited orientations was taking place in the context of a new and rapidly developing urban society that was generating a great deal of religious skepticism, as well as new modes of religious sensibility.

For the Buddha and the early Buddhists, the starting point which gave rise to their new orientation was a strong, existential sense of living in a world characterized by suffering. This sense was vividly depicted in the canonical accounts of the Buddha's famous visions of an old man, a sick man, and a corpse—visions which led to his renunciation of the household life, and his quest for release. And it was given the form of doctrine through its inclusion as the first of the four noble truths—the truth that all existence (*saṃsara*) is suffering. This association of existence and suffering was by no means unique to the Buddhists. But when it was transformed by the Buddha into a basic religious insight, it became a prime component in a new religious orientation.

What was, and still remains, distinctive about the new Buddhist orientation is the claim that the Buddha has discovered the origin of this *saṃsaric* world of suffering; that this origin is a process of dependent coorigination (*paṭiccasamuppāda*) of component psychophysical elements (*dhammas*); and that the "seeing" of this process of dependent coorigination is what is needed to halt the process and to eliminate the suffering[4]. Within the canonical tradition, the *saṃsaric* cosmogony of dependent coorigination is stated in a variety of different ways. Some modern Buddhologists have tried, with questionable success, to arrange the various canonical expressions of the doctrine into some sort of chronological order. But what is important for our purposes is not the chronology but the fact that, at some point before the end of the canonical period, Buddhists arrived at a "classical" formulation that the later tradition has accepted as authoritative. It goes as follows:

> When this exists, that exists or comes to be; on the arising of this, that arises. When this does not exist, that does not exist or come to be; on the cessation of this, that ceases. That is to say:
> on ignorance depend dispositions;
> on dispositions depends consciousness;
> on consciousness depend name and form;
> on name and form depend the six gateways;
> on the six gateways depends contact;
> on contact depends craving;
> on craving depends grasping;
> on grasping depends becoming;
> on becoming depends birth;
> on birth depend old age and death.
> In this manner there arises the mass of suffering [*saṃsara*].

Throughout the course of their history Buddhists have recognized that the *saṃsaric* cosmogony formulated in the doctrine of *paṭiccasamuppāda* (and the correlated conception of fundamental psychophysical components that are activated in accordance with that doctrine) is both central to the tradition and very difficult to comprehend. There has been much controversy, and many different interpretations have been proposed. For example, the Sarvastivadins have tended to emphasize the primacy of the introductory couplets ("When this exists, that exists . . ."), whereas the Theravadins have placed greater emphasis on the twelve-linked chain of causation ("on ignorance depend dispositions . . ."). Also there have been various levels of interpretation within the Theravada tradition itself. Some Theravada texts present *paṭiccasamuppāda* as a doctrine that affirms an explicitly nonsequential emergence of component elements. Other Theravada texts present the linkage of components as a kind of cosmogonic process. Certain treatises such as Buddhagosha's *Path of Purification* include discussions that relate the twelve-linked chain to this cosmogonic process as it occurs in and through three successive lives. Still other treatises such as the Vibhangappakarana (the second book of the Abhidhamma Pitaka) consider it with reference to the life of a single individual.[5]

Given these complexities and differences in interpretation of dependent coorigination, I cannot hope to present, in this context, anything like a full treatment. Instead, I will focus on two matters that are directly relevant to the present topic. The first concerns the emphasis which this *saṃsaric* cosmogony places on ethically valorized motivations and activity. The second concerns its specifically ethical content.

Within the Buddhist tradition the close connection between the doctrine of *paṭiccasamuppāda* and the importance that is given to ethically valorized motivation and activity is both pervasive and quite self-conscious. This is demonstrated by the fact that the ethical connection is almost always placed in the foreground when Buddhists argue against other cosmogonies or doctrines. Consider, for example, the Buddhist polemic used against eternalist and nihilist positions. Buddhist teachers have characteristically argued against eternalist alternatives (particularly, though not exclusively, those associated with Brahmanic teachings concerning the *atman*) by contending that the belief in an eternal god, world soul, or essential self results in the neglect of the concrete human individual who is the true ethical agent, and of the phenomenal world within which authentic ethical motivations and activity are necessarily situated. They have characteristically argued against the opposite, nihilistic alternatives by contending the nihilist denial of personal continuity and destiny removes the motivation necessary to stimulate and justify ethical activity. Or consider the traditional Buddhist polemics against determinist positions and positions that deny causality. Buddhist teachers have characteristically argued against determinist alternatives by maintaining that the rejection of human freedom eliminates the element of volition that is an essential component in any truly ethical action. They have characteristically argued against the opposite, fortuitous alternatives by claiming that the refusal to affirm the presence of causal connections leads, inevitably, to an ethical void.

When Buddhists have defended or explicated the doctrine of *paṭiccasamuppāda*, they have made the same point in a more positive way. They have contended that *paṭiccasamuppāda*, though it rules out any notion of eternal substance or essential self, nevertheless affirms the continuity of individual beings based on the series of causal connections that relate actions to their effects. They have contended that *paṭiccasamuppāda*, though it rules out any form of determinism, provides the grounding for a law of just reward and retribution (*kamma*) that both motivates and structures ethical activity. They have also contended that the doctrine, though it is antithetical to any view of reality in which chance or happenstance prevails, still establishes a basis for the kind of freedom that is essential for authentic ethical activity.[6]

The doctrine of *paṭiccasamuppāda*, in addition to emphasizing the importance of ethically valorized elements in the origin and structuring of phenomenal reality, identifies the specific, ethically valorized components that are involved. The first of these ethically valorized components is ignorance (*avijjā*). In the Western tradition there may be some doubt as to the ethical connotations associated with ignorance. But in Buddhism there is no doubt—ignorance (*avijjā*) is itself a vice; it is a vice that is especially supportive of other vices; and it is a vice that places a limit on the kind and extent of virtue that it is possible for human beings to cultivate.[7] The second ethically valorized component is craving (*tanhā*). For Buddhists it is a vice that is as fundamental as ignorance, and is directly related to it, both as a cause and as an effect. The third ethically valorized component is grasping (upādāna). The tradition, however, has never attributed to the vice of grasping the kind of primacy and centrality that it has attributed to the vices of ignorance and craving.

The substantive ethical point that Buddhists have drawn from the doctrine of *paṭiccasamuppāda* is quite clear. The vices of ignorance and craving are the vulnerable links in the chain of causation that is continually constituting and reconstituting the *saṃsaric* world and the suffering that characterizes it. By eliminating the vice of ignorance (through the realization that *paṭiccasamuppāda* is the truth which constitutes the middle way between doctrinal extremes such as eternalism and nihilism) and eliminating the vice of craving (through practicing the eight-fold path that is the middle way between self-indulgence and radical asceticism) an individual can bring his own involvement in the *saṃsaric* process to a halt. When an individual uproots these vices, and realizes the middle way both as insight and as path, his suffering is ended; his release has been achieved.

The Rūpic, Devolutionary Cosmogony[8]

From the Buddhist perspective the *saṃsaric* cosmos that is continually being generated and regenerated in accordance with the law of dependent coorigination is manifested in the form of cosmic world systems. In accordance with the Buddha's prohibition against speculation concerning an absolute "beginning," no attempt is made to determine when such world systems first made their appearance. But it is maintained that these world systems succeed one another in series, a vast number of these series proceeding simultaneously. Within each series, the destruction of one world system sets the stage for the devolution of the one that follows. In each case the devolutionary process proceeds according to a common pattern described in a mythic tradition that has played a central role in the life of the Buddhist community. This mythic tradition has a prominent place in the Pali canon, for example, in the Pāṭika and Aggañña Suttas which are found in the Digha Nikaya (*Dialogues of the Buddha*, Part III, 1921), in Buddhagosha's *Path of Purification* (Nanamoli 1956), and in the *Three Worlds According to King Ruang* (Reynolds and Reynolds 1982).

Though there are some variations, the basic structure of the myth remains remarkably intact through the course of the centuries. The story begins with a reference to the destruction of the previous world system in the series, and the rebirth of its worthiest inhabitants in a highly exalted realm of radiant *brahma* deities. This realm of radiance, though it is encompassed within the *saṃsaric* universe, is located above the series of world systems that come into being and pass out of existence. Following the destruction of the previous world system, after a very long time has passed, a new world system begins to devolve. At a certain point the life span or the merit of some of the *brahma* deities who have been dwelling in the realm of radiance is exhausted. At the same time, the water that had inundated the old world system begins to solidify and to take on color, odor, and taste. The *brahma* deities, who have a greedy disposition, taste the solidified "substance" and become consubstantial with it. In the classic scriptural account in the Aggañña Sutta, this is a one-step process that results quite directly in the descent of radiant, *brahma*-like beings to the earth where we ourselves presently live.

In the Pāṭika Sutta there are strong indications that even during the canonical period this segment of the myth was, in some contexts, more fully elaborated. The later versions contain an explicit account of the successive stages through which a whole series of cosmic levels come into being. As each level of the cosmos solidifies, beings from the realms above fall or are tempted into it, and become identified with it. After the regular number of *brahma* realms have appeared and been populated, the process continues with the devolution of the lower heavens and the gods called *devatā*. As the last of the *devatā* realms are established and populated, Mount Meru, the central cosmic mountain, comes into being; and then the surrounding mountains appear, along with the oceans and the islands that make up our own human realm. After the human realm appears, radiant deities are drawn to the earth and are reborn as the primal humans from whom humanity as we know it subsequently devolves.

The myth continues with an account of the devolution of human beings and the corresponding changes in the realm in which they live. As the self-radiant beings greedily consume the fruits of the earth, their materiality increases and their radiance declines. This leads to the emergence of immoral attitudes such as vanity and conceit. It also leads to a change in the source of illumination within the realm (as the beings lose their radiance, the merit they still possess produces a sun and a moon) and to changes in the character of the food that is available (it becomes more "substantial" and requires human cultivation). At a certain moment sexual differentiation takes place; this creates the need for privacy, and privacy is followed by the hoarding of goods and the claiming of property. Greed soon leads to stealing; and stealing generates censure and punishment. In the face of the increasing social disorder, the people consult together to determine how the process of disintegration can be halted. As a result of their deliberations they choose the man who is the handsomest and possesses the most merit to be the Mahāsammata, the Great Elect who will serve as their king. Thus the lineage of nobles (*khattiya*) is established. Some men and women, recognizing the immorality that is about them, take up meditation. Thus the lineage of *brahmins* is established. Some take up agriculture and trading, thus establishing the lineage of *vessas*. Others take up hunting, thus establishing the lineage of *suddas*.

The conclusion of the myth varies from text to text. For example, the Aggañña Sutta recounts that, as time goes on, certain dissatisfied members from each lineage withdraw from society. These individuals then constitute a company of mendicants within which—with the subsequent formation of the Buddhist order of *bhikkhus*—release from the *saṃsaric/rūpic* worlds can be attained. In the later Thai account the same basic point is made in a rather different way. This version of the myth makes no reference to the formation of the community of mendicants and the appearance of *bhikkhus*. Instead, the narrator comments on the continuing instability of the human condition. The lesson is then drawn that human beings should turn their attention to the attainment of the one reality that is truly stable and permanent—*nibbana* itself.

Some Western scholars, and some modern Buddhists as well, have questioned the centrality and significance of this *rūpic*, devolutionary cosmogony. Some have seen it as a fable addressed to those who are incapable of appreciating the real implications of the Buddha's insight and teaching. Others have seen it as a parody whose major purpose is to ridicule Brahmanic notions of creation by Brahma. From our perspective, however, this *rūpic* cosmogony is an essential element of the Buddhist *dhamma* that both conforms to the "original" or "primary" cosmogony set forth in the doctrine of dependent coorigination, and provides an essential supplement to it. It fulfills a fundamental need by formulating an ethically structured conception of the divine, the natural/physical, and the social aspects of reality. And in so doing it significantly extends the range of ethically valorized motivations and activities.

The *rūpic* cosmogony, in its account of the origins of the divine, the natural/physical, and social aspects of reality, draws heavily on preexisting and coexisting Indian traditions. But in each case the myth presents a distinctively Buddhist interpretation. Consider, for example, the deities who are involved in the devolutionary process and in the *rūpic* cosmos that results. The fact that these deities are called *brahma* and *devatā* clearly indicates their relationship to deities established within the Brahmanic tradition. However, they are, in very crucial respects, quite different from their Brahmanic namesakes. They are beings like ourselves who are similarly caught in the continuing *saṃsaric* process of birth and rebirth. They have become deities because of the *kammic* effects of their ethical activities in their previous lives. No one of the deities, nor any combination of them, is the creator of the world system. Nor do these deities determine the pattern according to which the world system devolves. To the contrary, they are themselves a part of the cosmic process that unfolds in accordance with the *dhamma* discovered and taught by the Buddha. Nor are these deities able to interfere in any way with the *kammic* process through which each of the other beings in the world system has created and is creating his or her destiny.[9] Thus deities are recognized, and they are given a significant cosmogonic and cosmological role. But this is accomplished within the context established by the doctrine of *paṭiccasamuppāda*.[10]

The conception of the natural/physical world that is formulated in the *rūpic* cosmogony has a similar relationship to preexisting and coexisting traditions. Many pan-Indian themes are apparent—for example, the hierarchy of heavenly realms, and the cosmography constituted by Mount Meru and the surrounding rings of mountain ranges and oceans. But in the devolutionary myth they are integrated within a distinctively Buddhist context in which ethically valorized action and the origins and character of the natural/physical world are closely correlated.[11] In the myth a clear correlation is drawn between the ethical devolution of deities and the formation and quality of the realms they come to inhabit.[12] The ethical decline of the primal humans is directly correlated with the transition from a period in which human beings are self-radiant and receive nourishing food from the earth without any effort, to a period in which human beings have lost their radiance and must work to bring forth the fruits of the earth. A direct causal connection is made between the merit acquired through *kammic* activity which humans, despite their ethical decline, still possess, and the appearance of the sun and the moon. And in the most developed version of the myth, a direct causal connection is made between the maintenance of proper morality in the subsequent human community and the proper functioning of the celestial bodies and the proper regulation of the seasonal cycle. In these ways the myth preserves the distinctive ethical emphasis which characterizes the doctrine of *paṭiccasamuppāda* at the same time that it provides Buddhists with a coherent conception of the natural/physical world in which they live.

The segment of the *rūpic*, devolutionary cosmogony that describes the origins of a functionally differentiated social order is the section of the myth that has the most obvious affiliation with the Brahmanic tradition. But once again, the myth presents a thoroughly Buddhist interpretation. According to the *rūpic* cosmogony the human social order is constituted by a fourfold hierarchy of lineages that has much in common with Brahmanic notions of *varna* (caste). However the fourfold hierarchy does not originate, as the Brahmanic myths would have it, through the creative activity and/or dismemberment of a deity.

Rather, it originates through human actions that are motivated by an ethical concern—the need to stem a primal tide of moral and social disintegration. And these actions are structured in accordance with an ethical norm—the *dhamma* in its role as cosmic and moral law. Established in this way, the four lineages do not constitute ontologically distinguished types of humanity. They are, on the contrary, groups of individuals sharing a common humanity that perform different social functions. The Buddhist myth also differs from its Brahmanic counterparts in that it reverses the order of precedence between the two highest lineages, the *brahmins* and the *khattiya*. It is the royal, *khattiya* lineage rather than the *brahmin* lineage that is first established, that is given the preeminent status, and that has the responsibility for regulating the functioning of the system as a whole. Thus, this very crucial aspect of the *rūpic* cosmogony maintains the distinctive ethical emphasis of the original *saṃsaric* cosmogony; it provides, in addition, an ethically structured theory of social differentiation that utilizes characteristically Indian themes; and it accomplishes this in a way that establishes the basis for a Buddhist society in which the preexisting Brahmanic priesthood is relegated to a secondary position.

The substantive ethical content of the *rūpic* cosmogony, like that of the doctrine of *paṭiccasamuppāda*, is heavily weighted toward the depiction of vices and their effects. However, the vices are somewhat different in character. They are not the cardinal vices of ignorance (*avijjā*) and craving (*tanhā*) whose uprooting is directly associated with the attainment of release. Rather, they are ordinary vices whose relative presence or absence determines the degree of suffering or pleasure that afflicts particular groups of beings and particular individuals. The myth gives an obvious precedence to the vice of greed (*lobha*); but it also refers to hatred (*dosa*) and delusion (*moha*), as well as vanity, laziness, and the like. In addition, the myth goes on to depict ethically valorized actions and to recount their effects. These include negatively valorized actions such as harming others, sexual misconduct, and stealing.[13] Such actions, which violate basic moral precepts, have a double effect. They lead to social disintegration and, at the same time, they generate personal destinies characterized by the grosser forms of suffering. On the positive side, the myth recounts the delineation and acceptance of social responsibilities in accord with a *dhammic* norm.

The Dhammic Cosmogony and Ritual Reenactment

The third kind of cosmos that Theravada Buddhists have recognized is a Buddhist cosmos in the full sense. This is the cosmos which

Buddhists continually regenerate in and through their ritual and ethical activity, the one in terms of which they—ideally at least—live their lives and seek their salvation. This cosmos is brought into being from within *saṃsara*, and from within the series of world systems; but it transcends them. As in the case of the *rūpic*, devolutionary cosmos, the *dhammic* cosmos is generated from time to time according to a definite pattern. Sometimes in an ordinary *rūpic* cosmos there is no *dhammic* cosmos that comes to fruition. In *rūpic* world systems such as the one in which we presently live, as many as five *dhammic* worlds reach fruition. In each case the *dhammic* cosmos is created or ordered by a Buddha through actions that he performs beginning with a moment at which he takes a vow to attain enlightenment for himself and to aid other beings in their quest. The cosmogonic process continues for eons and eons through a long series of his lives that culminate in a final life in which he discovers and preaches the *dhamma*. In so doing, he establishes a *dhammicly* ordered world.

From one point of view, everything Buddhists associate with the lives and teachings of the Buddha is a part of the *dhammic* cosmogony. However, there is a particular strand of the tradition in which this cosmogonic aspect is highlighted. At the level of the texts this strand is associated with rather dispersed accounts of various aspects of the Buddha's previous lives and his final, "historical" life as Gotama. The accounts of his previous lives, collected in a very famous and influential text called the Jātaka Commentary, describe the perfecting of the various virtues that prepared the *bodhisatta* (future Buddha) to be reborn as a Buddha.[14] The accounts of the cosmogonically crucial events in his final life include particularly those that describe his birth, his enlightenment, his first sermon, and his *parinibbāna* (his passing away at the end of his earthly career). The Buddha's birth is presented as a cosmic event in which his character as a *mahāpurisa* (a great being destined to become either a Buddha who orders and reigns over a soteriologically oriented cosmos, or a *cakkavatti* king who orders and reigns over a "secular" cosmos) is evident to those who have the eyes to see. His enlightenment is recounted in a narrative that clearly suggests a royally structured but supraroyal enthronement with cosmogonic significance. His first preaching is described as a setting in motion of the wheel of *dhamma*, a clearly cosmogonic act closely associated with a very archaic solar mythology. The account of his *parinibbāna* describes the transition from the primary form in which the *dhammic* cosmos was established during the historical career of Gotama, to the *dhammic* cosmos as it continues after his passing away. According to the Mahāparinibbāna Sutta, the monks become the inheritors of the *dhamma* (saving truth and cosmic law) which Gotama had discovered and taught. The laity, for their part, inherit his relics and enshrine them in *thūpas* like those erected to honor *cakkavatti* kings.

A more coordinated and succinct expression of this *dhammic* cosmogony can be found at the level of architecture and iconography, specifically in the architecture and iconography of the *thūpas*. It has long been recognized that well before the beginning of the Christian era *thūpas* were conceived and constructed not only as an iconic image of the Buddha but also as images of a *dhammicly* ordered cosmos complete with heavenly realms, a Mount Meru cosmography, and an effloresence of prosperity and affluence. What has been less generally recognized is the fact that the *thūpa* complex as a whole represents the Buddha and his *dhammicly* ordered cosmos as a cosmogonic process. This very basic cosmogonic dimension is made vividly apparent by the iconographic ornamentation that is, in fact, an integral element in the symbolism of the *thūpa*. This iconographic ornamentation includes, at the periphery of the *thūpa* complex, representations of stories depicting important events in the previous lives of Gotama. And around the monument which forms its central core, it includes representations of crucial events in the final life of Gotama. From this iconographic evidence we can conclude that the *thūpa* complex represents the generative activity of the Buddha that produced a *dhammically* ordered cosmos. We can also conclude that this generative activity encompasses both a *saṃsaric, rūpic* component (the kind of activity that characterized the Buddha's previous lives) and a path-centered, *nibbānic* component (the kind of activity that the Buddha discovered, exemplified, and taught in his final life as Gotama).

The importance of this *dhammic* cosmogony in the life of the Buddhist community is heavily underscored by the fact that it is, like many of the classic cosmogonies studied by historians of religion, ritually reenacted. This ritual reenactment is most evident in the great festivals associated with *thūpas*. But it is also implicit in all of the Buddhist rituals that involve activities of exchange between members of the laity and the monks. Laymen and laywomen, properly pure by virtue of their adherence to the five basic moral precepts (those against killing, stealing, lying, sexual misconduct, and the use of intoxicants), engage in activities that express and cultivate the virtues of the Buddha in his previous lives, particularly the virtue of selfless giving (*dāna*) which he perfected in his penultimate life as Mahāvessantara. The monks, properly pure by virtue of their adherence to the monastic rules attributed to the Buddha (especially the set of 227 rules known as the *paṭimokkha*), engage in activities such as meditation and preaching that express and cultivate the virtues of the Buddha in his last life as Gotama. The whole complex of ritually reenacted virtues and activities of the Buddha, organized in terms of a process of ethically charged exchange that is beneficial to each group and individual involved, results in the renewal of the *dhammicly* oriented cosmos. Continually renewed by this ritual/ethical activity, the *dhammicly* ordered cosmos provides a context that, in principle, encompasses the totality of Buddhist life, both social and individual.[15]

Just as some Western scholars and modern Buddhists have questioned the orthodoxy of the *rūpic* cosmogony, so too some interpreters of both types have questioned the orthodoxy of the *dhammic* cosmogony. In fact, however, this cosmogony and its correlated rituals are so structured that they neatly mesh with other Buddhist teachings, including both the doctrine of dependent coorgination and the myth of cosmic devolution. It is true that the creating figure in the *dhammic* cosmogony bears striking resemblances to certain Brahmanic deities. For example, the Buddha, like the Brahmanic Agni, generates, in and through his actions, a perfectly ordered cosmos. Like the actions of Agni in the Brahmanic *agnicayana* ritual, the Buddha's actions are represented by a monument constructed with bricks. (The monument representing the activity of Agni is a ritual altar; the monument representing the activity of the Buddha is a *thūpa*.) In both cases, this monument comes to serve as a substitute body animated by an appropriate symbol. (In the case of Agni, an image; in the case of the Buddha, a relic.) Other aspects of the Buddha's creative activity are strongly reminiscent of the creative activity of the Brahmanic Purusha. Like this Brahmanic deity, the Buddha—who was himself recognized as a Mahāpurisha (Skt. Mahāpurusha)—culminates his creation through a dispersion of his body into segments that become correlated with functionally differentiated segments of society. Purusha accomplished this through a dispersion of different portions of his substance into the four Brahmanic castes. The Buddha accomplishes it through the commitment of his *dhamma* body to the monastic order and the commitment of his *rūpa* body to the laity. But despite these similarities, the Buddha, as he is depicted within the *dhammic* cosmogony, remains firmly within the classical framework of Buddhist doctrine. The Buddha is not, like Agni, a deity closely affiliated with an "original" creator who first brought the phenomenal/*rūpic* world into being. Nor is he, like Purusha, a deity who generates the phenomenal/natural/social world out of his own substance. Nor is his creative activity, like that of both Agni and Purusha, directly associated with traditional forms of sacrifical ritual. On the contrary the Buddha, in accordance with the doctrine of dependent coorigination, comes to be within the *saṃsaric* and *rūpic* worlds where he is continously reborn, sometimes as an animal, sometimes (though very occasionally) as a god, and sometimes as a human being. Equally important, his cosmogonic activity is, in congruence with basic Buddhist teaching, thoroughly ethical in character.[16]

While the *dhammic* cosmogony and its correlated rituals mesh very neatly with the basic ethical formulations set forth in the *saṃsaric* and *rūpic* cosmogonies, they also add important new dimensions. For example, the notion of moral precepts is more fully developed. Five basic precepts are clearly delineated and associated explicitly with the laity; and a much more detailed and stringent set of 227 rules (the *paṭimokkha*) is set forth to guide monastic behavior. But what is more important than the addition of specific elements is the fact that the *dhammic* cosmogony and its ritual reenactments emphasize the realization of positive virtues and values rather than the mitigation or elimination of vices and immoral activities. In the new *dhammic* context, the taking of the five precepts (for the laity) and the confession of violations of the *paṭimokkha* (for the monks) serve primarily as starting points for more positively oriented ritual and ethical activity. This activity includes the expression and cultivation of lay virtues and monastic virtues that directly enhance the soteriological position of the actor. This activity also includes the exercise of lay responsibilities and monastic responsibilities which—though they do not have any immediate or primary impact on the soteriological position of the actor—serve to assure the maintenance and well-being of the community as a whole.

The Dhammic Cosmogony and the Coming of Metteya

The generation of a *dhammic* cosmos is not, for the Buddhist perspective, a once-and-for-all event. There are, even in the context of the present world system, a series of such occurrences. And there is one additional *dhammic* cosmogony that is directly relevant to the present situation—the one that will come to fruition in the far distant future when the Buddha Metteya descends from his present abode in the Tusita heaven and establishes, once again, a *dhammicly* ordered world.

The religious and ethical significance of this *dhammic* cosmogony that will come to fruition in the future derives from the fact that most Buddhists believe that the *dhammic* cosmos established by Gotama has been, from the moment of his death (or, in the view of some, from the time of King Asoka some three centuries after his death), in a process of continuous decline. Since many of these Buddhists have felt that the presently established *dhammic* cosmos no longer retains the kind of completeness and purity that is truly conducive to the salvation of those who live within it, they have turned their attention to the possibility of rebirth at the time of the coming of Metteya. Since Metteya will act and teach according to the same pattern as Gotama, and since the cosmos that he will establish will be structured in the same way as the one brought into being by Gotama, the recognition of its importance in Buddhist life adds little to an understanding of the basic principles of Buddhist ethics. However, such a recognition does enable us to understand the way in which many Buddhist practitioners relate the cultivation of worldly virtues and various merit-making practices in the present to the ultimate goal of entering the path and attaining *nibbāna*. The cultivation of the worldly virtues and other merit-making practices in the present will, they believe, lead to rebirth in the future *dhammic* cosmos to be brought into being by Metteya. In this Metteyan context they are convinced that the cultivation of the supraworldly virtues and in the attainment of *nibbāna* will, once again, become a realistic possibility.

Conclusion

In the preceding discussion I have focused attention on four cosmogonies that are basic to the Theravada tradition and have then proceeded in each case to analyze the distinctive ethical contents and implications of the cosmogony. If we combine these analyses, it is possible to identify four distinctive, though closely correlated and sometimes overlapping types of ethic, each relevant at two different levels of motivation and activity.

The first type of ethic, associated primarily with the *saṃsaric* and *rūpic* cosmogonies, is an ethic concerned with the mitigation and destruction of vices. At the *saṃsaric/rūpic* level, the mitigation and uprooting of ordinary vices such as greed, hatred, and delusion mitigates and eliminates the grosser forms of *saṃsaric* suffering. At the level of the path, the uprooting and destruction of the cardinal vices of ignorance and craving achieve the complete cessation of suffering that is identified as the attainment of *nibbāna*.

The second type of ethic, associated with the *rūpic* and *dhammic* cosmogonies, is an ethic of precepts and rules. At the *saṃsaric/rūpic* level represented by the laity, the primary emphasis is on the five basic precepts against killing, stealing, lying, sexual misconduct, and the use of intoxicants. Adherence to these precepts is prescribed as a way to prevent both social disharmony and an unfavorable personal destiny. At the level of the path symbolically represented by the monks, the primary emphasis is placed on the set of 227 rules called the *paṭimokkha*. Again at this level adherence is justified on the basis that it generates communal order at the same time that it fosters individual soteriological attainment.[17]

The third type of ethic, associated primarily with the *dhammic* cosmogonies, is an ethic concerned with the cultivation of virtues. At the *saṃsaric/rūpic* level the virtues that are commended for cultivation fall into two loosely differentiated groups. One of these groups includes virtues such as selfless giving (*dāna*) that are closely and directly associated with one's own soteriological advancement. The other includes virtues such as sympathy (*anukampā*) that are primarily directed toward the well-being of others. At the level of the path, the relevant virtues also fall into two categories. The virtues of insight and nonattachment are directly correlated with the attainment of one's own *nibbanic* realization. Other prominent virtues such as compassion are directed primarily toward the well-being of others.[18]

The fourth type of ethic, associated with the *rūpic* and *dhammic* cosmogonies, is an ethic of responsibility. As such, it is grounded in the responsibility to act in accordance with *dhammic* norms and does not relate, in any direct or immediate way, to the attainment of personal soteriological goals. At the *saṃsaric/rūpic* level this ethic of responsibility involves the performance of one's proper functions within the *dhammic* social order, including the maintenance of a proper protective and supportive relationship with the order of monks. At the level of the path symbolically represented by the monks, this ethic of responsibility involves the performance of proper monastic functions, including the maintenance of positive relationships with the laity.[19]

The recognition of the presence of these four cosmogonically grounded types of ethic does not exhaust the possibilities open to scholars interested in studying the ethical norms that have been formulated to guide Theravada behavior. The recognition of the presence and importance of these very different types of ethic does, however, provide a new kind of beginning and a firm basis for further research.

Notes

1. For the sake of convenience I have chosen to use the terms "Buddhism," "Buddhists," etc. to refer specifically to the early and subsequent Theravada traditions up to the time when the Theravadins encountered modern Western notions of cosmogony and cosmology.

2. In harmony with my focus on the Theravada tradition I will, throughout the essay, employ the Pali rather than the Sanskrit versions of Buddhist terms such as *paṭiccasamuppāda*.

3. I am convinced that the pattern of multiple cosmogonies that I have identified in Theravada Buddhism can be identified in other religions as well. For example, the Hindu Puranic tradition includes an "original" cosmogony (*prākṛtasarga*) associated with the birth of the god Brahma; a "secondary" cosmogony (*prātisarga*) associated with the awakening of Brahma each day of his life (this is the cosmogony through which the heavenly, natural, and social aspects of reality were formed); a series of semicosmogonic interventions by *avatāras* of the god Vishnu that precede the present; and a future cosmogony in which another *avatāra* of the god Vishnu will usher in a golden age. (For a full discussion of the system see Soifer 1978, especially chapters 4 and 8). A geographically more distant example of the same pattern can be found in the Christian context. Here we have an "original" cosmogony narrated in Genesis; a "secondary" and very historicized cosmogony that is narrated in Exodus; a new, soteriologically oriented cosmogony that is implemented in and through the career of Jesus; and a very closely related future cosmogony to be brought to fruition in and through Jesus' second coming. The presence of this common pattern of multiple cosmogonies, and the fact that in each case the variations in the pattern as well as the specific cosmogonies are replete with ethical implications, opens the way for a whole series of very interesting comparative studies. For the present, however, practical considerations have prevented any explicit exploration of comparative possibilities.

4. It is impossible to determine with absolute certainty whether the doctrine of *paṭiccasamuppāda* was an original discovery of the Buddha. The early Buddhist tradition itself raises questions in this regard when it affirms that the discovery of *paṭiccasamuppāda* is an achievement of all Buddhas, and makes a particular point of attributing it to an earlier Buddha named Vipassī. However this may be, the doctrine is a distinct and consistent feature of the Buddhist tradition founded by Gotama.

5. For further discussion of such matters, see Kalupahana (1975), pp. 141–46, and Woodward (1978).

6. The best discussion in a Western language is found in Silburn (1955), especially in chapter 6.

7. Though ignorance is the first in the list of links in the chain of causation, it is specifically not a "first cause." Just as the other links in the chain are causally dependent on ignorance, so ignorance is causally dependent on them.

8. I have chosen the phrase "*rūpic*, devolutionary cosmogony" advisedly. The justification for the choice is that the cosmogony is depicted as a devolutionary process that begins at the apex of what

will become a cosmos characterized by the presence of *rūpa* (form or materiality). This *rūpic* devolutionary cosmos is traditionally divided into two "worlds" (*bhūmi*), a higher one explicitly called the *rūpabhūmi* and a lower one called the *kāmabhūmi* (world of desire).

9. In certain strands of the early and Theravada traditions this last limitation has been stretched, and in some cases compromised. Such stretching and compromise has sometimes, though by no means always, been mediated through beliefs and practices that involve the transference of merit.

10. The character of the Buddhist deities obviously limits the kinds of relationship they may have to human ethical activity. At the same time, their character as temporary occupants of exalted cosmic positions open, in principle, to all endows them with one quite distinctive function vis-à-vis human ethical motivation. As a pantheon organized in terms of *kammicly* determined hierarchy, they present to the Buddhist community a colorful and appealing scenario of positive rebirth possibilities that can be actualized through ethical cultivation and action.

11. The exact character of the "correlation" between ethically valorized activity and the operation of natural/physical processes has been one of the more vexed questions in Buddhist scholastic thought. In several of the Hinayana schools there has been a tendency to speak in terms of ethical causation. Among Theravada scholastics, on the other hand, there has been a tendency to resist the idea of direct causal connections. (For the best available discussion see André Bareau [1957]). In my presentation I have tried to reflect the kind of connection suggested, in each particular instance of correlation, by the language used in the myth itself.

12. The various Buddhist schools, despite their differences regarding the relationship between ethical causation and natural/physical processes (see Bareau [1957]), were very careful to avoid any implication that the natural/physical world was, in and of itself, evil. The Theravada position on this subject is carefully explored in Y. Karunadasa (1967) especially in chapter 9 on "The Ethico-Philosophical Basis of the Buddhist Analysis of Matter."

13. In the mythic account (specifically the version contained in the Aggañña Sutta) the narrator takes pains to note that sexual misconduct had a different referent in the primal times than it does in subsequent human society. In the primal times, he implies, the fall from chastity was, in and of itself, inappropriate sexual behavior.

14. For an excellent analysis of the Jātaka Commentary that takes particular account of the relationship between ethically valorized action and the natural/physical world see Aronoff (1982).

15. This interpretation of the symbolism of the *thūpa* and its associated ritual, as well as the extension of the principles involved to the much broader complex of ritual exchange, represents a further development and adaptation of views presented in my earlier writing. See, for example, Reynolds (1978) and Reynolds and Clifford (1980).

16. For brilliant and extended discussions of many of the issues considered in this paragraph, see Mus (1935) and Mus again (1968).

17. For an excellent discussion of the ethical role and justification of the monastic rules, see Holt (1981).

18. An interesting analysis of several key virtues is provided in Aronson (1980). However, see also the review by Hallisey (1982).

19. Although this ethic of responsibility has been ignored by most Western interpreters of Theravada ethics, it has played a very important role in Theravada history. In fact, many of the most interesting ethical tensions that have arisen can best be understood in terms of conflicts that develop in concrete situations between a norm or norms set forth by another type of ethic, and modes of activity enjoined by the ethic of responsibility. For an interesting discussion of such tensions as they have arisen in the context of royal activity and its evaluation, see Clifford (1978).

References

Aronoff, Arnold
1982 "Contrasting Modes of Textual Classification: The Jātaka Commentary and Its Relationship to the Pali Canon." Ph.D. dissertation, University of Chicago.

Aronson, Harvey
1980 *Love and Sympathy in Theravada Buddhism*. Delhi: Motilal Barnasidass.

Bareau, André
1957 "Les relations entre la causalité du monde physique et la causalité du monde spirituel dans le Hinayana." In Kshitis Roy, ed., *Liebenthal Festschrift*, vol. 4, parts 3 and 4, of *Sino- Indian Studies*. Santiniketan, West Bengal: Visvabharati University.

Clifford, Regina
1978 "The *Dhammadipa* Tradition of Sri Lanka: Three Models within the Sinhalese Chronicles." In Bardwell Smith, ed., *Religion and the Legitimation of Power in Sri Lanka*. Chambersburg, Pennsylvania: Anima Books, 1978.

Hallisey, Charles
1982 Review of Aronson's *Love and Sympathy in Theravada Buddhism*. In *Journal of Asian Studies* 41 4 (August): 859–60.

Holt, John
1981 *Discipline: The Canonical Buddhism of the Vinayapiṭaka*. Delhi: Motilal Barnasidass.

Kalupahana, David
1975 *Causality: The Central Philosophy of Buddhism*. Honolulu: University Press of Hawaii.

Karunadasa, Y.
1967 *Buddhist Analysis of Matter*. Colombo: Department of Cultural Affairs.

Mus, Paul
1935 Preface. In *Barabadur: Esquisse d'une histoire du Bouddhisme fondée sur la critique archéologique des textes*. Hanoi: École Français Extreême Orient.

1968 "Où finit Purusha?" In *Mélanges D'Indianisme a la Mémoire de Louis Renou*. Publications de l'Institute de Civilization Indienne, Série IN–8, Fasicule 28. Paris: Editions E. De Boccard.

Nanamoli, Bhikkhu, trans.
1956 Buddhagosha, *Path of Purification*. Colomobo: R. Semage.

Reynolds, Frank
1978 "The Holy Emerald Jewel: Some Aspects of Buddhist Symbolism and Political Legitimation in Thailand and Laos." In Bardwell Smith, ed., *Religion and the Legitimation of Power in Thailand, Laos and Burma*. Chambersburg, Pennsylvania: Anima Books.

Reynolds, Frank, and Regina Clifford
1980 "Sangha, Society and the Struggle of National Integration: Burma and Thailand." In Frank Reynolds and Theodore Ludwig, eds., *Transitions and Transformations in the History of Religions: Essays in Honor of Joseph M. Kitagawa*. Studies in the History of Religions, 23. Leiden: E. J. Brill.

Reynolds, Frank, Phya Litai, and Mani Reynolds, eds. and trans.
1982 *Three Worlds According to King Ruang*. University of

California Buddhist Research Series, 4, Berkeley: Asia Humanities Press.

Rhys-Davids, T. W., trans.
1921 *Dialogues of the Buddha*. Sacred Books of the Buddhist, vol. 4, part III. London: Humphrey Milford, Oxford University Press.

Silburn, L.
1955 *Instant et Cause*. Librarie Philosophique. Paris: J. Vrin.

Soifer, Deborah
1978 "Beast and Priest: A Motific Study of the Narasimha and Vāmana Avatāras in Cosmological Perspective." Ph.D. dissertation, University of Chicago.

Woodward, Mark
1978 "Paradox and Causality: A Structural Study of Hinayana Philosophy." M.A. thesis, University of Illinois.

9 Completion in Continuity: Cosmogony and Ethics in Islam

Sheryl L. Burkhalter

The study of Islam has yet to carve for itself a comfortable niche within the history of religions discipline. Charles Adams has accounted for this tension in terms of the discipline's preoccupation with archaic religion:

> At the points where History of Religions has had some of its greatest successes, it has been irrelevant to the work of Islamicists, and Islamicists have in consequence exhibited very little interest in what historians of religions have to say. . . . For many Islamicists there is little of encouragement or stimulation in the quest of their colleagues to unlock the mystery of the archaic view of the world. Although it can easily be demonstrated that Islam has many archaic survivals in its developed ritual and thought forms, . . . yet there remains a vast segment of the Islamic experience that does not fall within the realm of the archaic. This segment is, furthermore, historically the most important portion of the Islamic tradition. (Adams 1967, 182, 186)

Although nearly two decades have passed since this statement was first published, its sentiments remain and recently found expression during a discussion of traditions during the symposium on cosmogony and ethical order that produced this book. Noting the conspicuous absence of a paper dealing with Islam, a participant suggested that the selection of cosmogony as a category of comparison had precluded the possibility of addressing Islamic materials. The comment intimates Adams's perspective, as the category of cosmogony has proved informative for scholarship recognizing the import of archaic traditions for the history of religions. The participant, therefore, was once again raising the more general question of whether a methodology which uses archaic traditions as a starting point can be helpful for gaining insight into the religious thought of Islam.

If the problem is to be addressed, one would be hard pressed to find a context more conducive to its discussion than that provided by "cosmogony and ethical order." While "cosmogony" may provide ample scope for analysis from a history of religions perspective, it would be difficult to conceive of a topic more appropriate for the study of Islam than "ethics." For, as Rahman writes of Islam, "Its ethics, indeed, is its essence" (1982, 154). A discussion of the relationship between cosmogony and ethics in Islam, therefore, provides an opportune test case for the impasse cited by Adams. Rather than dealing with archaic "survivals" at the tradition's periphery, an understanding of the relationship between the two categories requires delineation of the religious meaning cosmogony provides the ethical thought at the very heart of Islam.

The subject of this volume also lends a new perspective to the comparative task by emphasizing the need to derive categories for comparison from a holistic understanding of the tradition studied rather than from Western thought alone. Although this concern has informed studies of Islamic ethics in the past (see Izutsu 1959), the same considerations necessarily pertain to the scholar's working definition of "cosmogony." In both instances, materials from the tradition itself must be used to define the parameters of the comparative category.[1] It is largely here, I suggest, that the case of Islam has faltered.

To delimit the use of "cosmogony" to myths recounting the original creation of the world is to misrepresent the breadth of cosmogonic understanding in Islam. The tradition recognizes the creation of the world not only at the beginning of time, but through time; the creative process is understood not only through myths of beginning, but through myths of completion as well. An adequate appreciation of Islam's cosmogony thus requires recognition of both sets of myths which, while differentiated, are yet intricately related in their telling of one story: the creation of the world. This cosmogony not only provides insight into the structure of ethical understanding in Islam; it further suggests a framework for future dialogue between the Islamicist and the historian of religions which cannot easily be dismissed in the name of "archaic man."

Cosmogony Redefined

The Qur'ān's cosmogony is not a story of pristine perfection limited to an account of the initial creation and ordering of the world. To the contrary, the text describes an intentionally incomplete cosmogony and emphasizes the divine activity responsible for the continuous creation of the world through to the last day. Only then, with the final transformation and rearrangement of created order, will creation be complete. The study of cosmogony in Islam, therefore, requires that the narrow definition of the term, pertaining solely to the initial creation of the universe at the beginning of time, be set aside in favor of a more general definition which recognizes the continuation of cosmogonic activity through time.

Further, in describing the process of creation, Islam refers not only to the creation of the physical universe but to the creation of human activity as well. For man is understood to have accepted fundamental responsibility for the completion of the creative process. God, however, is not absent from the endeavor. Creator of the universe, he is also recognized as the maker of history. An adequate understanding of Islam's cosmogony requires consideration of both dimensions of creative activity: physical as well as historical.

Given this vast scope for divine creative activity, the question arises as to why it is meaningful to speak of "cosmogony" as differentiated from "cosmology" in Islam. Although the creation of the universe will be completed only with the end of time, Islam recognizes a period of perfected order achieved through the mission of the prophet Muḥammad and the community he established at Medina in accord with the divine guidance revealed through the Qur'ān. This perfect realization of the creative task characterizing the human situation fulfilled the incompletion described in the account of original creation: the myths of beginning told by the Qur'ān find their complement in the myths of the revelatory event itself as told by the community. It is through these myths, moreover, that the Muslim community determines the normative mode of action to which it is called.

Before discussing the cosmogonic dimensions characterizing the event of revelation as manifest in the mission of the Prophet, however, I will turn to the Qur'ān's references to the initial creation of the world. For it is these passages which reveal the structure of continuous creation and facilitate an understanding of man's place within it.

In the Beginning

Though the Qur'ān's account of the initial creation parallels the Genesis narrative in many respects, the differences are numerous and often significant. No single passage comprehensively describes the event; instead, the story of beginnings is variously referred to in selections found throughout the text. While one finds frequent mention of the disobedience of Adam and Eve, there is relatively little mention of the creation of the universe *ex nihilo*. The following verse, however, records the event:[2]

> Have not those who disbelieve known that the heavens and the earth were of one piece, then We parted them and We made every living thing of water? (21:30)

Beyond this, the Qur'ān mentions six days of creation during which the physical universe was created, each element therein taking form at the single command "Be!" (2:117; 6:73, etc.).[3] Unlike the biblical narrative, however, the Qur'ān makes no mention of the orders of creation which appear on each successive day. More significantly, there is also no mention of the seventh day of rest. In its stead one finds:

> Lo! Your Lord is Allah who created the heavens and the earth in six Days, then mounted He the Throne. He covereth the night with the day, which is in haste to follow it, and hath made the sun and the moon and the stars subservient by His command. His verily is all creation and commandment. (7:54)

Through this and similar verses describing the continual creation of the universe (10:4; 27:64; 29:19, etc.), the Qur'ān renders the import of the initial creation negligible except as it recognizes that creation as a product of and a context for God's ceaseless creative activity.

It is the perfection of the physical universe, rather than its origination, which is emphasized throughout the Qur'ān. Its verses repeatedly draw attention to the harmonious order found in nature:

> (Allah, the Beneficent, the Merciful) hath created seven heavens in harmony. Thou (Muḥammad) canst see no fault in the Beneficent One's creation; then look again. Canst thou see any rifts? Then look again and yet again, thy sight will return unto thee weakened and made dim. (67:3–4)

Elsewhere, the Qur'ān attributes this harmony of the universe to the creation and subsequent development of nature in accord with divine laws. (41:12). Endowed with a particular rule of behavior (*fiṭrah*), each element in the physical universe fulfills a prescribed role (*amr*) within the perfect order of the cosmos. As suggested by the following verse, this natural instinct is understood to result from the divine guidance provided all creation at its origin.

> And thy Lord inspired the bee, saying: Choose thou habitations in the hills and in the trees and in that which they thatch. (16:68)

With the bee, the entire universe submits to the divine laws which determine nature's development through time. From this perspective, the heavens and the earth can be meaningfully described as *muslim* (that which submits): through absolute submission to divine laws, they describe the unitary design of perfected order for participation in which all things have been created. Cosmic order thus stands witness to the *tauḥīd* (unity) of the Creator. Reflecting the perfection of his laws, nature further exemplifies Islam's understanding of divine justice (*ʿadl*), the principle assuring harmonious order wherein everything assumes its right and proper place.[4]

Numerous verses throughout the Qurʾān explicitly refer to the perfected order of nature as a "sign," a paradigm for the adoration due the Creator:

> Nowise did God create this but in truth and righteousness, (thus) doth He explain His signs in detail, for those who understand. (Y 10:5)

And elsewhere:

> Hast thou not seen that unto Allah payeth adoration whosoever is in the heavens and whosoever is in the earth, and the sun and the moon, and the stars, and the hills, and the trees, and the beasts, and many of mankind? (22:18)

The second passage reflects one qualification to an otherwise perfected cosmos: man. While the adoration of nature is universal, a mere "many of mankind" participate in the structured order of the universe. The Qurʾān's references to the creation of man explain the discord.

The last to be created, Adam and his progeny are distinguished from the rest of the created order from the very beginning. Several accounts in the Qurʾān provide a detailed description of the initial formation of man (32:7–9, etc.). Unlike all else which assumed form at God's single command "Be!", man was formed from previously created elements in nature, from "potter's clay of black mud altered" (15:28). Adam is thus "produced . . . as another creation" (23:14).

Once formed, he assumed a unique place within the created order when God breathed his spirit into the fashioned form. Two references to this event (15:26–29; 2:30–34), when taken together, associate this "breath of God" with the knowledge of "names" which God conferred on Adam.[5] As the recipient of these gifts, man was set above the rest of creation, even the angels (2:33).

This knowledge of "names" is interpreted as an understanding of everything's true nature, of its proper place within the cosmic order.[6] To understand the "names," therefore, is to recognize the divine laws which govern all creative activity, in accord with which all things manifest their potential. Granted this understanding, man stands with the rest of creation which likewise received divine inspiration at its origin: as nature is provided instinct (*amr*) as guidance for all action, so too is man's primordial nature imprinted with the knowledge of divine justice (91:7–8). Unlike the universe, however, man does not naturally act in accord with his true nature. Instead of a predetermined course of behavior, he is accorded the gift of "names."[7] In accepting this knowledge, therefore, man also assumed the burden of responsibility.

The Qurʾān describes this responsibility with a narration of man's selection as God's vicegerent (*khalīfah*) on earth (2:30). When God tells the angels that he is giving man this position, they ask, "Wilt Thou place therein one who will do harm therein and will shed blood, while we, we hymn Thy praise and sanctify Thee?" (Note the juxtaposition of injustice with justice.) God answers, "Surely I know that which ye know not." He then proceeds to teach Adam the "names" and instructs him to display his knowledge to the angels. Then, turning to them, God says, "Did I not tell you that I know the secret of the heavens and the earth?" (2:33). As suggested by the structure of the text, the gift of "names" dispels the threat of injustice which the angels feared would mar the earth.

This appointment of man as vicegerent is elsewhere referred to as an acceptance of the "trust" (*amānah*):

> We offered the trust to the heavens and the earth and the mountains, but they refused to carry it and were afraid of it, and man carried it. (Y 33:72)

Adam, alongside his progeny, assumed this responsibility through acceptance of what is referred to as the Primordial Covenant (*mīthāq*). The following verse describes the event:

> And when thy Lord drew forth from the Children of Adam—from their loins—their descendants, and made them testify concerning themselves (saying): "Am I not your Lord (who cherishes and sustains you)?" They said: "Yea! We do testify " (Y 7:172)

This primordial witness to God as Lord is understood to imply recognition of him as Creator and Sustainer of all that is and, thereby, the God to whom all are indebted.[8] In acknowledgment of this omnipotence, the appropriate response is singular: absolute submission to the divine laws governing creation. With the universe, man fulfills this obligation through recognition of his primordial nature (*fiṭrah*) and through action in accordance with the guidance (*amr*) his nature provides. Correlating the perfection of the universe with the perfection that man is called on to realize, the Qurʾān says:

> On the earth are signs for those of assured faith, as also in your own selves: will ye not then see? (Y 51:20–21)

As vicegerent, therefore, man assumes responsibility for the completion of cosmic order, a task realized through the establishment of divine justice within the human situation. Thus, the commandment:

> So set thou thy face steadily and truly to the Faith. (Establish) God's handiwork according to the pattern on which He has made mankind. No change (let there be) in the work (wrought) by God. That is the standard religion, but most among mankind understand not. (Y 30:30)

The myth of man's acceptance of the "trust," however, is always coupled with another myth. This one explains the lack of understanding on the part of "most among mankind" which foils the realization of divine justice, the "standard religion." It is the story of Iblis. When God granted Adam knowledge of the "names," he called on the angels to prostrate before Adam in recognition of the knowledge he shared with God (15:21ff.). All obeyed, save Iblis. As a result of his refusal, God cursed him and exiled him from heaven. The Qurʾān cites Iblis as saying in response:

> My Lord! Because Thou hast sent me astray, I verily shall adorn the path of error for them in the earth and shall mislead them every one, save such of them as are Thy perfectly devoted slaves. (15:39–40)

The devices of Iblis succeed in luring Adam and Eve to partake of fruit from the forbidden tree (7:19–23). This rebellion against God's explicit command leads to their expulsion from the Garden for a life on earth:

> We said: Fall down, one of you a foe unto the other! There shall be for you on earth a habitation and provision for a time. Then

> Adam received from his Lord words (of revelation) and He relented toward him. (2:36–37)

Though God forgives Adam, Iblis does not retract his threat. As all future generations had stood with Adam in their acceptance of the Primordial Covenant, so too would they, like Adam, be subjected to the enticements of Iblis.[9] Associated with the knowledge which set man above the rest of creation was both the responsibility for the vicegerency as well as the capacity to do evil. The "fall" thus initiated a subsequent history of incessant struggle against the powers of evil which cloud the primordial soul except where, through absolute submission, the faithful are fortified in their effort to fulfill the "trust."

As is evident from the Qur'ān's account of the original creation, its description of the "time in the beginning" is inextricably woven in with the cosmology it suggests. It is the story of nature's perfect creation not only at the beginning of time but throughout time; it tells of the responsibility of the Covenant accepted not only by Adam but by all his descendants; it describes the ways of Iblis who enticed not only the first couple but all succeeding generations as well. The narratives tell as much about the original creation as they do about divine creative activity through time. These texts, therefore, speak more of a cosmology than of a cosmogony. Yet, in the very telling of that cosmology as a cosmogony, the myths suggest a cosmogonic dimension which characterizes man's situation in the world.

At present, cosmogonic ordering is incomplete; its perfection will be realized only when divine laws describe a cosmos consistent within the natural order and the human situation alike. It is man's responsibility, as "vicegerent," to fulfill the purpose of creation through participation in the continuous process of creation. As indicated by the Adam narrative, however, it is ultimately God—through the mercy he shows in guiding those who submit to his will—who creates and orders. In history as in nature, his will is manifest, his dominion realized. Thus, a comprehensive appreciation of Islam's cosmogonic understanding requires that we move beyond the time "in the beginning" to consider creation as history.

Cosmogony as History

While Adam submitted in accord with the guidance provided him through divine revelation, the history of his progeny reveals the dark side of cosmogonic activity. Unlike nature, history is the fabric woven from human action. It narrates man's repeated failure to fulfill the creative task accepted with the "trust"; likewise, it mirrors his failure to use appropriately the gift of "names" so as to participate in the purpose of creation they describe. As with nature, however, the unfolding of history is governed at each moment by the Creator: ultimately it is his will which directs the course of events. The Qur'ān thus refers to history as "a detailed exposition of all things" (Y 12:111). With nature, it stands witness to the laws of justice with which the world is created. From this perspective, history reveals the paradox at the heart of Islam's understanding of man: true creativity implies utter submission to God's law in every aspect of one's existence; to act independently is merely to foment destruction. Though man is released from nature's determinism by being designated "vicegerent," it is only in becoming a "slave of God," to use a metaphor favored by the Qur'ān, that he is able to participate in the creation of the world and, thereby, in the perfection of its completion at the end of time.

Truly creative activity, therefore, is necessarily ethical. Solely through abiding by the rules for conduct (*ḥudūd Allah*, "bounds of God") with which he is created, can man fulfill his responsibility for participating in the continuous creation. The individual who thus acknowledges his place in the scheme of creation is a *sādiq*, one characterized by the "intention to be true to reality" (Izutsu 1959, 81). This intention is intrinsically related to *īmān* (faith) from which it necessarily arises; similarly, the intention, if pure, will naturally result in unquestioning obedience to the divine will. Or, as stated by Izutzu:

> Just as the shadow falls with the form, wherever there is *īmān* there is *sāliḥāt* or, "good works," so much so that we may rightly define the former in terms of the latter, and the latter in terms of the former. (1959,206)

The community of believers (*ummah*) is thus unified in its identity, function, and purpose. Through corporate action, it strives to establish an order which witnesses, with nature, divine justice and unity (*tauḥīd*). Solely through participation in this communal endeavor can the individual realize the potential he was created to manifest.

As told through the story of initial creation, this task entails incessant struggle against the blindness of evil which will give way only on the Day of the Appointed Time (15:38). In allowing the "seductions of Iblis" (7:29) to veil his primordial nature, man blinds himself to the "signs" in nature and history. Thus "deluded away from the truth" (6:98), he recognizes himself as self-sufficient; he fails to perceive the divine benevolence manifest in creation.

> Do you not see that God has subjected to your (use) all things in the heavens and on earth, and has made His bounties flow to you in exceeding measure? (Y 31:20)

The unbeliever, who is in arrogant disregard of this beneficence, is repeatedly referred to as a *kāfir* (one who is ungrateful) in the Qurʾān. His acts, no longer signifying submission to the One to whom he is indebted, fail to fulfill their purpose and ends.

> Nay, but man doth transgress all bounds, in that he looketh upon himself as self-sufficient. (96:6–7)

Transgressing divine law, man's acts disrupt, rather than establish, the order for which he is responsible. Consequently, he is also described as a *ẓālim* (one who is unjust). In failing to heed the creative laws, the *ẓālim* merely incurs his own destruction.

> Those who break God's Covenant after it is ratified and who sunder what God has ordered to be joined together, and who do mischief on [or: corrupt] the earth: these cause loss only to themselves. (Y 2:27)

While conversely,

> Prosperous is he who purifies (his soul). (Y 13:167)

The laws of divine justice which govern the entire process of creation can be recognized in the rise and fall of communities. The Qurʾān provides innumerable references to the manifestation of these laws in history.

> We wrongeth them not, but they did wrong themselves; and their gods on whom they call beside Allah availed them naught when came thy Lord's command; and they added naught save ruin. Even thus is the grasp of the Lord when he graspeth the communities while they are doing wrong. Lo! His grasp is painful, very strong. Sure in that is a sign for him who fears the chastisement of the world to come. (11:101–2)

As for the communities who are "putting things right," they will never be unjustly destroyed (11:117). For,

> Allah hath promised such of you as believe and do good works that He will surely make them succeed in the earth even as He caused those who were before them to succeed. (24:55)

Solely by acting in accord with the divine law which governs creation, is man able to participate in the positive unfolding of history. Ultimately, it is the choices he makes and the moral values implicit therein which shape the destiny of his future.[10]

True understanding of the scheme of creation entails participation in divine creative activity, continuous through history and consummated at the end of time, when

> The truthful will profit from their truth: theirs are Gardens, with rivers flowing beneath—their eternal home. (Y 5:119)

To "corrupt the earth" in service of ends other than those of the Creator, is to "bar the path of God and desire to have it crooked" (7:45). Unbelief entails participation in the destructive consequences of creative activity, both through time and at its end, when the perfection of completion will necessarily imply an absolute separation of good from evil. The *kāfir* is warned:

> And for those who disbelieve in their Lord there is the doom of hell, a hapless journey's end! When they are flung therein they hear its roaring as it boileth up, as it would burst with rage. (67:6–8)

It is the just law of God—inevitable in the unfolding of history (16:1) and perfectly manifest at the end of time—which judges a man worthy of participation in the consummation of creative activity. With this judgment is revealed the purpose of the world's creation:

> He it is who created the heavens and the earth in six days—and His Throne was over the waters—that he might try you, which of you is best in conduct. (Y 11:17)

To understand the beginning is to understand the end: acceptance of the Primordial Covenant implies participation in the creative endeavor, both in its continuity as well as in its completion.

Beyond the judgment of history, however, stands the Creator "oft-forgiving and most merciful" (4:25). His forgiveness counteracts the laws of history. While punishment is "easy" (4:169), because it results from the just laws running their course, the reward of the Garden is contingent on divine mercy as manifest in both guidance and forgiveness (Yusuf Ali 1977,233,246). The Qur'ān often refers to this expression of divine mercy with the metaphor of "turning":

> God doth wish to make clear to you and to show you the ordinances of those before you; and (He doth wish to) turn to you (in Mercy).
>
> God doth wish to turn to you, but the wish of those who follow their lusts is that ye should turn away (from Him)—far, far away. (Y 4:26–27)

This correlation between the "turning" on the part of man and God is further suggested by the Qur'ān's use of the same term (*taubah*) to refer both to the repentance of the believer and to the forgiveness extended him (Izutsu 1959, 106). Those who have "turned," who have submitted to the way of God in his creation of the world, are promised expression of this divine mercy when creation is transformed through the manifestation of absolute justice at the end of time.

> Verily, those who say, "Our Lord is God" and remain firm (on that Path)—on them shall be no fear, nor shall they grieve. . . . Such are they from whom We shall accept the best of their deeds and pass by their ill deeds; (they shall be) among the Companions of the Garden. (Y 46:13,16)

History is recognized as meaningful, therefore, inasmuch as it reveals the creative and destructive contours of cosmogonic activity. Its "signs" add depth to the "signs" in nature: while the perfection of the universe provides the paradigm for submission to the Creator, history provides its negative complement. It describes the shadow side in telling of man's illusion of independence. While the stories of communities destroyed narrate past history, for those "with understanding" they also prefigure the manifestation of absolute justice which will mark the end of time. It is in looking back to history in light of the future, that the Qur'ān calls for discernment of the ethical orientation demanded by the present. As one scholar observes:

> History in the Qur'ān becomes unified time. . . . The walls that separate the past, the present, and the future collapse and the three times comingle in a common destiny. Even the earth and heaven, temporal time and divine time, the story of Creation and the Day of Judgement . . . always meet in the present moment. (Khalīl, in Haddad 1982, 189)

It is through these dissolved chronological distinctions that the web of Islam's cosmology is spun.

While history provides "signs" which define the negative contours of the ethical paradigm, it also goes beyond nature in explicating the truths implicit in the perfect ordering of the physical universe: it describes the mission of the Prophet. This is the history to which Islam turns in its telling of the perfect realization of human potential, of the just ordering of the human situation called for by the Qur'ān's cosmogonic myths. Here the tradition finds a description of the first *ummah* established by Muḥammad in accord with divine revelation, a sociopolitical order which mirrored the "cosmos" of nature. For a time, the dual dimensions of creation—history and the universe—stood together in perfect witness to the *tauḥīd* (unity) of the Creator. In thus fulfilling the plan for creation, the first *ummah* is affirmed by the Qur'ān as "the best community . . . raised up for men" (3:110). The structure of cosmogonic understanding in Islam, therefore, suggests that we turn to this account in recognition of the tradition's primary cosmogonic myth.[11]

A Cosmogonic Myth: The Mission of the Prophet

With the "fall," man forfeited his right to the Garden. With Adam, however, he was promised the guidance necessary to recognize the "straight path" of return. This promise found fulfillment in a succession of prophets whose messages called on man to discern the directives of his true nature so as to acknowledge and fulfill the obligations of the Primordial Covenant.

While understood as historical figures, the prophets are significant inasmuch as their missions delineate the paradigm which informs, guides, and builds the community of believers. Their revelations provide no new truths; rather, as with the "signs" in nature, they explicate the truth implicit within man's primordial soul.

> Oh ye unto whom the Scripture hath been given! Believe in what we have revealed confirming that which ye possess. (4:47)

The Qur'ān thus refers to itself and the prophet Muḥammad as "reminders" or "The Reminder." For those "with understanding," they reveal nothing beyond a delineation of the truth evident within the realms of nature and history. Accordingly the revelations of the various prophets are understood to have repeatedly asserted the same truth:

> Say (O Muslims): We believe in Allah and that which is revealed unto us and that which was revealed unto Abraham, and Ishmael, and unto Isaac, and Jacob, and the tribes, and that which

> Moses and Jesus received, and that which the Prophets received from their Lord. We make no distinction between any of them. (2:136)

The succession of prophets was required not for the further revelation of divine will but for the restoration of the "true religion," which had been corrupted by the followers of previous revelations. Each prophet is recognized as recalling humanity to the Primordial Covenant and explicitly delineating an ethical order wherein the universal laws of creation might be manifest. As their message is constant, so too are the ethical norms they engender.

> (This was Our) way with the apostles We sent before thee (O Muḥammad): thou wilt find no change in Our ways [or:laws]. (Y 17:77)

As a reminder, the mission of the prophet is to offer clarification of the signs obscured through the "corruption of the earth." Awakened to primordial knowledge, the community is able to transform and order its life in accord with the divine will revealed through the recurrent revelation given to the prophets.

With Muḥammad, however, this cycle of prophecy ends. Called on to "follow the religion of Abraham, as one by nature upright" (16:117), Muḥammad's task was similar to that faced by the prophets before him: to revive the natural religion of pure monotheism. While in essence the same, however, the revelation given Muḥammad went beyond that of his predecessors as it further explicated the guidance they had received. As the earlier prophets had delineated the truths implicit in nature and history, so too the Qur'ān offered a "confirmation of (revelations) that went before it, and a fuller explanation of the Book . . . from the Lord of the Worlds" (Y 10:37). The guidance provided Muḥammad culminated a gradation of revelation initiated with the first creation of the world; with the injunctions of the Qur'ān directly ordering life in the community to which it was first given, the event of revelation perfectly actualized the ethical responsibility accepted with the Primordial Covenant. In Islam's cosmology, therefore, the period of the Prophet stands at the watershed of time. As told in the "Farewell Sermon" delivered by Muḥammad at the end of his mission:

> Time has completed its cycle and is as it was on the day that God created the heavens and the earth.(Ibn Isḥaq 1980,651)

History witnesses a second beginning that will find completion only with the final manifestation of divine justice on the Last Day. The significance of this beginning, distinguishing Muḥammad's mission from those of earlier prophets, can be discerned through a comparison of the three periods marking his prophetic career.

Muḥammad received the first surahs of the Qurʾān in Mecca. During this period, he called on the residents of the city to turn from polytheism and embrace the true faith of *tauḥīd*, the singular guidance of the Creator to whom all will return. He decried the cult that had corrupted the Ka'bah, "the first Sanctuary appointed for mankind" (3:96), which Abraham and Ishmael had built under God's direction (2:127). In explicit terms, he warned of the destruction individuals would face at the end of time if they failed to turn in submission to their Creator and blindly continued to oppress the weak and neglect the unfortunate, the widows, and the orphans.

Initially, Muḥammad's mission met with little support. Gradually, however, his following grew and came to pose a threat to the Meccan elite. Their opposition to the Prophet's call for reform intensified, severely limiting the possibility of establishing the moral order demanded by the Qurʾān. In 622 A.D. Muḥammad left Mecca with his followers and migrated to Yathrib. Eight years later, he returned to bring Mecca within the pale of his expanding authority. Tradition recounts the Prophet's entry into the Ka'bah, where he smashed all the idols, rendering the sanctuary suitable for the annual pilgrimage (*ḥajj*) called for by the Qurʾān. His followers performed the pilgrimage in the following two years; Muḥammad, however, is said to have waited until his power was sufficiently consolidated so that he could celebrate the *ḥajj* with no unbelievers present, in uncompromised accord with the prescribed rites first performed by Abraham (22:29). Not until 632, therefore, did the Prophet lead his followers on what tradition refers to as the Farewell Pilgrimage. Symbolically, the event expressed the intention of his mission: restoration of the true faith repeatedly called for by the prophets before him. In this, the Qurʾān determined him successful:

> This day I perfected your religion for you and completed My favor unto you and have chosen for you as religion al-Islām. (5:3)

This was the last verse revealed to the Prophet (Ali 1977, 237). Three months after his return to Medina, Muḥammad fell ill and passed away.

The unique dimension of the Prophet's mission, however, is found in the interlude between his original stay in Mecca and the subsequent reform of the Ka'bah. Unlike he did in the initial period in Mecca, the Prophet met with negligible resistance in the Yathrib community. As his followers accounted for most of the town's residents, Muḥammad was able to exact conformity not only on the level of personal piety but in his reorganization of the sociopolitical order as well.

Although in principle merely one of the faithful called on to recognize the responsibility and obligations implicit in affirming Allah as Creator and Lord, Muḥammad was clearly regarded as more (Hodgson 1975,690). It was he who conveyed the revelation of the Qur'ān as it provided continuous guidance for the establishment of the new moral order. Quite naturally, therefore, it was to the Prophet that the community looked for an interpretation of the divine directives: in his judgment, decisions, and actions, the faithful found guidance in matters not explicitly addressed by the Qur'ān. Consequently, Muḥammad assumed an authoritative role within the community and came to be recognized as the àrbiter of divine justice. In approval of this deference given the Prophet, the Qur'ān says:

> O mankind! the Apostle hath come to you in truth from God: believe in him; it is best for you. (Y 4:170)

> Ye have indeed in the Apostle of God a beautiful pattern (of conduct). (Y 33:21)

Through Muḥammad's leadership, therefore, the message he conveyed effected reform in both the personal and social spheres of community life. Appropriately, the name of the town was changed from Yathrib to *madīnat an-nabī* (Medina): the City of the Prophet.[12]

It is thus in Medina that Islam recognizes the perfection of the prophetic paradigm which elevates Muḥammad above his predecessors. The import of his mission extends beyond being a reminder of the Primordial Covenant, a call to reform acknowledged by a group of followers. Here, the divine guidance brought about a complete break with the past and established a sociopolitical order in perfect accord with the will of the Creator. Rather than the first revelation revealed in Mecca, therefore, it is this break as realized through Muḥammad's emigration (*hijrah*) to Medina, which takes Islam's cosmology back to the beginning and ushers in a new era of history. For here the story of initial creation finds its completion: the perfect actualization of the responsibility assumed through acceptance of the "trust." The community at Medina is understood to have exemplified in the same way as nature, the corporate endeavor which is at once the context for as well as the necessary result of absolute submission to the way of God (*sabīl Allah*) in the world. The "corruption of the earth" was eliminated and divine justice prevailed. A cosmos was created. The story of Muḥammad at Medina delineates the return to the Garden for subsequent history: humanity will no longer be sent a prophet as a reminder of the true religion, for history now stands witness to the perfect fulfillment of the Primordial Covenant.

A Cosmogony Remembered

With the death of the Prophet, the period of direct revelation illuminating and guiding the emergence of the *ummah* came to an end. The poetry with which Ibn Isḥāq closes his biography of the Prophet expresses well the loss experienced by the community at the death of their leader:

> By God, no woman has conceived and given birth
> To one like the apostle, the prophet, and guide of his people;
> Nor has God created among his creatures
> One more faithful to his sojourner or his promise
> Than he who was the source of our light,
> Blessed in his deeds, just, and upright. . . .
> O best of men, I was as it were in a river
> Without which I have become lonely in my thirst. (Ibn Isḥāq 1980,690)

The poem further serves as an indication of the sentiment from which emerged the myth of "the state of things at the beginning" (*al-amr al-awwal*), the myth fundamental to the *ummah*'s effort to compensate for their loss in the ongoing struggle to maintain the way of God in their community. Though the age of the Prophet had passed, the paradigm for true submission had been explicitly delineated. The memory of Medina under the Prophet, sustained through innumerable accounts (*hadīth*), assured living involvement with the truths manifested by Muḥammad's mission (Hodgson 1975, 324).

Knowledge of the Prophet's every word and deed became increasingly important for piety, for behavior consistent with the divine will. To acknowledge Muḥammad as the messenger of God was to recognize the normative character of the pattern of life he had established. His example was to be followed in all conduct, individual as well as social. As Rodinson writes of the Prophet:

> The smallest details of his behavior served as a model for all to follow. All courses, standards or ideals could be defended by

> attributing their origin to his words or notions. He had laid down the rules for divorce and government, just as he had shown how one should behave oneself at the table, blow one's nose and make love. (1971, 302)

This effort to maintain continuity with and sustain the vitality of the original community culminated in the codification of the *sharīʿah*, "the road leading to the water." In following this path, the "river" that had subsided with the death of the Prophet could be returned to, the "thirst" quenched.

While the *sharīʿah* is understood to provide norms for conduct consistent with the divine will, the initial cosmogony continues to describe the human situation: forces of evil continue to shroud the "signs" provided to man and lure him from the creative task to which he obligated himself in accepting the "trust." During the age of the Prophet, these forces had been counteracted in Medina through the community's perfect submission to the continuous revelations of the Qurʾān. Islam thus remembers this period of history as a "wholly different mode of time: a time made holy by divine activity, a 'time out of time' " (Graham 1977, 9). With the passing of the revelatory event, however, the older structure of chaos, fomented through blind "corruption of the earth," returned to threaten the order of the *ummah*. As Muḥammad's mission had provided consummate guidance, however, the threat of chaos can no longer be left to future prophetic activity. The mantle now falls on the *ummah* to "enjoin what is right and forbid what is wrong" (3:104), to restore the vision of the blind through a return to and manifestation of the potency of the creative moment, wherein the possibility of participation in the fulfillment of creation is realized. The creative task, therefore, continues to describe the Muslim's understanding of who he is within the community and before the Creator.

While the norms for this task are delineated by the event of revelation as remembered through the *sharīʿah*, the divine speech itself retains an active role within the *ummah*. A book to be carefully studied and memorized, the Qurʾān is also a text to be recited as it was by the Prophet when the surahs were first revealed to him. Through this recitation, the Qurʾān continues to engender the intention which informs the *ummah*'s dedication to the way of God. This ritual reenactment of the revelatory moment is understood as essential for the recovery of the prescribed past, of the "state of things in the beginning" (Naff 1981, 27). Of this return to the "time out of time" Hodgson writes:

> What one did with the Qurʾān was not to peruse it, but to worship by means of it; not to passively receive it but, in reciting

> it, to reaffirm it for oneself: the event of revelation was renewed every time one of the faithful, in an act of worship, relived the Qur'ānic affirmations. (Hodgson 1975, 367)

In reciting the Qur'ān the believer affirms the truth embodied in the divine speech. The act of worship is recognized as instrinsically related to the behavior it engenders; likewise, acts carry import only insofar as they express understanding of the Word which informs them (2:177). Thus, the initial event of revelation is recognized as both a book and the foundation of a community: to acknowledge the consummate guidance of the Qur'ān is to recognize the perfection it inspired in the community to which it was first revealed. One translator incorporates this understanding of the Medina ideal in his rendering of surah 2:143.

> God hath decided to raise you to the position of a model community, so that you might be an example unto others even as the Prophet hath been an example unto you. (Azad 1967, 58)

As two aspects of the creative event par excellence, the revelation of the Qur'ān and the ordering of the first *ummah* stand together in extending the possibility for true creative activity: for the first time, history witnesses the perfection of a "sign" consonant with the purpose of creation.

It is thus to the "state of things in the beginning," the complete integration of the Qur'ān as both text and event, that the Muslim community returns in its effort to sustain continuity with the order established at Medina. This concern informs the routine of daily life which prescribes five times for Qur'ānic recitation as part of ritual prayer (*ṣalat*). When order is threatened by serious illness or death, friends gather to recite the text many times through; similarly, when the promise of order is sought, as in the first habitation of a home or a child's entrance into school, a collective recitation of the Qur'ān is arranged. This search for order and renewal is evidenced on a quite different level with the *ummah*'s concern for the social fabric of the community itself. Again, it is to the initial revelatory event and the perfection it inspired that they return. An excerpt from a recent publication evidences how the Medina ideal, which has informed reform movements throughout the history of Islam, continues to provide Muslims with the promise of renewal in its explication of ethical responsibility.

> [T]hose who advocate an Islamic response to the crisis of the present want not only to recapture the potency of the past, but

> to affirm that that potency cannot be recaptured outside the prototype of the Medinan state as established during the lifetime of the Prophet. It is not only a spirit of the age that needs to be recaptured, but a whole social and ethical order that needs to be re-created and re-established. It is not an ideal or idealistic plan, but the only means to salvation of the world and the ascendancy of God's way in the world. (Haddad 1982, 137)

To discuss the tradition of Islam in light of this "return" would carry us far beyond the purpose of this paper, as it infuses almost every dimension of life that might be termed "Islamic." The two examples briefly discussed above, however, suggest the diversity of Islam's recognition of the promise inherent in the "return." They offer a sense of the broad spectrum of understanding which recognizes the perfection of creation realized through the mission of the Prophet as normative for the task described by the myths of initial creation. Despite the interpretation given to the event, however, a fundamental structure is invariably evident: in his effort to regain the harmony and order of the "state of things in the beginning," man in Islam turns not to the myth of an intentionally incomplete creation at the dawn of time, but to the myth of the Prophet. For it is this myth which describes completion in the continuous process of creation.

Conclusion

Charles Adams, in the article cited at the beginning of this essay, gives a succinct description of the character of Islam:

> Its principle mode of apprehending, responding to, and expressing the sacred has been ethical. Sunni Islam has been much more a matter of living out an ordained pattern of life than anything else. (1967, 183)

The accuracy of this statement is evidenced by the fact that it is the acceptance of the *sharīʿah* as an ideal to be acknowledged, if not a norm to be realized, which has accorded life within the *ummah muḥammadīyah* its unique character. This "principle mode of apprehending the sacred," however, cannot be adequately understood through a study of Islamic ethics alone. In search of the religious understanding which informs the "ordained pattern of life," one discovers a theme that provides the ethical norms with their meaning: cosmogony.

Through true understanding of the creation of the world, the Muslim recognizes who he is before the Creator. It is the Qurʾān's cosmogony which describes his participation in the Covenant made at the beginning of time, the creative task it implies through time, and the promise it holds for the end of time. These cosmogonic myths establish the framework through which man in Islam orients, synthesizes, and integrates the reality of his world. The beginning of the story delineates an ethical orientation in describing the human situation; the end of the story delineates the normative behavior required by that orientation in its telling of the fulfillment of the "trust" at Medina. In thus narrating the possibility of creation's perfection, the story's end also prefigures the end of creative activity which will be realized in history only with the complete manifestation of divine justice at the end of time.

These myths of creation do not belong to an archaic people. Islam does not seek to annul history; it embraces it as the arena within which the ethical and creative endeavor is to be realized. With archaic man, however, the Muslim finds meaning in that history not through its linear unfolding in time but through the paradigm for true submission it reveals: it is when chronological distinctions give way that the ethical demands of the present are recognized. While accepting history, therefore, the Muslim shares the archaic sense of "profane" time. He too seeks to return to the "state of things in the beginning" which promises continuity of "sacred" time though history; in the myth of beginnings, he too finds a paradigm for truly creative activity; it is through a return to the primary cosmogonic myth that he too finds the possibility of restructuring order amid the challenge of chaos. It might be said of man in Islam, therefore, as Eliade describes man in archaic traditions, that he "sees himself as real only to the extent that he ceases to be himself (for the modern observer) and is satisfied with imitating and repeating the gestures of another" (1974, 34). This similarity of response to the sacred suggests that a methodology which recognizes the importance of archaic traditions for discerning structures of religious experience may prove informative for understanding religious phenomena in Islam.

The potential dialogue it suggests, however, would not be one-sided. While the research of archaic traditions in the history of religions—in this case, Eliade's methodology in particular—may assist recognition of the structure of Islamic responses to the sacred in terms of a general typology of religious experience, the tradition of Islam, in turn, may lend a new perspective to the methodology itself. For example, particularly problematic for the case of Islam is Eliade's assessment of history (1974, 102–12). The categories of linear and cyclical time are discussed as mutually exclusive perceptions of time. Although Eliade does not discuss the case of Islam in detail, the tradition poses fundamental questions for his understanding of religions that recognize the importance of history. Does the value given history, in a recognition of linear time, necessarily transcend the traditional vision of cyclical time? Does divine intervention in history necessarily preclude the divine creation of archetypal gestures? Does the revelatory event remain limited to a moment in time? Does historical revelation necessarily imply new paradigms for action? Islam's understanding of history suggests that these questions be answered in the negative. The importance it gives history may be characterized as at once linear and cyclical, the one complementing, rather than excluding, the other.

While Islam's perception of history poses questions for a methodology which acknowledges the import of archaic traditions for the historian of religion's task, it also becomes problematic for the Islamicist who sees "the historical stuff of Islamic religiousness as extraordinarily, one may say almost perversely, impervious to significant analysis along the lines which the majority of historians of religions have followed and are following" (Adams 1967, 181–82). Recognizing Islam's perception of history, scholarship using archaic traditions will be concerned not with the significance of pre-Islamic "survivals," but with the experience which informs the ethical response central to the tradition: the understanding of a cosmogonic myth which at once expresses and engenders the *ummah*'s vision of a perfect order through which the completion of creation is realized.

Notes

1. For discussion of the need for redefining "cosmogony," see, in this volume, Reynolds, chapter 8, pp. 203–24, and Sturm, chapter 14, pp. 353–80.

2. In support of their understanding of the eternity of the world, Muslim philosophers have rightly pointed out that there is no explicit mention of creation *ex nihilo* in the Qurʾān. On this point, and throughout the essay, I have limited the discussion to Sunni interpretations.

I have used M. Pickthall's translation of the Qurʾān except where otherwise indicated with a "Y" before the sura reference, indicating the use of A. Yusuf Ali's translation.

3. Though the Qurʾān frequently mentions six days of creation, it elsewhere equates one of God's days with fifty thousand years (70:4).

4. Explaining the particular meaning of the concept of justice in Islam, Gibb writes, "Justice is not used in the same sense of political

or judicial application of manmade laws, but as a principle of order and wholeness: all elements, endowments and activities of life shall be in harmonious relation with each other, each fulfilling its proper purpose and ends in a divinely-appointed system of interlocking obligations and rights" (1964, xi).

5. Sura 15:28–31

And when thy Lord said unto the angels: Lo! I am creating a mortal out of potter's clay of black mud altered,	ANNOUNCEMENT OF CREATION OF MAN
So, when I have made him and have breathed into him of My spirit, do ye fall down, prostrating yourselves unto him.	SPIRIT OF GOD BREATHED INTO MAN
So the angels fell prostrate, all of them together. Save Iblis.	PROSTRATION OF ANGELS

Sura 2: 30, 31, 34

And when thy Lord said unto the angels: Lo! I am about to place a viceroy in the earth. . . .	ANNOUNCEMENT OF CREATION OF MAN
And He taught Adam all the names, then showed them to the angels, saying: Inform me of the names if ye are truthful. . . .	NAMES TAUGHT TO MAN
And when we said unto the angels: Prostrate yourselves before Adam, they fell prostrate, all save Iblis.	PROSTRATION OF ANGELS

6. Rahman appropriately refers to this knowledge as "creative," for it implies a correlation between God's creating/ordering the universe and man's ability to create/order a social order which reflects the justice of the universe (1980, pp. 10, 18).

7. Space does not permit a detailed discussion of the free will debate which has colored theological developments throughout the history of Islam. The problem, however, does not affect the primary thesis under consideration. Both arguments recognize human action as participating in the continuous creation of the world; they disagree over the extent of man's freedom within that process of creation. Those who recognize absolute determinism claim that "each event

that happens or fails to happen is the result of a particular creative act on God's part," while their opponents assert that "man creates his own acts and so becomes the cause of his own bliss or damnation" (Goldhizer 1981, 81, 112).

This discussion, however, presumes the position current in many circles, as represented by Rahman's statement: "To hold that the Qur'ān believes in absolute determinism of human behavior denying free choice on man's part, is not only to deny almost the entire content of the Qur'ān but to undercut its very basis" (1980, 20).

8. Although the word "rabb" is usually translated into English as "Lord," the Arabic term carries broader meaning than does the English. As al-Attas writes of the word as it appears in this verse, "To acknowledge God as Lord means to acknowledge Him as Absolute King, Possessor and Owner, Ruler, Governor, Master, Creator, Cherisher, Sustainer—since all these meanings denote the connotations inherent in the concept of Lord" (al-Attas 1978, 49).

9. Islam has no concept of original sin which recognizes man as inherently sinful; quite the contrary, man is understood to be inherently good and potentially better. He is "sinful" only inasmuch as he fails to perceive the directives of his primordial nature. The Qur'ān thus warns, "O ye Children of Adam! Let not Satan seduce you in the same manner He got your parents out of the Garden." (Y 7:27).

10. The promise of prosperity for those who submit to the divine will is sometimes qualified, as by the following verse: "Before thee We sent (Apostles) to many nations and We afflicted the nations with suffering and adversity, that they might learn humility" (6:42). Ultimately, however, the believers who endure through the trial are understood to prosper, as evidenced by the early persecution of the Prophet and his followers in Mecca which preceded the flourishing of the community in Medina.

11. Unfortunately, the term "myth" continues to be burdened with the meaning of "an untrue narrative." As the term is often offensive to Muslim sensibilities, I use it with hesitation, though in appreciation of its reference to the ultimate truths revealed through accounts of divine participation in the human situation, accounts which reveal the true nature of the cosmos.

12. As Rodinson points out, scholars have differed with Islamic tradition regarding this etymology: "The Jewish name for Yathrib was the Aramaic *medīntā*, which means simply 'the city'; this became in Arabic *al-medīna*, from which we get Medina. That the Koran itself calls it by this name is proof that Yathrib did not, as has often been claimed, take its second name from the phrase *madīnat an-nabī*, 'the city of the Prophet' " (1971, 139).

References

Adams, Charles
1967 "The History of Religions and the Study of Islam." In *The History of Religions: Essays on the Problem of Understanding*, ed. Joseph Kitagawa, pp. 177–93. Chicago: University of Chicago Press.

Ali, Yusuf A.
1977 *The Holy Qur'ān: Translation and Commentary*. Indianapolis: American Trust Publications.

al-Attas, S. M. al- Naquib
1978 "Religion and the Foundation of Ethics and Morality." In *The Challenge of Islam*, ed. A. Gauhar, pp. 33–67. London: Islamic Council of Europe.

Azad, Abul Kalam
1967 *The Tarjumām al-Qur'ān*. Vol. 2. Ed. and trans. S. A. Latif, Bombay: Asia Publishing House.

Eliade, Mircea
1974 *The Myth of the Eternal Return*. Princeton: Princeton University Press.

Goldziher, Ignaz
1981 *Introduction to Islamic Theology and Law*. Ed. B. Lewis, trans. A. and R. Hamori. Princeton: Princeton University Press.

Gibb, Hamilton
1964 Preface. In Sayyid H. Nasr, *An Introduction to Islamic Cosmological Doctrines*, pp. xiii–xvii. Cambridge: Harvard University Press.

Graham, William
1977 *Divine Word and Prophetic Word in Early Islam*. The Hague: Mouton.

Haddad, Yazbeck
1982 *Contemporary Islam and the Challenge of History*. Albany: State University of New York Press.

Hodgson, Marshall
1974 *The Venture of Islam*. Vol. 1. Chicago: University of Chicago Press.

Ibn Isḥāq, Muḥammad
1980 *The Life of Muḥammad: A Translation of Ibn Isḥāq's Sīrat Rasūl Allah*. Trans. A. Guillaume. Karachi: Oxford University Press.

Izutsu, Toshihiko
1959 *The Structure of Ethical Terms in the Koran*. Tokyo: Keio Institute of Philological Studies.

Naff, Thomas
1981 "Towards a Muslim Theory of History." In *Islam and Power,* ed. A. Cudsi and A. Bessouki, 24–36. Baltimore: Johns Hopkins University Press.

Rahman, Fazlur
1980 *Major Themes of the Qur'ān*. Minneapolis: Bibliotheca Islamica.
1982 *Islam and Modernity: Transformation of an Intellectual Tradition*. Chicago: University of Chicago Press.

Rodinson, Maxime
1971 *Muḥammad*. Trans. A. Carter. New York: Pantheon Books.

10 *Creation Narratives and the Moral Order: Implications of Multiple Models in Highland Guatemala*

Kay Barbara Warren

While creation mythology may once have been a more elaborate, less cryptic body of oral narratives for the Mayan Trixano Indians of San Andrés Semetabaj, this was not the case in the early 1970s on the eve of a collapse of traditionalist religion in this western Guatemalan town.[1] At that point ritual guides (*camol beij*) recounted world creations in striking but fragmentary images, not infrequently including first-person editorial comments and asides. The line between myth, descriptions of "ancient times," and exegesis was not clearly drawn or apparently relevant. This verbal informality contrasted with the guides' careful, formalized recitation of ritual prayers and discourses. The guides see the decline of knowledge about traditional beliefs (*costumbre*) as part of a pattern of social change, as they point out in the invitational prayers at major rituals:

> [The guide] has come to bring us, to continue the *costumbre* that our dear past fathers invented, in procession behind the saints. Perhaps we cannot fulfill this *costumbre* to perfection, but although it be only a sign that we make, although it be only a little, we are doing it. The *costumbre* does not extinguish itself, but rather we ourselves are those who leave behind our *costumbre* in this land, in our town.

> Perhaps there was a word, a better form of expression that our father ancestors used, but now they have returned in the hands and at the feet of God. They have taken their words and expressions; they returned with all of this. Now we remain.

And so it is reasonable at the outset of this analysis to ask if mythic narratives have become so eroded by social change that they have lost their semantic load, their expression and systematization of fundamental conceptions of the Trixano worldview. In fact in studying the thematic interrelation of oral narratives one finds cohesiveness and subtlety in the portrayals of the creation of the natural and social orders. Moreover, the thematic connections of the oral narratives to ritual prayers and discourses form a broader meaning-system with complex messages about cosmogony, world order, and ethics.[2]

In this analysis, I examine mythic narratives and ritual discourses which describe (a) the general creation of the world order, (b) the Spanish conquest and transformation of the Indian community of San Andrés, and (c) the blessing and protection of Indian communities after they were given by Jesucristo to the saints, and (d) the role of Indian ancestors in creating Trixano *costumbre*. My argument here is that these narratives express similar principles of creation, yet they describe distinctive social and ethical orders. Of particular interest are the different images of moral action. Central to the conquest myths is a brutal social world in which the preeminent models of moral action—as pictured in the initial, generalized world creations—are unrealizable for the conquered Indians. Trixano worldview presents multiple cosmogonies and ethical orders. The moral conflicts and ambiguities of the creation-conquest cosmologies are resolved through another set of narratives and further articulated in Trixano ritual which describes a universe not cynical about moral action or hostile to the wants and needs of the Indian community.

The goal of my analysis will be to present fine-grained anthropological examination of Indian narratives as examples of multiple cosmogonies and ethical orders. In addition, I will go on to consider some broader conceptual issues involved in the comparative study of meaning systems.

World Creation: A Process of Cyclic Refinement

Trixano narratives tell of a series of creations and destructions through which the world was formed and ordered and the Eternal Father's direction was taken over by other sacred intermediaries, first his Son, Jesucristo, and later the saints. The chronology of creation forms interlocking stages: the epoch of the Eternal Father, the life and crucifixion of Jesucristo, and the division of the earth's population into towns governed by the saints. Each stage is structurally marked by the addition of new intermediaries for the sacred realm and the division of the world into more specific domains.

Trixanos conceptualize creation in a vocabulary rich in agricultural metaphors. Initially all living beings were periodically "harvested" through a world destruction commanded by God. Later, as a consequence of Jesucristo's coming to earth, individuals were allowed independent life-cycles and individualized "harvests" of death. In both cases death is thought to be necessary for a broader continuity and fertility.

During the epoch of the Eternal Father, human beings were limited in intelligence and knowledge. While God monitored people's behavior, moral action was not an individualized issue in the face of a world order that was recreated through periodic total destructions:

> The Eternal Father lived in the sky, in heaven, and only came in person here on earth for a time, a moment. He had to come to be aware of the good and the bad that people did and for the blessing of the crops. In ancient times there were no ideas like [those that] now exist in [the minds of the] people, but rather they needed the direction of someone superior in idea and intelligence.

Not only were humans wanting in intelligence, but these "humble and obedient" creations are said to have become increasingly disobedient over time. The growth of disobedience is pictured as a natural tendency, and Trixanos offer no further explanation of it. The Eternal Father allowed humans extremely long lives, but as evil increased God periodically judged and condemned the world to destruction by a devastating flood:

> When there was a holy judgment, all the people in the world died. All that was in the sea and the birds in the mountains, all died. The seed of different animals and of a few families was kept. I do not know who the person was who kept all the seed; some say it was Noah, but I cannot be sure. Then those who did not help to collect the seed, all died, and the man who kept the seed made a kind of box in which to keep all the animals. On top of all the animals went this person with his family. When the seas covered all the world, these boxes floated. And when the seas receded, all the animals went in different directions.

God's cataclysmic judgment of the world contains another agricultural parallel. The world was God's field in which he harvested humanity and used a few people as seed for the next planting, which gave rise to new generations.

Creation narratives repeatedly contrast the ambiguously structured universe of the Eternal Father with the subsequent neatly categorized and delineated world order of Jesucristo. According to Trixano belief, the Eternal Father left the world physically unfinished in the sense that the seas, land, and sky were not clearly bounded or differentiated. Finally, he is said to have become "bored" with his labor in overseeing creation and returned permanently to the heavens, an act that spatially separated them from the nature of life on earth. His Son was sent at that point to continue the process of world creation. He began by further defining the natural environment:

> From the time that Jesucristo came, he ordered the world. He separated the waters in some places and put the earth in other places. In ancient times in the epoch of the Eternal Father all was united. There was very little land. One could not walk much because one went along falling into the pools of water.

What becomes clear in the analysis of creation narratives is how closely the transformations of the physical universe parallel changes in its inhabitants. During the epoch of the Eternal Father, the human beings, who lived on an earth indistinct from the waters, were not clearly separated from animals, in the sense that all living things communicated through "spoken word" and lived in peace. In his role as second creator Jesucristo is associated with the division of human beings from animals:

> The animals in the epoch of the Eternal Father were not the enemies of the people, because they spoke and if they needed something they asked for it. The change happened in the epoch of Jesucristo. The Eternal Father watched over the people of his creation and did not want the people whom he had created to suffer. When Jesucristo came, all the animals were deprived of speech because there was much evil in the people. Because of the danger that [wild] animals present, one remembers that there is a God when one comes face to face with an animal.

In separating for once and for all the interpenetrating levels and categories of the world order, Jesucristo ordained the uniqueness of humans in the realm of living things.

Humanity was at this point allowed to separate into different races:

> Before, there was no difference in race because only one kind of person existed. There were neither Jews nor Spaniards, but rather one class of persons created by the Eternal Father. God permitted the races to differentiate at the time of Jesucristo because it was not a bad thing that people were succeeding to know more of the world. The rich were beginning to know and populate different places; and when they came to primitive places, the mixed race arose, a mixture of foreigners and natives.

In creation narratives, foreigners who had knowledge of God and the world through exploration are contrasted with the native race associated with the wilds:

> There existed then native people who might not have known of the existence of God. Here one still sees the [archeological] remains of the first town of San Andrés which was called Chutinamit ["small town"] and was located to the north of the present town. The primary reason why this place was abandoned is not known, but I believe that it was found to be secluded and was the object of many dangers from the steep gullies and wilds.

Jesucristo appears to have been sent both to take the Eternal Father's place in world creation and to stand in for humanity in the cycle of regeneration through periodic destructions. Trixanos say that Jesucristo was born and crucified so that more human beings would die and more would be born. Rather than causing all to die so that more would be born, Jesucristo determined an individualistic cycle:

> If a man dies in the morning another is born in the afternoon; if a person is born now, later someone will die.
>
> We die so that there will not be much evil, since the bad are taken back more rapidly. If it were the contrary, evil would grow. We would kill each other every moment on earth so that we could feed ourselves and get what we need.

Along with the creation of individual life-spans, Jesucristo is associated with the creation of the moral order. This order, which calls on the individual to choose between alternatives, resulted from a treaty between Jesucristo and his chief apostle, who tried to take over Jesucristo's creative powers by imitating and taking advantage of him. When Jesucristo asked San Gabriel to punish the disobedient follower, the apostle was transformed into a being with human and animal characteristics, into Satanás. By refusing to obey the new order, Satanás came to mirror aspects of the old undifferentiated order in which humans were not totally distinct from animals. Thereafter, Jesucristo's domain was the heavens, mountaintops, and the town where his followers lived; and Satanás's domain, the wilds and the underworld.

Trixanos believe that this agreement meant that each individual must choose between paths: one very rugged, narrow, and full of spines (i.e., sufferings) that leads to God; the other wide and overflowing, with bright flowers (i.e., riches) that leads to Satanás. Those who follow the path to gain riches without suffering through hard work will be punished because this way leads to the wilds and the underground home of Satanás. There individuals are transformed into human-animals, into beasts of burden who must work eternally, without rest, stoking the volcano's fires. Individuals as moral beings must choose between the new order, with a spiritual existence after death, and the old order in which they become anomalous beings after death. Both Satanás and those who make pacts with him assume the form of human-animals, rejecting the clear separation of living things associated with the ordering of Jesucristo.

Trixano creation narratives are the syncretistic product of preconquest Mayan belief, sixteenth-century Spanish missionization, nineteenth-century state consolidation, and twentieth-century Catholic revitalization through recent pro-orthodoxy movements. Despite these diverse sources of symbolism, there is a coherence in the cosmogony which reflects peasant Indians' resynthesis of beliefs into a system through which they might make sense of their changing world.

First, the cosmogony defines the fundamental importance of choice to moral action. Creation narratives do not provide exemplars for moral action or detailed codes for conduct, rather they present a symbolic choice of allegiances between Jesucristo and Satanás, the town and the wilds, with differential rewards and punishments for either path.

Second, creation demonstrates a fundamental process of refinement, clarification, and the differentiation of key categories of the world order. Just as humanity with its capacity for moral choice is marked off from other living beings, so behavior that is radically immoral is clothed in a variant of premoral symbolism: as part-human/part-animal. This symbolism is repeated elsewhere in Trixano folklore. Individuals who commit incest, Indian women who sleep with non-Indian men, ritual coparents who are sexually attracted to each other, and those individuals who make pacts with Satanás are all transformed into animals who, as they suffer in special hells, betray their human origins. In each case these individuals can be shown to have violated crucial distinctions of the world order. They are abominations to Mayan classifications of the social world (Douglas 1966).

Third, in a fundamental sense creation is left hanging in these stories of general world ordering. The races are allowed to differentiate and to possess the intelligence to explore the world; individuals become moral beings who determine the form of their afterlives by allegiances and actions in the world; and the Indian town Chutinamit is left to be discovered in the wilds. What happens when these aspects of creation catch up with each other is the subject of another set of narratives which portray the conquest of Chutinamit by the Spaniards.

The Conquest: An Extension of the Cycles of Creation

Conquest narratives take the separate themes of world creation and ultimately show the problematic nature of moral action in a society composed of different races. As was previously mentioned, Trixanos believe that their society was in some sense left behind in the ordering of the world during the epoch of Jesucristo. Secluded in the mountains, Indian "tribesmen" wore animal skins and feathers and worshiped "deities of stone and roots." They were unaware of the existence of other races in the world, although other tribes lived in the region, practicing agriculture and waging wars to extend their dominions. Within the kingdom people lived peacefully without major divisions of labor or disparities in wealth. It is said that the king, along with other officials, earned respect because he did not allow social differences and presented himself to his people as an equal.

What is notable about these descriptions, of course, is the similarity in symbolism used to characterize the epoch of the Eternal Father and to picture Indian society before the Spanish conquest. To this social order—incompletely separated from nature—came the Spaniards. In effect, the Spaniards through their destruction and recreation of Indian society brought Indians into the mainstrean of Jesucristo's creations. Indians were divided from nature and resettled at the present site of San Andrés; their clothing was changed to the town-specific cotton and wool peasant dress of today. Indians are said to have been "civilized" by being exposed to knowledge through a newly established school and a new religion which centered on the worship of God and the saints and called for humans to choose between good and evil.

Interestingly, narrations of the Spanish transformation of Indian belief consistently portray the conquerors as attempting to fool the

Indians and benefit commercially from the religious practices they introduced:

> Long ago there were no images of the saints, only false gods in the form of stones with carved figures or faces. Then the Spaniards came and they tried to change these traditions, bringing religion and images of the saints to take the place of the stones and roots. They explained that only one God was creator of the world, and they founded religious brotherhoods with their grand devotions. For each image that came, they had a great ceremony. To promote the consumption of their commercial products, the Spaniards made people believe that they had to celebrate these fiestas. People did this by taking incense and candles to where the image had "miraculously" appeared at the foot of a mountain, and they set off firecrackers and had great feasts. Afterwards they moved the image to the church and then created the religious brotherhoods. It was a great ceremony with marimba, fireworks, flowers—everyone brought something.

Spanish trickery in introducing images of the saints is a common theme, as is the notion that, although earlier gods may not have been totally efficacious, the tribal people initially resisted the loss of their traditions. Nevertheless, the colonists were successful in putting "the name of the image on the town."

Individuals who were "baptized" into the new colonial social order as Indian laborers (*mozos*) for the new Spanish masters (*patrones*) experienced a psychologically complex transition:

> Before, there was one kind of primitive people. When the Spaniards came, they brought mirrors, earrings, and razors. They cut the hair of one of the primitive men and shaved him. When he felt clean, he felt a great change in his personality, because the people of the tribe had been accustomed to long hair and beards and it was difficult to tell the men from the women. They also bathed him. He felt like a Spaniard. However, there was a problem, because the rest of the tribe did not like the way he looked and he was forced to work alone for the Spaniards.

Significantly, the Indian initiate "felt like a Spaniard" but did not become one, though he clearly underwent a dramatic shift in identification, work, and cultural practices. The core image of the new order was a massive stone and adobe church which Indians were forced to build in the center of the new settlement by working as "beasts of burden." The relationship of the Spaniards as masters to

Indians as subordinate laborers continued as the colonists organized plantations and exploited Indians as a source of labor.

The Spaniards are remembered in commentaries as innovators who brought religion and education. Yet Trixanos stress that in the process of civilizing the tribes, the Spaniards made people suffer and sacrifice themselves for the colonial social and religious orders. Furthermore, they stripped the town of wealth and took over Indian agricultural lands, impoverishing the population. Later, it is said, the Spaniards returned to their native land, leaving their descendants, called Ladinos, to continue the bi-ethnic social order.

The conquest narratives show how Indians were finally incorporated into the earlier patterns of world creation and how the present bi-ethnic town of San Andrés came into existence. When brought together for a broader thematic analysis, the narratives reveal just how ironic this process turned out to be. In effect the Spaniards are portrayed as introducing the religious concept of the individual as a moral actor and then making it impossible for Indians to actualize this concept of the individual. After the conquest, individuals were not allowed moral autonomy (that is, to choose between paths and allegiances) because Spanish racism and domination submerged individual identities into their ethnic origins. By virtue of impoverishment and racial subordination, Indians would now automatically merit heaven where they would "receive their rest because one has already come from suffering here on earth."

For Spaniards, who are said to have enjoyed an easy life directing the labors of Indians, ethnic identity also superseded the issue of moral choices of the individual. Thus, according to Trixano elders, when the Spaniards died, "they had to go to a special place inside the wilds where there is a city for them. But they have to work not as they worked here on earth [managing the labors of others], but rather the work there is physically more arduous." By subordinating Indians, Spaniards gained riches without suffering through hard work and thus followed the path leading to a special hell. In fact Spaniards and their descendants are associated in Trixano folklore and ritual with Satanás and the Lord of the Wilds, all transformations of the divinity who challenged and opposed Jesucristo's moral order.

Creation's denouement at the conquest results in a world order with important unresolved questions. To use Yearly's distinctions (cf. pp. 384–85 below), Trixano cosmogony has important explanatory and comforting functions for the victims of 450 years of poverty and racism in a bi-ethnic society where they are the structural subordinates. The logical tension between the moral orders of Jesucristo and of the Spaniards helps to highlight the injustice of the conquest. The myths, however, largely lack a guiding function. As Geertz would phrase it, the oral narratives provide a sequence of "models of" social reality while remaining laconic about "models for" action within that culturally constructed reality (1973). Suffering is now forced on Indian society; it is not a consequence of moral decision-making. In the wake of the paradoxes of the conquest, the Trixanos' concept of the individual is stripped of its ethical dimension. Cosmogony appears to imprison Trixanos in a deterministic world where Indians are cast as passive victims in an unchanging social order.

Once again, however, other narratives complicate the picture; for, in addition to believing that the Spaniards created the town of San Andrés, the Trixanos hold that the town was given by Jesucristo to its patron saint of the same name. The saints were intermediaries sent by God to finish the task of creation after Jesucristo's crucifixion and return to the heavens:

> Holy Week is a remembrance of the passion of Jesucristo, a remembrance of a time of suffering. He died to win the world; His death is the most important. The Apostles were the witnesses to the acts of his suffering; they had to be the witnesses of all that, giving testimony. Jesucristo told each one of them that they would have a piece of earth, and for that reason there are distinct names for the towns, following the names of the Apostles.

Each saint protects and ensures the fertility of his town—his fields—and is celebrated by Indian religious brotherhoods. The saints are not autonomous but rather embedded in a hierarchy of divinities:

> The saints command also. They make their miracles and blessings. But God has given them these powers, this permission. For example, if you have a chief and he asks you to complete the orders. In the same way God commands what the saints do. Thus God has given San Bernadino powers. He has the work of illuminating; he has more important work like a first secretary or a first official who has a higher grade or better work.

The apostles are also thought to be particularly powerful because they accompanied Jesucristo. Most significantly, in this context, the saints' association with the town is understood to define it as an Indian community, not a bi-ethnic community. What this alternative ending to the creation process suggests is the possibility of a moral community independent of the conquest. In Trixano culture, this possibility is pursued in the ceremonies of the Indian religious brotherhoods where distinctive values and concepts of moral action are celebrated.

There are many additional questions to be asked as we move beyond multiple cosmogonies to other symbolic systems. In ritual, do Trixanos elaborate what is begun in mythology: the articulation of multiple models of the ethical order? Are codes for conduct and a fuller conceptualization of the individual issues elaborated elsewhere in the Trixano worldview? To what extent does the image of the individual, whose actions cannot really make a difference, overflow from the cosmogony of conquest to everyday life? To what extent does this cultural system provide room for what Cua (1978) and Burridge (1979) would call individual moral creativity within an overall deterministic and fatalistic world order?

Trixano Ritual: Another Order Echoed in a Stronger Voice

For Trixanos, creation mythology is the backdrop for a cultural stage on which ritual is the preeminent genre for cultural expression and organization. As has been argued in this analysis, mythology outlines the devastating moral paradox of the conquest and suggests a circumvention. Brotherhood rituals and their exegesis provide us with another voice in the expression of the Trixano world view and ethos, a voice that situationally puts aside the conquest to pursue a distinctive sense of moral community, origin, and Indianness:[3]

> May God be with your souls under San Bernadino, the sun. May you have received the gift of life, of good health, in the hands and at the feet of Christ, Señor San Salvador. Or perhaps not, perhaps we are not aware of our souls, the souls of our fellow men, our family, our child, our grandchild, of one of our elders bent over with white hair, a father, a mother who still lives—if they are well, their souls full of life.
>
> And this is what we should believe: He who invented the past days, the past hours, Señor Jesucristo and the angel San Bernadino, who saw above, below the sky, below the world, and on the world, who became a circle—he saw and found out what happened, where there was a punishment, where there was a pain, where one of our fellow men was being counseled and punished, where he is paying his punishment. We don't become aware of this because there is no way to compare our vision with the flame of pine pitch, with the lamp of the angel San Bernadino.

> Perhaps from there, from this angel, came the generation of our first fathers, our first mothers who lived the first lights of day.

Of their traditional religious practices, Trixanos say: "If we didn't do these activities, it would be as if there were no law or [system of] order in force." They are part of *costumbre*, of the traditions invented by the first Indian ancestors to guide their lives and worship God and the saints. *Costumbre* recognizes a tempered reciprocity as an important dimension of moral action tying divinities to humans:

> If one promises something to a saint or to God and doesn't complete the promise, then he is punished with sickness, or a bad harvest. For example, among us we may promise something that later we don't do, and the other person will be bothered. Perhaps we have our reasons, but the other person does not understand this. Even more so, this is true for the saints.
>
> God punishes each town when he sees an evil and for this reason the rain doesn't come. Then the people beg with a mass or a procession. As when a father punishes to a child; but afterwards he repents giving this punishment, and this is when the rains come.

Tradition also designates the ancestors as exemplars of correct behavior, which Trixanos believe was rewarded by the fertilities of a long life—many children and successful plantings and harvests. *Costumbre* unites that which is separated: bilateral kin with their ancestors as well as individuals with their fellow Indians in the town. San Andrés, the patron saint of the settlement, directs his children "to unite, love each other, and express mutual understanding" by joining together to serve the community. The commandment is fulfilled through cooperation and common effort among members of the town's Indian civil-religious hierarchy (including the religious brotherhoods, church caretakers, and the municipal government).

Ritual discourses present contrasts between valued and devalued postures toward key cultural images. For example, one major discourse describes the respect ancestors showed for their own elders, a respect which was rewarded in the past by a long life and the attainment of the status of "elder." As the discourse begins, the ancestor addresses an elder on the path of life:

> "Excuse me father, excuse me mother. Permit me to pass before you, to pass behind you," they said, and this was the indication of their words. And it is possible that their day grew

> and their birth grew because they did not push aside or expel their elder father or mother on this path, on this earth. They asked their pardon and their peace. For this reason their day and their birth grew. They were given more life by God; they became white-haired; they became bent over and walked with canes on the earth.
>
> They multiplied with offspring and inhabited these lands for much more time. And thus they did their service, their work for two or three times in the holy church; they did their service, their work for two or three times in the municipal government; they did their work, their service in the religious brotherhoods. Why? Because their behavior was good. They came together in a good way with good ideas, and from there we came to exist.
>
> Perhaps for us, the elder father or mother is not important when we meet them at the crossroads of a path, of life. Perhaps we encounter an elder father who comes with his cane and we pass by, pushing him aside because of our careers, our pressures, our strength. Perhaps we mimic him and imitate his way of walking. This elder feels sad because of our criticism. And so he might say, "You are my child, my neighbor, my grandchild, and because of your strength, this is what you do to me. You mimic me, push me, and imitate my way of walking." When we see the angel San Bernadino leave, he will kneel and ask our punishment. For this reason our day arrives or does not arrive, the hour of our return. Why, if not because of our bad behavior of imitating, mistreating, criticizing our elder neighbor, because of our careers, our strength, our drinking. And this is not what our Señor Jesucristo has commanded, but rather that we should unite and love each other. And this was how our first ancestors behaved.

Elders say that by the very enactment of *costumbre*, it is evident that they are adhering to the model of the ancestors. In following their heed, one shows respect by acknowledging the presence of elders and by acting with reference to them in daily life. Such behavior is rewarded so that the respectful become respected, supported by canes as they become weakened with age. In contrast, disrespect is shown by casting elders off the "path" (that is, by denying their importance as reference points) and by refusing to empathize with their sufferings or deaths.

In addition to dealing with respect, discourses point out the interdependence of the generations. Individuals are asked to identify with the life cycle of interdependent generations through the image of an old man, bent over with age and supported by a cane, walking the path of life. Trixano interpretations of this image note that the elders are guides because of their knowledge and ideas acquired through a life of suffering for their families and of sacrificing to leave their children land. In their waning years the retired elders are supported by canes, their descendants, who work under their direction. The youth are referred to as the elders' "canes" and their "flowers, buds, and fruit." They support the older generation and are the product of the fertility of the elders who are their fathers and mothers. To deny the elders is to reject the connected continuity of Indian descent, land, and *costumbre* as inherited from past generations.

For Trixanos, moral standards are taught through the discourses at shrine rituals and enacted by participation in the civil-religious hierarchy. Responsibility to the community involves an individual's identification with others, with their health, sufferings, worries, hardships, and deaths. Thus, at the close of another part of the discourses, the death of an individual is described as the loss of the entire community:

> When we hear the ringing of the holy bells on earth, the dominions of the Apostle San Andrés, we might say, "Ah Dios, who might this mother be, who might this father be who goes back, who returns before God? God, that you pardon him; God, that you take care of him." This is what we would say if we were good Christian souls. But perhaps we do not have good souls, because ours is another generation. Perhaps we are happy, we laugh at the death of our fellow man, the return of our father, of our mother. But this is not what God commands. He has told us to come together, to love each other like brothers. We lack the presence of our first fathers, our first mothers of the past day, of the past hours, who had better words, better expressions and behavior before a justice, before a guardian of the church, before a brotherhood.

I would argue that Trixano values as expressed in ritual discourses do not take the form of a detailed set of rules to guide behavior. Rather the discourses summarize key social connections between individuals and present striking images through which Trixanos should orient themselves to others. Through the image of the elder on the path of life, Trixanos define respect and the interdependence of generations; by recalling the halting, mournful ringing of the holy bells high on a hill above town, Trixanos represent the importance of identifying with others, a theme common throughout the discourses.

For the Trixanos, *costumbre* as celebrated in shrine rituals represents a notion of cultural origins independent of the conquest. *Costumbre* stresses continuity from the world creations of Jesucristo to the division of the earth's population into towns governed by the saints to the ancient times of the first Indian ancestors. Brotherhood ritual celebrates a distinctive set of world orders, ultimate values, and rewards and punishments for behavior. In this alternative, solely Indian order, fertility (a longer life, many children, and good harvests) is God's blessing to those who "adore God, behave correctly and love their fellow townspeople." Those who fail this social order may be condemned by their elders and taken back early by God. Ritual emphasizes a world order in which individual actions involve choices with ethical implications for the person and the Indian community.

Trixano mythic narratives describe multiple cosmogonies which are foundational to the emergence of ethics and society. The connectedness of cosmogony to ethics becomes more explicit over time as the creation of the physical universe grades into sociogony. Creation involves a process of clarification in an unstable world where the growth of evil inevitably calls for new cycles of recreation through destruction.

Multiple cosmogonies allow Trixanos to discuss the emergence of ethical choice, its wholesale compromise with culture contact, and yet its continued existence via the saints and ancestors. The two cosmogonic endings and their ritual elaboration have become a way for Trixanos to come to grips with the significance of being a conquered people, but also to assert the "radical imaginings' of a self-defined identity, social order, and ethical system. *Costumbre* as anchored in ritual discourses, builds on the separatist creation of the town by the saints, giving the mythic alternative an ethical content and broadening the Trixano range of ethically valorized motivations and activities (cf. Frank Reynolds, chapter 8, this volume).

One last aspect of the Trixano worldview which should be explored at this point is the cultural conception of the individual as it shapes commitment to the world orders portrayed in these symbolic systems.

The Individual and Moral Creativity

At the heart of Trixano conceptions of the individual is the notion of *voluntad* ("will"). Stubbornly within the control of the individual, *voluntad* is the locus of each person's unique perspective on the world and is generally knowable by others primarily by its final product, action. *Voluntad* is exercised by an individual's ranking of valued alternatives in everyday behavior. *Voluntad,* however, does not involve the kinds of moral choices spelled out in creation mythology, of allegiances to God or Satanás. Rather, this aspect of the person reflects the Trixano belief that each person "thinks in a different way" about the set of alternatives embodied in *costumbre*.

The goal of the civil-religious hierarchy is not to challenge different *voluntades* or marshall consensus but rather to create a setting in which individuals with different *voluntades* can "unify and understand each other." Trixano religion does not demand strict compliance to a set of highly detailed obligations. Instead it presents a broad range of alternatives through which commitments to *costumbre* can be expressed. Motives for actions are thought to be diverse and difficult if not impossible to influence. For example, all those serving in a brotherhood would be said to have the *voluntad* to do so, though their reasons for serving would be assumed to be various, ranging from the fulfillment of personal promises and special devotion to a particular saint, on the one hand, to the economic advantages of access to communal lands offered to the landless, or the avoidance of more demanding duties in the civil government, on the other hand. *Voluntad* emphasizes the contrast between action and nonaction, leaving aside the specification of precise motives for action.

Just as Trixanos assume a diversity of motivations for behavior, there is also an accepted pluralism of belief within *costumbre*. There are great differences in levels of ritual knowledge, differing opinions over which divinities are most efficacious, and variations in individual Mayan Catholic syncretism. In celebrating a diversity that is unified, the brotherhoods incorporate variations in belief within the broad rubric of shrine rituals. As one head of a brotherhood observed: "There are many who invoke only the World-Earth, others only God. For this reason, in the discourses, both are mentioned. The discourse is not to divide people; it says something for each person."

At this point, then, we are able to address the issue of individual moral creativity within an overall deterministic world order. *Voluntad* and the assumed pluralism of belief call on individuals to develop their own personal motivations for religious participation and commitments to divinities. Like Lévi-Strauss's *bricoleur* (1966), Trixanos are free to arrange the components of a cultural set in a way that reflects their own perceptions and problems. The wider system makes room for them to do this without directly soliciting or questioning the results. Rather, the ritual guides act to show that *costumbre* and the blessings of the saints overarch the full set of recognized alternatives. The guides themselves are considered exemplars of the positive direction of individual wills toward the unifying principles of *costumbre* and the perpetuation of the Indian community.

There are, however, culturally defined limits to *voluntad* demarcated by the contrasting concept of *suerte* (luck-destiny). Trixanos believe that each person is born with an unchangeable *suerte* which may cause a person to pursue goals outside the moral universe of *voluntad* and *costumbre*. Such goals include (a) becoming a diviner-sorcerer with the power to bring misfortune to people who may not deserve it, (b) making pacts with Satanás for wealth in this life at the expense of suffering in an underground hell after death, and (c) pursuing nontraditional occupations involving ladinoized schooling or the exploitation of fellow Indians as laborers. None of these are felt to be issues of choice for Trixanos, and all are felt in some fundamental way to invert the moral understandings of *costumbre*. With the concept of *suerte* Trixanos define the limits of moral agency and suggest that the violation of those limits is outside both individual and communal control. As I have argued elsewhere (1978), Trixanos have used their concept of *suerte* to define social change—particularly individualism, materialism, and ladinoization—and its moral implications. It is particularly interesting to note the available idioms that Trixanos have *not* selected for understanding change: the variation of *voluntades* or the free choice of allegiance to God or Satanás.

Suerte can be seen as dealing with unresolved issues generated in both the separatist and the subordinate images of Trixano community. On the one hand, separatist discourses stress unity and the importance of the individual's identification with the community. Through the concept of *suerte*, competition between individuals, envy, and curses meant to undercut another's fertility are dealt with outside the religious brotherhoods in the realm of the diviner-sorcerers. On the other hand, the subordinate image of identity is built on an unambiguous ethnic division of exploiter and victim. Indians, who through training, occupation, or sudden wealth, blur the ethnic distinction are also dealt with outside *costumbre*. This latter case parallels earlier cosmogonic myth in that those who have the *suerte* to become like ladinos in this life are thought to suffer a special retribution in the next. They are condemned to an underground hell, as were the conquerors and their descendants, where the Ladino Lord of the Wilds, another guise of Satanás, oversees their labors.

In some ways my findings echo Cua's framework for moral creativity (1978). Rather than presenting rules to govern behavior, Trixano myths and brotherhood discourses elaborate diffuse orientations, identifications, and reciprocities associated with *costumbre*. Individuals are called upon in this worldview to be creative actors in order to bridge the gaps between the key scenarios of the discourses, the flexibility of *costumbre*, and their own motivations for behavior. Trixano notions of the individual—particularly their concepts of *voluntad* and *suerte*—channel this creativity within cultural bounds and deal with issues that are problematic for Indian culture: control over the individual and social change.

Interpretation and Comparison

In this concluding section I want to identify several generic problems reflected in the diverse ways that disciplines such as anthropology, history of religion, and philosophy approach the analysis of cosmogony and ethics: differences in theories of the production and organization of meaning, contrasting conventions for the bounding of our inquiries, and, finally, distinctive disciplinary cosmogonies.

In interdisciplinary forums for the study of cosmogony and ethics, anthropologists are pulled in two directions. One is in the familiar direction of analyzing cultural systems in their own terms. Concretely, this means locating the centers of gravity of belief in a particular tradition; examining the interconnections of cultural thematics and preoccupations; and tracing the interrelations among such genres as myth and ritual. One reservation with this approach is the fear that highly disciplined relativism will produce results that are ethnographically rich but difficult to use for broadly comparative purposes. Behind this concern lies the very live issue in anthropology and in history of religion of the theoretical diversities hidden by the label "describing systems in their own terms." Our various modes of description—with unstated but implicitly contrasting theories of meaning—may be significant creators of the intriguing differences and similarities we find among cultural systems.[4]

A second, less common direction for anthropologists leads to experiments with questions and perspectives developed by scholars in philosophy and comparative religions. In this case anthropologists encounter new organizing questions, generated far outside the discipline's tradition. There is a great deal worth critically examining here: Green's notion of the internal logic of belief as generating new problems to be solved by religious reason (1978); Little and Twiss's framework for examining action guides in legal, religious, and ritual domains (1978); Cua's (1978) and Burridge's (1979) models for the study of moral creativity; and the central questions phrased by Frank Reynolds and others concerned with multiple cosmogonies in this volume.

There are, however, important interpretive tensions which anthropologists encounter in undertaking this kind of interdisciplinary analysis. In moving past Lévi-Straussian structuralism to the more process-oriented analysis of Geertz (1973; 1983) and Bourdieu (1977), anthropology has edged away from narrowly formalized examinations of belief as an isolated cultural domain of logic and taxonomy. Increasingly, the field is concerned with meaning in action, with the forging and manipulation of belief in everyday life, with the ongoing experience of moral reasoning for individuals and groups in other cultures. Comparativists from philosophy and ethics with Kantian perspectives argue that we need not get mired down in the specifics of context, that cosmogonies and ethics can be usefully studied independently of their communities of practitioners. The problem for anthropologists is that it is impossible to bracket or to factor out context because the historical, political, and individual details of particular communities at particular moments breathe life into cultural categories and lend new substance to old cultural constructions. Anthropologists are centrally concerned with understanding how ideas give form to action and social environments, and how social environments and action shape ideas. Comparative ethical inquiry concentrates on the ways in which cultural understandings of the universe and society make action meaningful and valued. Anthropology adds to this a concern with systems of belief and behavior *in action.*

By way of demonstrating how context and social process transform meaning, I want to draw attention now to the nature of the contexts that need to be examined for a full presentation of Trixano belief. First, of course, is the historical reality of the three-hundred-year Spanish conquest and colonization of Guatemala which both created the problem of a bi-ethnic society and introduced a religious language designed to perpetuate the colonial social order. In syncretizing their own version of Mayan and colonial Spanish beliefs, Indian populations managed to transform colonial religious language and to present their own understanding of culture contact. In the Trixano case cultural reworking has occurred in several ways: (a) the appropriation and recombination of religious symbols infused, in the process, with new meanings (e.g., Satanás, the Lord of the Wilds, and Judas all appear as non-Indian Ladinos in myth, folklore, and ritual); (b) the use of multiple cosmogonies, models of the social order, and definitions of community to develop a contrast between the acceptance of an imposed system and the development of distinctive understandings of the world as one's culture is being radically reshaped; and (c) the reorientation of religious brotherhoods, initially founded by Spanish missionaries, to serve as the mainsprings of a separatist Indian culture with its own ritual language and valued images for behavior.

Reintroducing the issues of social process and context opens up important new areas of inquiry, including most notably the politics of symbolism. Sturm and Lincoln have suggested that we need to examine the social and political processes through which symbolic systems are imposed or eliminated as relevant options by the broader political system.[5] The Trixano analysis adds another political dimension: how people use, test, and perhaps subvert imposed orders in their everyday lives. Subversion is not a single act but rather a patterned social and historical process of responses to power, and so its successes and failures must also be seen in the wider context of possibilities and constraints.

The limits of subversion are clear when we examine brotherhood separatism in community-wide celebrations. Ritual discourses and prayers are embedded in rounds of other religious activities: processions, liminal dances, and divinatory rites. As I have discussed elsewhere (Warren 1978), separatist discourses often give way outside brotherhood shrines to the symbolism of the conquest and the realities of rigid ethnic hierarchies in the bi-ethnic town. In processions, the transition from one symbolism to the other is marked by a shift in the carriers of the saints from members of the brotherhoods to Indians under the direction of Ladino authorities. Each shift brings to the foreground one of the available set of meanings for the saints and the origins of *costumbre*. During the course of every major celebration there are oscillations between separatist and subordinate images of creation and the moral order.

Another contemporary aspect of context which needs to be emphasized is the fact that Indian world view has been neither static nor pervasive in San Andrés. In fact the town experienced a decline and, by the mid-1970s, a near collapse of traditionalist religious organizations.[6] Actively seeking the end of brotherhood religion were new pro-orthodoxy Catholic and evangelical groups. The resulting clash of different religious groups threw into relief certain aspects of the beliefs of each. For example, these pressures forced traditionalist brotherhood officials to distinguish between *costumbre* as convention and as a set of ideals, resulting in a new public discussion and evaluation of motivations and commitments to belief. Religious contact and the consequent threat to brotherhood religion led to a serious questioning of the scope of individual moral and religious creativity in a way that had not happened previously because of Trixano notions of individual *voluntad* and *costumbre*. For their part, the new religious groups increasingly defined their beliefs with reference to how they contrasted with traditionalist religion. They have even pursued an analysis of the political implications of traditionalist belief, criticizing the limitations of Indian separatism in effectively challenging Ladino domination (cf. Warren 1978). The point here is that religious and ethical systems are more fluid and phenomenologically and historically shaped than the noncontextual analysis of cultural logics would suggest.

If interdisciplinary studies are to pursue more explicitly comparative work, what frames of analysis can anthropology contribute? One possibility is to treat systems outside cosmogony, world order, and ethics as the independent variables in our comparisons. This might lead us to compare those systems arising from state-based societies with each other, peasant societies with each other, and tribal societies with each other. Implicit in this case would be the notion that social systems (systems of power, hierarchy, social organization and production) vary significantly, and comparisons within broad categories of social systems might turn out to be particularly appropriate and fruitful. The major problems that need to be avoided with this comparative strategy include evolutionary modes of thinking, which underlie the typology, and the mechanical assumption that symbolic systems are superstructures directly derived from nonsymbolic aspects of society.[7]

Alternatively one might turn to a historical time frame and compare systems in contact to ask if the resulting political and ideological interplay leads to different, but in some sense jointly shaped, unresolved issues or focal points for the traditions involved.[8] In contact (and specifically colonial) situations where there are power differentials, one might pursue the cosmogonic and ethical dimensions of the ways in which "otherness" is conceptualized by those who control the state, particularly when it seems to justify a not fully human status for some categories of people who are being economically and politically integrated into a transformed society.[9]

The choice of comparative frameworks is made difficult in part by the rootedness of modes of analysis in what I would call "disciplinary cosmogonies." For instance, anthropology justly prides itself for having gone beyond its nineteenth-century evolutionary origins, when it ranked societies from barbaric to civilized with moral perfection located at the individualized, Western apex of human progress (cf. Tylor 1958; Hatch 1983). Clearly, the field initially made ethnocentric and overdetermined connections between its theory of cultural origins and ethical development (and at least some of these connections persist today in American worldviews). Anthropologists must continue to look for evolutionary echoes in the language of their comparative frameworks, particularly in those which incorporate materialist or psychological approaches that share these nineteenth-century origins.[10] The critical awareness of historical and cosmogonic roots, and how they influence the questions we pursue and the interpretations we find most satisfying, is another generic problem for the disciplines. So, too, is the need to come to an understanding of the ethical assumptions implicit in our phrasings of central questions for comparative research.

This analysis and the work of others such as Frank Reynolds and Douglas Sturm in this volume suggest that the comparison of cultural systems with multiple models of cosmogony and ethical orders will continue to be a challenging line of inquiry, one that successfully criticizes neo-Kantian notions of the universal structuring of moral reasoning. The Trixano culture is a good example of what Reynolds and Lovin term in their introductory essay a "problem-posing/problem-solving" relation of alternative cosmologies. What is notable about the Trixano belief is the *interlocking* resolution in cosmogonic myth, in ritual, and in concepts of the person of the problems generated by a hostilely conceived world. In religious terms, multiple models allow the Trixano traditionalists both to express and to seek to resolve a fundamental moral paradox. In political terms, the models define the nature of racial oppression, while perhaps masking its scope in arguing that Indian cultural separatism is possible. That the view which fosters a measure of cultural autonomy and community may have also perpetuated Indian subordination is an ironic—though not an inevitable or necessarily long-enduring—consequence of the multiple models of cosmogony and ethics in this case. For an important moment in modern Trixano history this patterned fiction—that separatism is distinct from subordination—has been central to Indian concepts of cosmogony, ethics, and world order.

Notes

1. This analysis has benefited greatly from the interdisciplinary discussions at the "Cosmogony and Ethical Order" conferences at the University of Chicago. I am grateful to Frank Reynolds, John Reeder, John Carman, Wendy O'Flaherty, Paul Williams, Bruce Grelle, and other conference participants for helpful comments and questions. Fieldwork for this research was undertaken in the western highlands of Guatemala for thirteen months in 1970 and 1971 with a return visit in 1974. The texts of the myths and ethnographic materials presented in this essay are my own translations of data gathered during these periods of fieldwork. Some of these translations have

been drawn from Warren (1978) which provides, in addition, a fuller treatment of community politics, economics, and social change.

The community of San Andrés includes a nodal town of 1,000 inhabitants and outlying hamlets and plantations that bring the total municipal population to approximately 3,500 people. In the western highlands, traditional religion, social organization, and Indian identity focus on the *municipio* (the municipal unit or country) making this a logical unit for study of religion.

2. Redfield also noted the fragmentary nature of origin myths, but apparently failed to see significant connections with other genres: "The present-day Maya that I know do not, however, much concern themselves with the origins of things; they have origin myths, but these are unsystematized" (1953, 98–99).

For other studies of highland Indian worldviews, see especially Bricker (1981), Bunzel (1952), Guiteras-Holmes (1961), Mendelson (1958), Nash (1967/68), and Tedlock (1982).

3. The disjuncture between the image of a separatist *costumbre* and the realities of subordination is dealt with at length in Warren (1978).

4. At this point my sense is that the findings of singular or multiple cosmogonic or ethical models in a given tradition may be more an artifact of our analytical approaches than of the nature of the tradition in question. In this volume, Atkins gives us historical reasons for the focus on a singular model in the Greek case; Sullivan's perspective explores symbolic congruencies, rather than symbolic variations, in an Andean worldview. These are good examples of very different sources of singularity. One suspects that multiple models may be very common given the general polysemic nature of symbols, the liminality and communitas of rituals (Turner 1969), and the symbolic problems posed by social change (Geertz 1965). Nash's pathbreaking contribution (1979) analyzes the nonproblematic coexistence of radically different forms of discourse among Bolivian tin miners in the context of severe exploitation and rapid social change.

5. Sturm, Lincoln, and others raised this important issue in the conference discussions.

6. Bloch's analysis (1975) of the limitations of highly formalized oratorical styles for coping with change and responding to variations in political interests describes the Trixano ritual discourses very accurately.

7. Sahlins (1976) explores the conceptual difficulties in the symbolic/nonsymbolic distinction implicit in this approach, while Taussig (1980) makes the case for strong interconnections.

8. Revitalization approaches to colonial contact and change have been very productive. See Wallace (1970) for an example with interesting parallels to the Trixano material.

9. Bruce Lincoln drew our attention to the issue of otherness during the conference proceedings.

10. See papers by Sturm on Marx and Yearly on Freud in this volume. Clearly we need to know more about the relation of historical process to our disciplinary lenses. For a good example of "evolutionary echoes" in scholarship which does not see itself as evolutionary, see Kohlberg (1981).

References

Bloch, Maurice
1975 *Primitive Language and Oratory in Traditional Society.* New York: Academic Press.

Bourdieu, Pierre
1977 *Outline of a Theory of Practice.* Cambridge: Cambridge University Press.

Bricker, Victoria R.
1981 *The Indian Christ, the Indian King.* Austin: University of Texas Press.

Bunzel, Ruth
1952 *Chichicastenango, a Guatemalan Village.* Publication of the American Ethnological Society, No. 22. Locus Valley, New York: J. S. Augustin.

Burridge, Kenelm
1979 *Someone, No One.* Princeton: Princeton University Press.

Cua, A. S.
1978 *Dimensions of Moral Creativity.* University Park: Pennsylvania State University Press.

Douglas, Mary
1966 *Purity and Danger.* London: Routledge and Kegan Paul.

Geertz, Clifford
1965 *The Social History of an Indonesian Town.* Cambridge, Mass.: MIT Press.

1973 *Interpretation of Cultures*. New York: Basic Books.

1983 *Local Knowledge; Further Essays in Interpretive Anthropology*. New York: Basic Books.

Green, Ronald M.
1978 *Religious Reason*. New York: Oxford University Press.

Guiteras-Holmes, Calixta
1961 *Perils of the Soul: The World Views of a Tzotizil Indian*. New York: The Free Press.

Hatch, Elvin
1983 *Culture and Morality; The Relativity of Values*. New York: Columbia University Press.

Kohlberg, Lawrence
1981 *The Philosophy of Moral Development*. San Francisco: Harper and Row.

Lévi-Strauss, Claude
1966 *The Savage Mind*. London: Weidenfeld and Nicolson.

Little, David, and Sumner B. Twiss
1978 *Comparative Religious Ethics*. San Francisco: Harper and Row.

Mendelson, E. Michael
1958 "The King, the Traitor, and the Cross: An Interpretation of a Highland Maya Religious Conflict." *Diogenes* 21: 1–10.

Nash, June
1967/68 "The Passion Play in Maya Indian Communities". *Comparative Studies in Society and History 10:318–27.*

1979 *We Eat the Mines and the Mines Eat Us*. New York: Columbia University Press.

Redfield, Robert
1953 *The Primitive World and Its Transformations*. Ithaca: Cornell University Press.

Sahlins, Marshall
1976 *Culture and Practical Reason*. Chicago: University of Chicago Press.

Taussig, Michael
1980 *The Devil and Commodity Fetishism in South America*. Chapel Hill: University of North Carolina.

Tedlock, Barbara
1982 *Time and the Highland Maya*. Albuquerque: University of New Mexico Press.

Turner, Victor
1969 *The Ritual Process*. Chicago: Aldine.

Tylor, Edward Burnett
1958 *Primitive Culture*. New York: Harper and Row.

Wallace, Anthony F. C.
1970 *Death and Rebirth of the Seneca*. New York: Knopf.

Warren, Kay B.
1978 *The Symbolism of Subordination*. Austin: University of Texas Press.

Part IV

Cosmogony, Philosophy, and Ethics

11 Ethics and the Breakdown of the Cosmogony in Ancient Greece

Arthur W. H. Adkins

It has been suggested[1] that an autonomous ethic appears in a culture when the traditional cosmogony breaks down. In this paper I shall trace the development of such an ethic in ancient Greece. The first necessity is to explain "autonomous" and "breaks down." "Autonomous" here denotes an ethic in which sufficing reasons for moral action can be given without invoking divine sanctions or sanctions from any other aspect of the cosmogony. That is to say, unjust behavior between mortals neither provokes Zeus to send a bad harvest nor causes a bad harvest to occur simply because the cosmos is organized in such a way that it occurs, without any divine action, in response to injustice. (For a belief similar to the latter, see Homer, *Odyssey* 19. 107–14.) I emphasize "sufficing reasons." Such an autonomous ethic may adduce additional reasons for moral action from the traditional cosmogony, or from a new cosmogony, for those who are prepared to accept them; but it must claim that moral action is adequately motivated without them. An ethic autonomous in this sense evidently need not be autonomous in other senses. Indeed, this essay will argue that all ancient Greek ethics are heteronomous in the Kantian sense, and necessarily so. "Breaks down" does not mean that all the members of the society find it impossible to accept the cosmogony, but that a significant proportion, reckoned in terms of numbers, prominence, or both, no longer accept it.

I assume that readers of this essay have already read my "Cosmogony and Order in Ancient Greece," which appears earlier in this volume. Here I shall briefly trace the history and development of the Hesiodic world of values to the point, in the second half of the fifth century, where a breakdown in the cosmogony occurred. I shall discuss the reasons for the breakdown and its consequences. The remainder of the essay will give examples of the first solutions attempted in the later fifth and early fourth centuries, draw some conclusions about the relationship of possible and actual systems of autonomous ethics in Greece to the values enshrined in the cosmogony, and hazard a more general speculation.

Hesiod's *Theogony* created a universe consisting of and inhabited by divine powers, and portrayed their relationships and the manner in which they evaluated them: *arete, philotes,* and *moirai* of *time* were of paramount importance, while justice was less valued. Hesiod's *Works and Days* is a handbook of advice to mortals who inhabit this universe. Their goal is to succeed as well as possible, to maximize their *arete* and their *moirai* of *time.* Justice is one of a number of routes to this goal, none of which, whether taken singly or together, absolutely guarantees success. If the gods will, the feckless farmer and the unjust may prosper more than the prudent and just, sometimes even at the expense of the prudent and just.

For Hesiod and his contemporaries, and for most who came after them for many generations, such occurrences do not shake belief in the gods' existence. The action of deity is seen in every springtime and harvest, in every sunrise, thunderstorm, plague, earthquake, and flood. The gods are believed to reward and punish by sending such "natural" phenomena; but it is not believed that the gods are omniscient, omnipotent, or reliably benevolent and just. The Greek of this period does not say "If such events are not governed by justice, there can be no gods." Some later Greeks have a different attitude (below, pp. 290–91, 292).

There are, however, other problems. If there is to be stable society, its members must have some reliance on one another's justice; they must believe that their neighbors will for the most part cooperate with them, or at least refrain from attacking or plotting against them; and that if they do not, the powers that be, whether divine or human, will dispense even-handed justice. But if success is the goal of one's neighbors, and of oneself, and the powers that be are not always just, justice is not a reliable means to the desired end; and, as Hesiod himself says, *W*270–272, "Now may neither I nor my son be just among men, since it is a bad thing for a man to be just, if the more unjust is to come off better."

No one doubted that the gods existed; their works were everywhere to be seen, and all success and failure were ascribed to them. But if being just was not a reliable route to success, why be just? Given Hesiod's general attitude in the *Works and Days*, it should suffice him that justice is a *more* reliable route to success than injustice. In that case, the existence of some who are unjust but prosperous does not cancel the choiceworthiness of justice.

If the *Works and Days* is understood in this way, Hesiod is making moderate demands of the cosmogony, demands which it may be able to bear, provided the proportion of the successful unjust does not rise too high. If it does, or if others demand that injustice should never prosper, the cosmogony is tested to the breaking point. In the dominant belief in ancient Greece, all good and ill fortune—interpreted as reward and punishment, or rather as proof of the favor or hostility of the gods (Adkins 1972a, 16–17; 1972b, 78–87)—must come in this life: the wronged or unsuccessful cannot reflect that the unjust or overproud will suffer after death for his offences.[2] The belief that injustice does not bring success in this life is always capable of empirical falsification; and it would be a fortunate society in which it was not falsified quite frequently.

Falsified, that is to say, in the person of the wrongdoer; but where the kinship group is treated as being more important than the individual, a solution to the problem is possible: punishment reliably comes in this life, but it may come upon the children, grandchildren, or even more remote descendants of the wrongdoer. (The belief is invoked by Herodotus [1.91] to explain the fall of King Kroisos in the fifth generation after his ancestor Gyges usurped the throne of Lydia.) In the early sixth century the Athenian Solon wrote (Solon 13, ll. 25–32, in West, II, 128):

> Such is the punishment [*tisis*][3] of Zeus. He is not swift to anger at each [person? deed?] as a mortal is; but it does not escape his notice for ever who has a guilty mind. . . . No; one man pays the penalty [*tinein*] at once, another later; and if they themselves escape . . . , their children or family, guiltless [*anaitioi*], thereafter pay for [*tinein*] their deeds.

Solon records this belief without complaining. The sentence would record a complaint only if Solon additionally believed it wrong for Zeus to punish those who were not guilty or causally responsible.[4] That the gods punish the innocent *together with* the guilty, or those not causally responsible with those who are, is a belief found already in Homer and Hesiod. It is indeed an empirical datum in their eyes: the gods send famine, plague, and similar occurrences, and these affect innocent and guilty alike.[5] The belief that children may be punished for the wrongdoing of their parents doubtless long antedates Solon, and may well have aroused no complaint. Property consisted primarily in the—inalienable—family holding, which remained while generations came and went; and divine reward and punishment were likely to consist in abundance or dearth of the products of that continuing holding. The punishment may have appeared appropriate and just.

Provided the kinship group is regarded as more important than the individual, justice may be linked indissolubly with prosperity, injustice with disaster. The belief covers all possibilities and cannot be empirically falsified: if one is just and prospers or unjust and meets with disaster, there is no problem; and now if one is unjust and prospers, one's descendants are certain to suffer (and few families will not encounter disaster of some kind in a generation or two); while if one is just and meets with disaster, one is merely paying for the misdeeds of one's ancestors (and few families could not find some appropriate misdeeds). The theory will enable the moral observer to explain the success and woes of others; and if a moral agent is sufficiently convinced that the kinship group is more important, and more significant in the eyes of deity, than he himself is, he may be constrained to behave justly.

The emphasis given to this theory in the sixth century may reflect increased demands on the cosmogony: some have demanded from heaven a guarantee that injustice *never* prospers. The theory furnishes an adequate explanation. It will depend on other factors whether it is an ethically acceptable one.

For the belief reflected in Solon 13 is found in Greece at the point where the individual is beginning to feel himself more important; and a writer of the Theognid corpus, probably also in the sixth century, responds to it in a very different manner (731–52):

> Father Zeus, would that it might become dear to the gods that *hubris* should be pleasing to the wicked. . . and that whoever. . . committed harsh deeds, without care for the gods, should later pay for the harm [*kaka*] he has done, and that the wicked deeds of the father should not in later days become a source of woe [*kakon*] to the children; and that the children of an unjust father whose thoughts and deeds are just, O Son of Kronos, in fear of your wrath, and who love justice from the

> beginning among their fellow-citizens, should not pay for the transgressions of their fathers in their stead [*antitinein*]. Would that this were dear to the blessed gods; but as it is the one man does wrong and escapes, and another suffers woe [*kakon*] in the future. And how is *this* just, king of the immortals, that a man who is clear of unjust deeds, and has committed no transgressions nor wicked false oath, but is just, should not experience what is just? What other mortal, looking at this one, would then fear the immortals. . . , when a man who is unjust and wicked, and takes no steps to avoid the wrath of man or of the immortals, commits *hubris* sated with wealth, while the just are overwhelmed, under the burden of grievous poverty?

Those who cause woes (*kaka*) to their fellow-men should pay for it (*tinein, antitinein*) by suffering *kaka* in return. Where their fellow mortals are unable or unwilling to ensure that the guilty pay the penalty, Zeus or the other gods should do so; and they should punish the wrongdoer, not his children. But they do not; and the *tisis* of Zeus is here explicitly censured as unjust.

The theory continues to explain all the phenomena; but the individual regards such "justice" of Zeus as flagrant injustice. Even now the possibility that one's own injustice might harm one's descendants could serve as a check, both from human affection and in the light of the belief that one's well-being after death might depend on the existence of descendants to make funeral offerings to one's shade. But the thought that one might behave justly all one's life, foregoing all chance of unjust success, and *also* suffer divinely sent disaster for the wrongdoing of one's ancestors might well rankle with anyone conscious of his own individuality; and it evidently rankles with this poet.

But note that the poet does not deny the existence of Zeus or the other gods. He is trying to recall them, and Zeus in particular, to a proper sense of their own responsibilities. The implication is clear: if the gods do not punish the unjust, mortals will not behave justly where they can behave unjustly without fear of human reprisals.

We need not conclude that they will cease to have any dealings with the gods: they may pile up unjust gains and give a share to the gods in sacrifice, a more reliable method of securing their good will. In Plato's *Republic* (362b7–c6), written in the fourth century, Glaukon says that some believe that the successful unjust man gets the better of his enemies

> and becomes rich and benefits his friends and harms his enemies, and makes sacrifices and offerings to the gods in a splendid

> manner, and serves the gods much better than does the just man. . . so that in all probability he becomes more dear to the gods than does the just man.

This is not the cynicism of an unbeliever, but a reasonable deduction from much of early Greek religious belief. To obtain a maximum of *eudaimonia* in a world inhabited by Greek deities, a coherent life-plan is needed; but justice need not play an important role in it.

The moralist must continue to proclaim that injustice does not pay; and that it does not pay the wrongdoer himself. Nothing else will make justice choiceworthy. In Aeschylus's *Eumenides* (produced in 458 B.C.) the Furies, in fear that the matricide Orestes may be acquitted by the jury and snatched from their punishment, sing (508–25):

> Nor let anyone call out when struck by disaster [i.e. when someone else harms him]. . . "O Justice, O thrones of the Furies." Perhaps a father or mother, new in pain, might lament thus, for the house of Justice is falling.
>
> There is a place where the *deinon* [what is terrible] is beneficial: it should remain enthroned as a watcher over the mind. It is profitable to *sophronein* [be prudent, self-controlled] under constraint. What man who fears nothing in the light of his heart[?] would continue to revere Justice in like manner as in the past?

When Orestes is acquitted, the Furies suppose themselves to have lost *time*, and since the jury of Athenian citizens gave the verdict, the Furies threaten Athens with disaster (778–92, etc). Athena, by a combination of threats (826–28) and promises (881–900), wins them over, and they are established in Athens as Eumenides, the Kindly Ones. The name is a euphemism. Athena says of them (927–37):

> These things I do with kindly intent for my citizens, having found an abode here for deities who are mighty and hard to please. For they obtained by lot [as their *time*] the management of all human affairs. He who finds them a heavy burden does not know from where come the blows that strike his life [or livelihood: *bios*]. For the misdeeds of former generations [?] bring him to these deities, and a silent doom levels him with the dust, even though he boasts loudly. . . .

The Eumenides are still to punish the wrongdoer. Only so will justice be valued, for only so will injustice not be a possible road to success, *olbos* and *eudaimonia*. Indeed, Athena seems to say that they punish later generations for the misdeeds of earlier ones. Some scholars have denied this, but there is no reason to suppose that the belief was dead, even if its usefulness to the moralist had diminished. If "boasts loudly" is the correct translation, Aeschylus is suggesting that the person punished is giving some offense to deity himself; but it is not unjust to boast. A little later in the *Republic,* Adeimantos speaks of "wizards and soothsayers" who claim to be able to "heal" any injustice committed by oneself or one's ancestors (364b–c). The concern about ancestral injustice may not be altruistic.

The vehemence of the Eumenides, and Aeschylus, is great. The argument is not overtly against disbelief, against those who maintain that the gods do not punish, for they do not exist; it is against one god, Apollo, who is endeavoring to save Orestes from the vengeance of other gods. In the first half of the fifth century, no Greek so far as is known had yet openly denied the existence of deities, nor yet that they sometimes punish some persons for some actions, even if Presocratic philosophy furnished the premises for such a conclusion (below, p. 287).

But human self-confidence was growing. The Persian Wars were a decisive period. In the generation before the *Eumenides,* a Persian force had attacked Attica, and had been defeated on the plains of Marathon (490). A decade later, a much larger land and naval force had been defeated in a naval battle at Salamis (480) and a land battle at Plataea (479); and in a sustained series of campaigns the Greeks, soon under the leadership of the Athenians, had swept the Persians from the shores and islands of the Aegean.

It is still apparent in the extant texts written after the event that the Greeks had not expected to win. The odds were not as heavily against them as they supposed; but they were seriously outnumbered. The immediate response of the Greeks, once the dust had cleared and they found themselves the victors, was to give credit to deity. The Persians, and the Persian King Xerxes in particular, had committed *hubris,* had aspired too high for mortal man, and had been brought low by jealous deities.[6] So the English inscribed on the Armada medal: "God blew, and they were scattered."

But both in Elizabethan England and in the fifth-century Aegean, the victors were to experience a rarely equalled surge of energy and confidence. Less than twenty years after the Eumenides we find in Sophocles' *Antigone* the famous ode (332–75):

> Many are the things that are *deinon,* and none is more *deinon* than man: he crosses the grey sea, driven by the stormy south wind, making his path beneath swelling waves that tower around

> him; and Earth, the most august of the gods, Earth the immortal, the unwearied, does he wear, as the ploughs go to and fro from year to year, cultivating with the offspring of horses.
>
> And he, cunning man, traps in woven nets and leads captive the race of light-hearted birds, the tribes of wild beasts and the creatures that are produced in the sea; and he overcomes by his devices the mountain-roaming wild beast that dwells in the wilds and tames the shaggy-necked horse, putting the yoke upon its neck, and the bull of the mountain that does not grow weary.
>
> And he has taught himself speech, and thought swift as the wind, and dispositions suited to living in cities and how to escape the darts of the frost that strike from a clear sky, and the darts of the rain. He is all-resourceful; nothing that is to come does he meet without resource; only from death will he contrive no means of flight: from baffling diseases he has devised means of escape.
>
> Having, in his contriving skill, something *sophon* [clever] beyond expectation, at one time he comes to *kakon* [woe] at another to *esthlon* [what is beneficial for him]. When he observes the *nomoi* [laws and customs] of the land and the justice which he has sworn by the gods to uphold, high in his city; but that man who consorts with what is not *kalon* [honorable] by reason of his daring—he has no city. May he who does these things never share my hearth, and may I never be of one mind with him.

In the *Eumenides* the Furies are the source of the *deinon* which alone ensures that men behave justly. In the *Antigone* ode, nothing is more *deinon* than man. This is not Sophocles' "message" in the *Antigone,* as indeed the last stanza of the ode itself indicates; but the ode evidently reflects one contemporary attitude.

Another challenge to the traditional cosmogony had begun to develop earlier. Hesiod's *Theogony* is the only surviving example of a set of early Greek cosmogonies; but though in a sense those cosmogonies competed with one another, they invoked similar types of explanation, depending on personal or personalized deities. Thales, however, who flourished about a hundred years before the Persian Wars, in the early sixth century, was the first of a long line of Greek thinkers, known collectively as "Presocratics," who produced a different type of account of the universe. "Presocratics" is a modern term. The thinkers (Thales, Anaximander, Anaximenes, Pythagoras, Parmenides, Heraclitus, Empedocles, and many others) differed in many ways from each other, and few generalizations would fit all of them. Their accounts may be termed cosmologies rather than cosmogonies; but not to distinguish them in any simple sense as secular rather than religious. Thales is reputed to have said "All things are full of gods" (Aristotle, *De Anima* 411a7), apparently because the universe contains motion, and motion is imparted by life. Similarly, Empedocles employs *Neikos* (Strife) and *Philia* (Love or Friendship) as what we should term ordering principles of his universe (B16 etc. D–K); and when similar abstract qualities occur in Hesiod, they are treated as deities and personalized (*T*224, 229).

Whether or not a particular Presocratic regarded such principles as divine is too large a question for this essay. It is also irrelevant. Any Presocratic who explicitly banished deity from the cosmos was explicitly challenging the cosmogony. Those who did not were challenging the cosmogony nonetheless; for their deities were not personalized or anthropomorphic. They were not concerned about their *moirai* of *timē*;[7] they did not expect sacrifice and prayer; and they did not reward justice or punish injustice. Human behavior was of no interest to them.

In the second half of the fifth century there appeared another type of thinker, then and now termed "sophists." At this time, the word "sophist" was not pejorative: it denoted anyone who was wise or clever in any field.[8] The sophists in question were professional—and expensive—itinerant teachers, who would teach anything for which there was a demand. If cosmology was asked for, cosmology would be taught, whether the sophist had anything original to say or not. The sophists must have served to disseminate cosmological ideas more widely; for the Presocratics were not primarily teachers. Some sophists may have made contributions to cosmology; but they are best known for their thoughts on ethics, politics, and rhetoric. Earlier Greece had been a traditional, custom-bound society. Some sophists used the distinction between *phusis* (nature) and *nomos* (custom, convention, law) to develop a view of the "real" man. Since *agathos* denotes and commends a good specimen of human being, it is not surprising that the "real" man proved to be the traditional *agathos* with his competitive excellence, using his courage, strength, skill, and resources to maintain and increase his well-being, and caring nothing for mere *nomos*.[9] The sophistic education was largely devoted to increasing the skills of the *agathos*, to enable him to succeed in politics and the courts (Adkins 1960a, 195–219). Since Athens in the second half of the fifth century was a direct democracy in which, in theory, any citizen might address the assembly and every citizen who engaged in litigation had to conduct his own case, in Athens and similar cities rhetoric played a prominent role in the New Education.

The cosmogony was under great strain. Demands were being made on it which it was not designed to bear; and thinkers were challenging it in a number of ways. I have roughly distinguished between Presocratics and sophists; and some scholars distinguish between sophists and rhetoricians. The distinctions have a limited explanatory use; but in the later fifth century the important opposition in the minds of contemporaries is between "the good old ways" and "dangerous new-fangled nonsense" or "exciting new truths" and "old-fashioned childishness," according to taste. All of the new thought is perceived as one movement by those who are opposed to it, and all the old ways are subject to attack by the new thinkers.

The Attic Old Comedy took as its themes matters of public interest. The sole surviving dramatist is Aristophanes, whose extant comedies are spread over the years 425 B.C. (*Acharnians*) to 388 B.C. (*Plutus*). In 423 B.C. Arisitophanes produced the *Clouds*, devoted to the New Education. The *Clouds* is a difficult work for the literary critic: what we have is a second draft, the first having been unsuccessful in performance (Dover, lxxxi-xcviii). Significantly, we are told that the play was a failure because Aristophanes appeared insufficiently disapproving of the New Education. Here, my concern is with the judgments passed in the play, and with juxtapositions of ideas.

In the *Clouds*, Strepsiades—the name means something like "Twister"—is in debt as a result of the extravagances of his wife and son. He wishes to cheat his creditors, and endeavors to persuade his son Pheidippides, a young man whose chief passion is horses, to go to Sokrates' "Thinking-Shop" (*Phrontisterion*) to learn the Unjust Argument. Pheidippides refuses, and Strepsiades goes himself. He is introduced to the Clouds, the goddesses of the *Phrontisterion*, but is too stupid to learn anything. Pheidippides is then persuaded to try. He learns all too successfully. Strepsiades scores a victory over his creditors with his son's help; but then Pheidippides beats him and adduces arguments in justification, to which Strepsiades has no reply. When Pheidippides proposes to beat his mother too, Strepsiades in rage sets fire to the *Phrontisterion*, and the comedy ends in confusion and rancor.

When Strepsiades first mentions Sokrates' school, he immediately juxtaposes cosmology and injustice (94–99):

> This is the *Phrontisterion* of *sophos* [wise, clever] minds. There dwell men who in their arguments try to persuade us that the heavens are an oven which is all around us, and that we are the coals. If one gives them money, these people teach how to win both just and unjust victories by one's oratory.

When Sokrates appears, he is sitting in a basket suspended in the air. His remarks would convict him of *hubris* in the eyes of a traditional believer (223–27):

> *So:* Why do you address me, you creature of a day (i.e. mere mortal)?
> *Str:* First, I beg you, tell me what you are doing.
> *So:* I am walking on air and contemplating [*periphronein*] the sun.
> *Str:* So from a basket you are looking down on [*huperphronein*] the gods, and not from earth?

Sokrates gives him a "Presocratic" answer, which draws on the kinship between mind and air.

Sokrates tells Strepsiades that the members of the *Phrontisterion* do not believe in the gods (247), and offers to tell him the facts about "divine matters" (250). The Clouds are true deities (252–53); and Sokrates now invokes them (264–66):

> O master and lord, Air the immeasurable, who holdest the earth in space, and bright Aether and Clouds, august goddesses who cause thunder and lightning, appear, lady goddesses, and manifest yourselves on high to your Thinker.

Later, Sokrates terms the Clouds (316–19)

> Clouds of heaven, mighty goddesses for lazy men, who furnish us with intellect, discourse, mind, verbal marvels, circumlocution, cheating, and quickness of apprehension.

Cosmology is associated both with cleverness and with dishonest use of language.

It has already been hinted that the Clouds, not Zeus, cause the weather. Sokrates becomes more explicit (365–71):

> *So:* These alone are deities. The rest are only nonsense.
> *Str:* But Zeus. . .the Olympian, is he not a god?
> *So:* Zeus indeed! Do stop babbling. Zeus does not even exist.
> *Str:* What's that? But who rains? Tell me that first of all.
> *So:* Why, the Clouds do. I'll prove it to you conclusively. Did you ever see it raining without clouds? But if what you claim were true, Zeus could rain from a clear sky. . .

A naturalistic explanation of thunder follows (375–81):

> *So:* The clouds thunder as they roll along.
> *Str:* How, you daring man?
> *So:* When they are filled with water and are compelled [*anankazein*] to move, hung up on high full of water, under the influence of necessity [*ananke*], then they bump heavily into each other, break and rumble.
> *Str:* But who compels them [*anankazein*] to move along? Is it not Zeus?
> *So:* Not at all, but [a] Vortex [*Dinos*] of the aether.
> *Str:* Vortex? I had not realized that Zeus was no more, and that Vortex ruled in his stead.

Ananke is the ordinary Greek word for "compulsion." It is also used by the Presocratics for a quasi-scientific "necessity." *Anankazein* is the ordinary Greek work for "compel," in nonphilosophical writers usually with a personal subject. Strepsiades, till now a believer in personal deities, asks *who* compels the clouds to move; and naturally interprets Vortex in personal terms. If he accepted the Hesiodic cosmogony, Strepsiades knew that already Kronos had ousted Ouranos, and Zeus, Kronos; and if he remembered the myth of Aeschylus's *Prometheus Bound,* Strepsiades knew that the overthrow of Zeus had almost occurred in the past. What more natural than that Strepsiades should suppose that Zeus had been ousted by another deity, particularly as *Dinos* (Vortex) sounds so like the oblique cases of Zeus (*Dia, Dios, Dii*)? Aristophanes makes his point most neatly at 828: Vortex (*Dinos*) rules, having driven out Zeus (*Dia*).

In the *Clouds,* of course, the Clouds appear on the stage in personal form, and Sokrates himself prays to them, and to the Air and Aether. Aristophanes needs the Clouds as his chorus, and to supply his denouement; and their presence on stage renders far more vivid the displacement of the deities of the cosmogony.

After more naturalistic explanation, Strepsiades asks (395–403):

> But tell me this: where does the thunderbolt come from. . . ? Clearly Zeus hurls *that* against perjurers.
> *So:* You fool, you remnant from the days of Kronos, you antediluvian nitwit! If Zeus strikes perjurers, why didn't he burn up Simon or Kleonymos or Theoros, who are certainly perjurers, rather than striking his own temple, and Cape Sounion and great oak trees. Why should he do that? An oak tree doesn't perjure itself.
> *Str:* I don't know.

And well might Strepsiades be puzzled, or anyone else who expected the gods to punish all wrongdoers and to punish them promptly and them only. If the gods do not punish in this way, the conclusion may now be drawn that the gods do not exist. Sokrates then offers Strepsiades a "Presocratic" explanation of thunderbolts. The Clouds promise Strepsiades that he will be *eudaimon* if he works hard and takes as his goal (as most *agathon*) "the goal that is reasonable for a clever man to take, victory in action and counsel and fighting with the tongue" (412–19). Sokrates calls on him to reject the other gods and believe only in "Chaos, the Clouds and the Tongue." Strepsiades does so, and the Clouds promise him success as a democratic politician; to which he replies that success in the courts is enough for him.

Socrates fails to teach the stupid Strepsiades. While Pheidippides is being taught, the Just Argument and the Unjust Argument contend in person with each other (889–1104). The Just Argument commends a moral—and traditional—way of life; the Unjust, accepting all the pejorative terms used of him by the Just Argument as compliments (907–12), argues that skill in rhetoric will enable one to behave as one chooses and get away with it (1080–82): "If you are caught in adultery, you will reply to your accuser that you have done no wrong. Take Zeus as your example: he was overcome by love of women: how could you, a mere mortal, resist temptation when a god could not?" Earlier, the Unjust Argument had claimed that Justice (a deity) does not exist; and when the Just Argument, in reply, said with Hesiod (*W*256ff.) that 'Justice dwells with the gods' (903), asked "Why, if Justice exists, did not Zeus perish when he had bound his father Kronos?"

In the play—and particularly in the second draft—Aristophanes can ensure that Right prevails. Strepsiades pays for his folly when Pheidippides, taught in the *Phrontisterion,* beats him up and justifies the action; the Clouds, unlike the principles in a Presocratic cosmogony, reveal that they see to it that wrongdoers suffer "so that they may learn to fear the gods" (1458–61); and Strepsiades burns down the *Phrontisterion,* and possibly Sokrates in it. A stupid or unthinking member of the audience might have supposed that all the nasty, newfangled notions of Presocratics, sophists, and rhetoricians had been refuted by the arsonist's torch; as, some twenty-five years later, he might have believed that a few ounces of hemlock had eliminated Sokrates' disturbing ideas together with Sokrates.

But Aristophanes was an intelligent man; and though his political views can be endlessly debated, he certainly preferred peace to war and justice to injustice. Even had he not been intelligent, he would surely have realized that justice was underwritten by the belief that the gods punish injustice, if not invariably then frequently. Being intelligent, he must have realized that the arguments of Sokrates and the Unjust Argument were unanswerable: the thunderbolt did strike at random, the cosmogony was not constructed on the principle of justice, and important deities did behave unjustly to each other. It is not surprising that the *Clouds* is a tense and uneasy play (Adkins 1970b, 13–24).

There are indications that demands on the cosmogony are increasing at this period, as it becomes less and less able to sustain them. In Aeschylus's *Oresteia,* produced in 458 B.C., Orestes is acquitted in court in the last play of the trilogy, the *Eumenides;* but many considerations besides strict justice are thrown into the balance (Adkins 1982a, 229–31). In the second play, the *Choephori,* Orestes and Elektra offer many traditional inducements besides the justice of their cause to persuade Zeus and the shade of Agamemnon to help them (246–63, 305–478). In Euripides' *Electra,* produced about 413 B.C., these elements are reduced to a minimum, and Orestes says (583–84) "For if injustice is to get the better of justice, we must no longer believe that the gods exist." All of the traditional beliefs continue to be held in the society (Adkins 1960a, 131–52). The punishment of the unjust is only one of the gods' relationships with mortals; but in the minds of some articulate Greeks it is now the crucial one.

A very different document gives similar testimony. A speech from a satyr-play by Kritias runs as follows (D-K 25B: the rest of the play is lost):

> There was a time when the life of man was without order, beast-like and under the dominance of strength; when there was no prize for the *esthloi* nor punishment for the *kakoi.* And then, I think men made themselves *nomoi* [laws] as punishers, that *dike* [justice] might be tyrant and have *hubris* as her slave, and that anyone who erred might be punished. Then, when the *nomoi* restrained them from committing open and violent crimes, they committed crimes by stealth, and then I think that some *sophos* [shrewd and clever] man devised for mortals fear of the gods, that the *kakoi* might have some fear even if it was by stealth that they were doing, saying, or devising something. For this reason, then, he introduced the idea of the divine, maintaining that there is a deity flourishing with a life that is immortal, hearing and seeing with his mind, whose thoughts are exceedingly great and whose attention is directed to mankind, who has a divine *phusis* [nature] and will hear all that is said among mortals and will be able to see all that is done. And if one silently plots some *kakon*

> [harm], this will not escape the notice of the gods; for their intelligence is too powerful. By uttering these words he introduced the most pleasant of teachings, hiding the truth by a false claim. He asserted that the gods dwell in the region that was certain to cause most terror to man, the region from which he knew that fears come to mortals and also benefits for their wretched lives, from the circling heavens above us, where he saw the flashes of lightning and the dread rumble of thunder, and the starry body of the sky. . .and from where come the bright rays of the sun and the wet rain descends to earth. Such were the fears he set around mortals, and by their means in his words he gave the deity a home, *kalos* and in a fit place, and quenched anarchy [*anomia*] by laws [*nomos*]. . . .Thus, first, I think, someone persuaded mortals to believe that there exists a race of gods.

The speaker affects to be on the side of law and order. But when belief that the gods punish injustice is a necessary condition of the desirability of justice for anyone who feels free of human constraint, to proclaim that the gods are a convenient fiction is to destroy the cosmogony and the claims of the cooperative values—completely, this time. It is not merely that the gods do not concern themselves with reward and punishment; they are inventions, given a fictional abode in a Presocratic cosmos which goes on its heedless way.

Any Athenian citizen could attend the dramatic festivals at which the plays of Aristophanes and Kritias were produced. Not all shared the views of Aristophanes' Sokrates, the Unjust Argument, or Kritias's speaker, either before the performance or afterwards. Not all well-known Athenians did so. In 413 B.C. Nikias, a prominent politician, one of the richest men in Athens and one of the Athenian generals before Syracuse, was induced by the occurrence of a lunar eclipse—the moon is, of course, a goddess in the cosmogony (*T* 19, 371)—to delay retreat for "thrice nine days," with disastrous results for himself and his army; and, says Thucydides, the greater part of the army agreed with him (7.50.4). But for anyone who thought seriously about it, the cosmogony was now in ruins.

Those who accepted Presocratic cosmologies in general now believed[10] either in no gods or in gods who did not reward or punish; and the traditional cosmogony could not withstand the strains put upon it by those who demanded just deities reliably punishing individuals in their own person in this life. It was not that kind of cosmogony. The choiceworthiness of justice depended on its role as a means to success and well-being. Where human restraint of the unjust was unavailing, it rested with the gods to guarantee that one could not be unjust and prosper. That guarantee was now cancelled.

To appreciate the seriousness of the problem, one must recall the nature of Greek values. The goal is well-being, termed both *olbos* and (by the fifth century) *eudaimonia*. The words do not mean "prosperity," but "well-being" is usually understood primarily in material terms. Health, a flourishing family, and social position are usually also components (Adkins 1972b, 79–81); but Medea can say, in a conscious paradox criticizing the overmaterialistic usage, "May I never have a *eudaimon*—i.e. prosperous—life that is painful to me, nor an *olbos* which grieves my mind" (Euripides, *Medea* 598–99). She would of course have termed *eudaimonia* the way of life she herself regarded as most desirable. *Arete* is, or the plural *aretai* are, the sum of the qualities belived to produce or maintain *eudaimonia:* usually courage, fighting ability, and the skills conducive to political success. *Eudaimonia* is desirable, and the qualities which produce it, the *aretai,* are not surprisingly desirable too, particularly as the exercise of *arete* requires material goods, so that the status of the *agathos* qua *agathos* is more desirable than that of the *kakos.* The *arete* of the *agathos* defends the well-being not only of the *agathos* but also the well-being and indeed the existence of the other members of his group (household, political faction, or city); they value his *arete* too. Hitherto, justice has not been observed to make sufficient contribution to the well-being of household or city to warrant its being termed *arete:* even the *kakos* values the justice of the *agathos* less than his *arete* (Adkins 1960a, passim). The *agathos* is evidently drawn by his values to pursue a maximum of *eudaimonia* for himself and his friends, employing his *arete* and any other appropriate qualities to attain his goal. He will be just only if justice is conducive to *eudaimonia;* if, that is, his ill-gotten gains might be greater, but would not last, whereas well-gotten gains, if fewer, persist (Solon 13, ll. 7–32, in West, II, 128). Of that the gods have been till now the—unreliable—guarantors; but now articulate doubts are heard. In any society, the breakdown of the cosmogony might have led to the pursuit of success by any effective means; in fifth-century Greece, when divine sanctions were absent, the *agathos* was encouraged by the most powerful values of his society to pursue success, for himself and his group, by any effective means.

The following quotations from Platonic dialogues show these values in action. Kriton, an old friend of Sokrates, was a prosperous and respectable citizen of Athens. In Plato's *Crito,* Kriton comes to prison to urge the condemned Sokrates to escape. He thus attempts to persuade him (45c5–46a4):

> Then again, Sokrates, you seem to me to be attempting something that is not even just [*dikaion*], betraying [*prodidonai*] yourself

> when you might be saved. You are eager to suffer the kind of fate that your enemies would be eager for you to suffer in their desire to destroy you. In addition, you seem to me to be betraying [*prodidonai*] your sons too, whom you will leave and abandon, though you could bring them up and educate them. So far as you are concerned they will have to fare as best they may; and their experiences are likely to be the customary ones of orphans in their orphanhood. One ought either not to have children or take one's due share of the toil of bringing them up and educating them; but you seem to me to be choosing the easiest way out. You ought to choose what an *agathos* and brave [*andreios*] man would choose, seeing that you claim that you have been practising *arete* all your life. For I am ashamed [*aischunesthai,* compare *aischron*] on your behalf and on behalf of us your friends lest the whole business should appear to have come about through some cowardice [*anandria*] of ours,—both the coming of the case into court when it need not have done so, and the manner in which the case was handled, and this final event, the supreme absurdity of the whole business, that you should seem to have departed from us as a result of some *kakia* and cowardice [*anandria*] of ours, since we did not save you and you did not save yourself, though it was perfectly possible if we had been of any benefit to you at all. Take care, Sokrates: this may appear to be not merely *kakon* [harmful] to you and to us, but *aischron* too.

Arete is displayed in defending one's own well-being and that of one's family and friends. Sokrates is failing to protect himself or his sons; which is betrayal, and shows cowardice. *Andreios* (brave) and *anandria* (cowardice) are formed from the Greek word for "man." Kriton and Sokrates' other friends are failing to save Sokrates. The loss of Sokrates will not merely be a great woe (*kakon*) for them, but *aischron* too, for they are failing to live up to the demands of the most powerful values of their society.

Kriton is a decent Athenian gentleman. Kallikles, who appears in Plato's *Gorgias,* and Thrasymachos, in Plato's *Republic,* are usually termed "immoralists." In terms of Greek values, the matter is not quite so simple. Kallikles says (492a) that it is those who lack the capacity to get their own way who say that licentiousness (*akolasia*) is *aischron,* in order to enslave those who are more *agathoi;* and they praise self-control and justice because of their own *anandria.* If anyone inherited a position of power or gained it by his own efforts (492b),

> What would be more truly *aischron* and more *kakon* [harmful to its possessor] than self-control and justice for these men? They could enjoy good things [*agatha*] and no one could prevent them;

> but they would then bring in as a master [*despotes*] over themselves the law and words and blame of mankind. How could they not be wretched [*athlios*, the opposite of *eudaimon*] as a result of the *kalon* of justice and self-control, if they gave no more to their friends than to their enemies, when they themselves are ruling in their own city? In truth, Sokrates—and it is truth that you say is your goal—the matter stands thus: luxury and license and liberty [*eleutheria*], if they have the resources to gain their ends, are *arete* and *eudaimonia;* the rest are merely trimmings, human contracts which are contrary to the nature of things, worthless nonsense.

Similarly, Thrasymachos, *Republic* I, 344c5–6, assures Sokrates that "injustice on a sufficient scale is something stronger, and more worthy of a free man and a master over slaves, than justice." Since the role of *arete* was to preserve one's freedom and prosperity, and all Greek *agathoi* were masters over slaves, it is not surprising that Thrasymachos maintains that injustice, not justice, is the *arete* (348c5–10). He is using the traditional criteria of *arete.* He is not rejecting the most powerful values of his society, but drawing from them conclusions which were very difficult for a Greek of the period to counter.

For Kriton, Kallikles, and Thrasymachos *arete* is displayed by helping one's friends against one's enemies, a task in which it is *aischron* to fail, *kalon* to succeed. The difference between them is that Kriton is more concerned with defending what he has, Kallikles and Thrasymachos with acquiring as many *agatha* as possible. Recall also that justice is choiceworthy whenever in naturalistic terms, too, it is likely to produce more *eudaimonia* than injustice. Doubtless Kriton usually found that justice paid; Kallikles (492a) implied that this is the situation of the great majority of mankind, and he indicated that they have their own way of using *kalon* and *aischron.* The usage is a recent development (Adkins 1960a, 172–94). But for outstanding *agathoi* like Kallikles and Thrasymachos, what could be more shameful than to gain fewer *agatha* than they have the power to do? The free man, the man of *eleutheria,* should have no *despotes,* for a *despotes* is a master of slaves, and he is free: he should not have as his master the law and blame and words of the mass of mankind. To a god the mortal must submit, for gods are more powerful, and if they did not punish his injustice as injustice, they might resent excessive prosperity. But the heavens are empty: "some shrewd and clever man" in Kritias's words, pretended that there were gods up there. To a Kallikles or a Thrasymachos, to submit to the mere words of mere mankind would be *aischron*, for it would be unfitting for a truly free man.

These are not problems for the study or lecture-room only. By 399 B.C. the Peloponnesian War, which had begun some thirty years earlier, and had been marked by steadily increasing brutality, had ended in utter—and unexpected—defeat for the Athenians; the Thirty Tyrants, who included Kritias, had come to power and, after a reign of terror constrained by no thought of law or morals, had been put down by the restored democracy; and under the democracy Sokrates had been put on trial for his life, and executed, as a "corrupter of the young." Among his pupils had been Kritias, Charmides, another of the Thirty, and Alkibiades, the latter the most talented and unscrupulous of politicians who had held elected office. Kritias and Charmides were close relatives of Plato; and he was sincerely appalled by their actions. Plato believed Sokrates to have been the finest person—the most *agathos* person—had ever met.[11] Since Sokrates had been poor all his life, had defended himself unsuccessfully in court, and had died a condemned criminal—all marks of *kakia*—Plato's contemporaries would not be easily persuaded. For many reasons, Plato might be moved to serious thought about Greek values, now that the cosmogony had collapsed.

The urgent need was to make justice choiceworthy even for those who were too strong for human constraints and did not believe in the punishment of heaven.[12] Anyone who believed that justice was an *arete* should of course pursue justice, for traditional *aretai* are preferable to traditional *kakiai:* it is better to be a success than a failure. Sokrates must have thus valued justice, and evidently acted on his evaluation. Other examples are found in the later years of the fifth century. In Euripides' *Electra*, produced about 413 B.C., Elektra, though a princess, has been married off to a poor farmer so that her children will be no danger to their mother Klytemnestra and Klytemnestra's lover, the usurper Aigisthos. The farmer, from both self-control (45–46, 52–53) and fear of reprisals should Orestes, Elektra's brother, return (260), has not attempted to consummate the marriage. Orestes is amazed, and begins to reflect upon the nature of *euandria*, the qualities of the *agathos* man (380–90):

> For such men administer well both their cities and their own households, whereas those who are nothing but senseless lumps of muscle are mere ornaments of the market-place, for a strong arm does not even endure a spear-thrust any better than a weak one. No; such ability lies in a man's nature and in his excellence of spirit. For this man, who neither has a high position among

> the Argives, nor is puffed up by the fame deriving from noble lineage, but is a man of the people, has proved to be *aristos*. Will you not come to your senses, you who wander about full of empty opinions, and in future judge men by their mode of life, and hold those to be noble, who lead noble lives?

Orestes argues that self-control and justice are aspects of *arete*, on the grounds that the just and self-controlled are better at administering cities and their own households, which consequently fare better than they would without these qualities. This speech enrolls self-control and justice as *aretai* on the same terms as the traditional ones. Since *aretai* are desirable per se, no divine sanctions are needed. Indeed, no human sanctions should be needed.

Orestes, of course, is praising the farmer for something he has already done, not trying to persuade him to do something that he is reluctant to do. (Nor were these the farmer's own reasons for doing it.)[13] His argument is persuasive only to anyone who accepts that justice and self-control do improve the administrative efficiency of one's household and city. Again, traditionally the farmer's *arete* would not be impaired by forcing his attentions on Elektra; but traditional *arete* did not require him to do so.

Kriton is in a different situation. Traditional *arete* does demand that he help Sokrates to escape from prison, and he wishes to do so. In reply, Sokrates asks him, among other questions (48b4–c2):

> *So:* Consider also whether we are still agreed on this, that it is not living on which one should set the most store, but living well [*eu*].
> *Kri:* We are.
> *So:* And are we still agreed that living well [*eu*], living honorably [*kalōs*] and living justly [*dikaiōs*] are the same thing?
> *Kri:* We are.
> *So:* Then in the light of our agreements we must consider whether it is just or unjust for me to try to escape from prison without the consent of the Athenians; and if it seems just, let us make the attempt; but if not, let us abandon it.

The force of this argument evaporates in translation. In English, "well," "honorably," and "justly" are vague terms whose meanings overlap; and it makes sense to ask with respect to any of them whether it is to one's advantage to live in that way. In Greek, to live *dikaiōs* has no initial attraction unless so to live brings advantage. To live *kalōs* is traditionally to live according to the requirements of *arete*, an admired and successful existence of value to one's community. *Arete* may require one to die defending that community; but the life of *arete* is desirable, and "better death than defeat" and "better death than slavery" are central tenets of *arete*-ethics. To live *eu* is to live with *eudaimonia;* and such a life is the goal, as Sokrates' first question indicates. To subsume living justly under living *kalōs* and living *eu* is to render the just life choiceworthy *in all circumstances,* for living *eu* must be advantageous. Once Kriton has accepted the premises, the conclusion follows: if to live justly is to live the life of *eudaimonia,* one will of course choose the just course of action.

Kriton apparently accepted the equivalence of the terms in the past; but his long speech (45c5–46a8) makes it clear that he continued to treat the traditional *aretai* as *aretai.* No Greek polis could do without the courage of its citizen-soldiers; and the *agathos* had to defend his own household in many ways, unfamiliar to a modern head of household in a settled society, which demanded courage and initiative. Kriton is in a situation in which courage and justice seem to demand different courses of action; and not surprisingly he feels more strongly the demands of the traditional competitive *aretai.*

Kriton is an ordinary decent Athenian of his day. Anyone who is prepared to agree, as he is, that justice (*dikaiosune*) is (an) *arete,* or that to live justly (*dikaiōs*) is to live *kalōs* and *eu,* should be just even though the cosmogony has collapsed about his ears, for the just life is—in some way not specified—more advantageous for him than the unjust life, and will lead to the desired *eudaimonia* as the unjust life will not.

Not all of Sokrates' companions in argument are so obliging. In Plato's *Gorgias* Polos, a pupil of the great rhetorician of the title, holds that committing injustice (*adikein*) is more shameful (*aischron*) than suffering injustice (*adikeisthai*), but that suffering injustice is more harmful (*kakon*) for the sufferer than committing it. Common sense seems to us to be on Polos's side, for Polos's values now seem more familiar than the traditional values of Greece; and since the Greek in pursuit of *eudaimonia,* the possession of good things (*agatha*) however defined, should pursue the greater *agathon* or the lesser *kakon,* and suffering injustice is a greater *kakon* than committing it, it is evidently preferable to commit injustice rather than suffer it, if Polos's evaluations are correct.

Polos's use of language reflects the recent broadening of the use of *arete* and *kalon,* to which Kriton assented. Traditionally, it is common sense to prefer *arete* and what is *kalon* to *kakia* and what is *aischron,* for *arete* and *kalon* commend prosperity and success and the qualities which promote them, while *kakia* and *aischron* are similarly linked with poverty and failure. Kriton, in accepting the new extended sense of *kalon* and *arete*[14] which includes justice, accepted the "logic" of *kalon* and *arete:* if justice is more *kalon* it is also better, more advantageous, for its possessor. Polos accepts the extended use of *kalon* and *aischron,* but rejects the "logic": injustice may be more *aischron* without being less advantageous.

Polos now acknowledges two kinds of quality denoted by *kalon;* one is advantageous to its possessor, the other is not. A more clear headed person than Polos might have wondered whether it was appropriate that both should be termed *kalon.* The woolly-minded Polos rapidly falls prey to Sokrates' wily attack.

Kalon is applied not only to behavior but to bodies, colors, shapes, and sounds, in a usage which we—inaccurately—translate as "beautiful." (As Sokrates' first question below indicates, a *kalon* body may be one serviceable for its task.) Sokrates asks Polos whether there is any reason why all these things are termed *kalon (Gorgias* 474d–475b):

> *So:* For example, do you not call *kalon* bodies *kalon* either for their usefulness, with respect to the purpose for which each is useful, or in respect of some pleasure, if they cause pleasure to those who look at them? Can you say anything other than this about the *kallos* of a body? *Po:* No. *So:* And do you call all shapes and colors too *kala* either on account of the pleasure or benefit they provide, or for both reasons? *Po:* I do. *So:* And sounds and everything associated with music? *Po:* Yes. *So:* Now consider *kala* laws and practices. Surely these do not fall outside these criteria, of being either beneficial or pleasant or both? *Po:* I don't think so. *So:* And the *kallos* of subjects of learning also? *Po:* Very much so; and *now* you are defining the *kalon* in a *kalon* manner, Sokrates, in defining it in terms of pleasure and *agathon* [what is good for its possessor]. *So:* And in defining the *aischron* in terms of the opposite, pain, *kakon* [what is bad for its possessor]? *Po:* Necessarily. *So:* Then when of two *kala* one is more *kalon,* it is more *kalon* because it exceeds in either or both of these characteristics, pleasure or benefit? *Po:* Certainly. *So:* And when of two *aischra* one is more *aischron,* it will be more *aischron* by exceeding in pain or *kakon?* Is that not necessary? *Po:* Yes.

Sokrates now moves in for the kill (475b–c):

> But what was being said just now about committing and suffering injustice? Weren't you saying that suffering injustice was more *kakon* [for the sufferer], committing it more *aischron? Po:* Yes. *So:* Then if committing injustice is more *aischron* than suffering it, it is either more painful, and more *aischron* by exceeding in pain, or in harm, or both? Mustn't that be so? *Po:* Of course.

> *So:* Let us consider first whether committing injustice exceeds suffering it in painfulness, and whether the unjust suffer more pain than those to whom injustice is done. *Po:* They do not, Sokrates. *So:* So committing injustice does not exceed in pain? *Po:* No. *So:* And if not in pain, it could not exceed in both harm and pain? *Po:* Apparently not. *So:* Then it must exceed in the other. *Po:* Yes. *So:* In harm. *Po:* Apparently. *So:* Then if it exceeds in harm, committing injustice must be more harmful [*kakon*] for the agent than suffering it. *Po:* Clearly.

Polos goes down almost without a fight. The reason is evident. He has not ceased to use *kalon* and *aischron* in their traditional contexts, in which *kalon* is attractive because it is linked with pleasure and benefit to its possessor. Sokrates draws these traditional usages to Polos's attention, and invites him to agree that the "logic" of the new range of *kalon* and *aischron* must be the same. Once Polos agrees, he is defeated. He should have replied that that was what Sokrates was supposed to be proving, and that he, Polos, saw no prima facie good reason to believe it.

For the present discussion, it is important to note that it is the beneficial which is pursued. The general tenor of Sokrates' argument with Polos (*Gorgias* 461b3–481b5) shows that anyone who evaluates the situation as he does will pursue the benefit of committing injustice, and pay little heed to the *aischron* involved. For him, indeed, *kalon* and *aischron* should henceforth be words of little emotive power. But, as the argument discussed here itself shows, it is very difficult to change the emotive power of a word; it is much easier to find a word with the appropriate emotive power and change its range of application. Sokrates and those who agree with him have done this in using *arete* and *kalon* to commend cooperative excellences, wishing to carry the "logic" of *arete* and *kalon* over to the new contexts; but Polos really should have termed committing injustice *kalon,* suffering it *aischron,* since that accords with the "logic" and emotive power he wishes the words to possess. (Similarly, the speaker in the Kritias-fragment quoted above, pp. 292–93, uses *esthlos* to commend cooperative excellences; but he evidently does not believe that the cooperative excellences pay, and is acknowledging rather than endorsing this use of *esthlos*.)

Kallikles, Thrasymachos, and their like will not succumb to such arguments. What they term *arete* and *kalon* is what they believe to be advantageous, *agathon,* for themselves and conducive to *eudaimonia.* Argument is possible, for they are not nihilists: they have strongly held values. Nor are they "immoralists," in the sense of having turned their backs on the most powerful values of their society. Qualities termed *arete* and acts termed *kalon* have always been commended by these terms on the grounds that they made an evident contribution to the *eudaimonia* of their possessor and his group. Justice was less valued because its instrumental contribution to *eudaimonia* was less evident. Kallikles and Thrasymachos deny that, for a powerful *agathos*, justice makes any instrumental contribution at all: it is injustice that increases the *eudaimonia* of the *agathos* and his group. The last three words are important: these men are not invoking a Hobbesian *bellum omnium contra omnes* as an ideal. Their group is their own household and other similar *agathoi* with their households: a group smaller than the city, a political faction. The Thirty Tyrants did not set out to be unjust to one another.

These men rest their claim to be preeminently *agathoi* on their strength, their courage, their intelligence, and their resources; and these are the traditional criteria. Traditionally, the deterrent upon the use of injustice as a means to greater *eudaimonia* was the belief that it was not an effective means: the gods would see to that. Plato cannot use such arguments, for these men do not hold the belief. Instead, he argues that just behavior is more like skillful, intelligent behavior than is unjust behavior (*Republic* I, 348d–350c); and when Kallikles claims both that the *agathos* is the *phronimos*, the practically intelligent man, and that the *agathon*, and hence *eudaimonia*, is pleasure, Plato argues that there are situations in which the fool feels more pleasure than the *phronimos* (*Gorgias* 487e–498c).

The arguments contain numerous fallacies: but that is not the important point for the present essay. It is important to note that even if the arguments were logically impeccable, they would not be psychologically cogent. Accustomed to a vigorous competitive *arete* exercised to secure success in public life and to a materialistic view of *eudaimonia*, how could Plato's contemporaries have readily accepted that anyone who was just but unsuccessful was more *agathos* and more *eudaimon* than the unjust but successful man? Even if they were to accept Plato's claim that Sokrates was a just man unjustly condemned, how could Plato convince them by verbal dexterity of this kind that Sokrates, poor, unpracticed in the skills of public life, a miserable failure when called upon to defend himself in court, a condemned and executed criminal, was *agathos* and *eudaimon*, and indeed outstandingly *agathos* and *eudaimon?*

After the argument with Thrasymachos in *Republic* I, Plato devotes the remaining nine books of the dialogue to providing more psychological cogency. At the beginning of Book II, Glaukon and Adeimantos survey the arguments about justice and injustice current in their day.

People say that justice is pursued by those who do value it solely because they are not strong enough to commit injustice with impunity: anyone who could commit injustice and get off scot-free would do so. And those who do praise justice praise it for the prosperity it brings, whether naturalistically or from the gods. Adeimantos also enumerates all the other nonmoral methods of winning divine favor, all of which are available to the prosperous and unjust.

In sum, everyone praises either justice or injustice for their—extrinsic—consequences. Glaukon and Adeimantos challenge Sokrates to demonstrate that, no matter how great the material well-being of the unjust man or the ill fortune of the just, the just is still *eudaimon* and the unjust not *eudaimon*.

In reply, Sokrates argues that a *polis* is larger than an individual, and that justice should be easier to see there. He then claims that there are four kinds of human beings, with different kinds of *psuchai* ("souls"; the presence of *psuche* is what gives life to a living creature): gold, silver, bronze, and iron. He takes four *aretai* as cardinal: political wisdom, courage, justice, and self-control. The gold are capable of all four, the silver of all but political wisdom, the bronze and iron of only justice and self control. Sokrates then delineates a *polis* of three classes: those with gold *psuchai* rule when older, having served as soldiers earlier; those with silver *psuchai* in their youth and prime serve as soldiers, and are then presumably retired; and all the rest take no part in public life, but perform each the trade or profession for which he is best suited by nature. The wisdom of the *polis* is clearly to be found in the rulers, the courage in the soldiers. All must show self-control by acknowledging that it is appropriate for the rulers to rule, the soldiers to defend, and the remainder to perform their own tasks; and justice consists in everyone performing his own task. Since it is taken as a datum that each person is fitted by his *psuche* for a particular task, it follows that when each performs his own task the *polis* is most efficient and enjoys most well-being, *eudaimonia;* and this efficiency depends on the *aretai* thus defined, so that the most just *polis* enjoys most *eudaimonia*.

Sokrates now turns to the individual *psuche,* and discovers in it a rational element, a "spirited" element, and an appetitive element. Using the analogy of the *polis* to guide the search, he "discovers" wisdom in the rational element, courage in the spirited element; while self-control consists in the acknowledgment by each element that it should perform its own task, and justice consists in every element performing its own task. Since a *psuche* so organized will evidently "run more smoothly" than a *psuche* which is not so organized, it evidently enjoys well-being, *eudaimonia,* irrespective of any ill fortune; and it is the *aretai,* thus defined, which bring about such *eudaimonia.*

This account of Sokrates' argument is sketchy. I offer a fuller analysis elsewhere (Adkins 1960a, 282–93). It is not my purpose here to discuss the myriad difficulties of the argument, or to consider whether Sokrates has met the challenge of Glaukon and Adeimantos. It is the nature of the challenge and the response that is important. Glaukon and Adeimantos complained that the rewards of justice—or injustice—were extrinsic. Glaukon said, "I want to hear what power [intrinsic effects] justice and injustice themselves have when present in the *psuche,* and to set aside their rewards and other [extrinsic] consequences" (358b5–7); and asked also that the just man should have the reputation for injustice till he dies and vice versa so that when the one has come to the limits of justice, the other of injustice, we may judge which is the more *eudaimon* (361d2–3). Justice is to be valued as an *arete,* if at all, for the intrinsic contribution which it makes to a nonmaterialistic *eudaimonia.*

Recall the development of Greek values, from Homer and Hesiod to the *Republic* of Plato. In Homer and Hesiod, and in general, it is necessary to be prosperous in order to be *agathos* and possess *arete;* and one employs one's *arete* to protect and, if possible, increase one's prosperity (whether termed *time, olbos,* or *eudaimonia*) and that of one's group (*oikos* or *polis*). Justice is less valued. No one likes to suffer injustice; but in early Greece it is apparent to all, *kakos* as well as *agathos,* that the *arete,* the competitive excellence, of the *agathos,* makes more contribution to the well-being of all than does his justice. At the beginning of the *Odyssey,* Telemachos is an estimable young man if one employs the criteria of the cooperative excellences. But the urgent problem is to drive away or kill the suitors; and Telemachos' cooperative excellences cannot solve the problem.

The Greek cosmogony as represented by Hesiod employs the same values. The regime of Zeus, too, is a "political" arrangement resting on *arete, moirai* of *time,* and *philotes.* It is based on power, not justice or any superiority in terms of cooperative excellences. In their relationship with human beings, gods may sometimes take note of justice and injustice between one mortal and another; but, as the *Works and Days* and other works of Greek literature show, they care for many other things too: sacrifice, observance of due season, hard work, none of which, singly or in combination, guarantees, in Hesiod's view, the success to which they are a hoped-for means. The relative modesty of Hesiod's claims upon the cosmogony renders them less susceptible to empirical falsification. An attempt to render the cosmogony proof against such falsification foundered on the increasing importance of the individual vis-à-vis the kinship group; and thereafter the more the demand for divine justice grew, the less able was the cosmogony to satisfy it. (The demand was not universal among believers: as the words of Glaukon and Adeimantos in *Republic* I, 358b–367e show, all the traditional modes of belief in deity remain prevalent in the early fourth century.) The Presocratics and the sophists in different ways made their contributions to the breakdown of the cosmogony, and a considerable proportion of articulate Greeks found it no longer possible to believe in divine retribution. Alternatively, one might say that the Presocratics replaced the traditional cosmogony (or rather cosmogonies) with a cosmogony (or rather cosmogonies) of their own, which demanded a different ground for the desirability of justice.[15]

An ethic which made no demands on deity was urgently needed. But evidently Plato, or any other Greek, had not a free choice of starting points. One can persuade others only by offering them what they want. Greeks from Homer onwards had valued qualities as *aretai* if they made evident contributions to the well-being of one's *oikos* or, later, *polis*. The goal is *eudaimonia*, the important means are *aretai*. These values take precedence over all others. Since *aretai* render their possessor a good and effective specimen of human being, any quality termed *arete* requires no divine sanction to render it desirable. The well-being may be rendered entirely intrinsic, the self-interest as enlightened as may be; but the resulting ethic must be eudaemonistic, not deontic, if the latter be defined as an ethic in which the concept of "ought" or "duty" is primary. In Greek, "Why?" is always a sensible reply to "You ought to do that"; and the only answer that is final is "Because it will make you *eudaimon*, or more *eudaimon* [or less wretched] than any other course of action available to you." (See, for example, Plato, *Republic* I 352d2, discussed in Adkins 1960a, 253.) To the problem of the breakdown of the cosmogony there are other answers besides Plato's; but the totality of Greek values from Homer and Hesiod, the status of justice and the relationship of values to cosmogony and cosmogony to values, ensure that they are all eudaemonistic answers.

Let me remind my readers of the goal of this essay; to argue that, in ancient Greece, when the cosmogony broke down, the cosmogony-dependent ethic was replaced by an ethic autonomous in the sense that it made no demands on divine sanctions. The Platonic ethic, and the other philosophic ethics which subsequently appeared in Greece, satisfy this requirement.[16] In his myths Plato depicts a system of divine sanctions too, but these are additional inducements: justice and the other cooperative excellences have been shown to be choiceworthy without invoking any such sanctions, in Plato's view (*Republic* 612a8–614a8). In addition, the metaphysic of the Forms plays an essential role in his philosophy, at all events at the period of the *Republic;* but though the Forms are eternal, they neither reward nor punish; and though it is presumably they, or rather some of the most important Forms, that render justice and the other cooperative excellences intrinsically good for their possessor, the argument from the structure of *polis* and *psuche* precedes the introduction of the Forms and philosopher-rulers into the *Republic,* and does not depend on them in any way. In addition, earlier Sokratic arguments of the kind discussed here do not rest upon any metaphysic. In the sense defined, then, the Sokratic/Platonic ethic is autonomous, and a response to the collapse of the cosmogony.

A final speculative paragraph. I have tried to show both that an autonomous ethic developed in ancient Greece when the cosmogony broke down, and that many characteristics of that ethic were closely linked with those of the cosmogony-dependent ethic which preceded it. I do not claim to have demonstrated a general truth; only extensive empirical research can reveal whether the Greek experience was an instance of a general rule. But suppose that this is the case. We may then inquire under what conditions a deontic ethic may develop. The Greek gods are more powerful than mortals, but not ethically superior; fear of reprisals motivates obedience, and in Greece no deontic ethic seems possible. Is it not likely that a necessary condition[17] for a deontic ethic's developing in the ruins of a cosmogony is that at least part of the motivation for obeying deity was that deity was believed to be more just and righteous than mere mortals could hope to be, and justice was underwritten not only by divine power but also by divine authority?[18]

Notes

1. By John Carman and Robin Lovin on several occasions during the conferences that gave rise to this volume.

2. Some Greeks believed in different treatment after death for different groups of the dead; but initiation per se seems to have been most frequently the ticket to bliss, rather than just behavior (Adkins 1960a, 140–48). Plato uses the belief as an additional inducement to good behavior, as in the myths of the *Gorgias* (523a–527a) and *Republic* (614b–621d); but the Platonic ethic claims that justice is choiceworthy whether or not the myths are believed (Adkins 1960a, 259–315).

3. *Tisis* and *tinein* denote transfers of *time* (Adkins 1960b). The punished individual or family loses *time* when a bad harvest or some other woe sent by the gods reduces his material prosperity (Adkins 1972a, 3–11). *Olbos* is now used of material prosperity, but the use of *tisis* and *tinein* remains.

4. In *Iliad* 19, 85–90, Agamemnon says that not he but three deities were *aitioi,* causally responsible for his depriving Achilles of Briseis. (LSJ interprets such usages as meaning "guilty," and cites no instance of *aitios* in the sense of "causally responsible" earlier than Herodotus; but "causally responsible" is a much less tendentious translation in passages such as this, including the lines of Solon discussed here.) But since Achilles has lost *time* in losing Briseis, Briseis must be restored together with a large amount of additional *time* to placate Achilles. If the father in Solon's poem has unjustly acquired more goods than he should have, and Zeus sees to it that the children lose them, the children are *anaitioi* in a different sense from Agamemnon; but unless a writer explicitly complains, there is no reason to suppose that he finds the practice unjust.

5. In *Iliad* I, Agamemnon refused to return Chryseis to her father, though the army wished him to do so. At the father's request, Apollo sent a plague on the army, which was guiltless in the matter.

6. Aeschylus, *Persae* 808, 821, Herodotus, 7.16, 8.77.1 (an oracle). Xerxes' scourging of the (divine) Hellespont and the accompanying speech display *hubris,* 7.35. For jealous deities, e.g. 7.46, Aeschylus, *Persae* 362.

7. *Dike* is sometimes used as an explanatory concept, e.g., Anaximander, D-K B1, explains growth and decay in terms of *dike, tisis,* and *adikia;* and Heraclitus, D-K B94, says "The sun will not go beyond its measures; for otherwise the Furies, the allies of Dike, will find it out." But there is no suggestion that these cosmological principles concern themselves with human transgressions.

8. For ancient discussions, see D–K vol. 2, 252–53. There is a convenient modern discussion in Guthrie, 27–54.

9. For *phusis* and *nomos,* see Heinimann passim, Adkins 1960, 1970a, 1972b, Indexes s.vv. *Nomos* and *Phusis,* and Guthrie, Index s.v. "*Nomos-physis* antithesis."

10. More accurately, "logically should have believed"; but those—if there were any—who attempted to combine cosmology with the gods of the cosmogony were in a difficult intellectual position.

11. And the most just, *dikaios, Phaedo* 118a17. Even those who conceded Sokrates' justice would not have granted his outstanding *arete.*

12. In general, there is an attempt towards the end of the fifth century to replace value-words which require divine sanctions with value-words which do not (Adkins 1976, 301–27).

13. In addition to self control (45–46, 52–53) and fear of Orestes' anger (260), he thought himself socially unworthy of Elektra (43–46).

14. Since *kalon* and *aischron* are applied to individual actions, *arete* to characteristics of persons, the extension of the usage of *kalon* and *aischron* may have been easier (Adkins 1960a, 179–81).

15. The distinction between cosmogony and cosmology calls attention to some important differences; but both cosmology and cosmogony furnish a frame of reference for ethics.

16. Aristotle, the Stoics, and the Epicureans each have their own answers, and their own philosophical cosmologies.

17. The condition is doubtless not sufficient. It seems likely that the deity must, like *Yahweh,* express his moral requirements in *explicit* commands to his worshipers.

18. Since writing the above, I discover that MacIntyre has interesting suggestions on the rise of deontic ethics, or rather deontic ethics in the Western tradition. But neither he nor I nor anyone else has engaged in the empirical research, both far-ranging and detailed, necessary to render our suggestions more than a speculative hypothesis.

References

Adkins, A. W. H.

1960a *Merit and Responsibility: A Study in Greek Values.* Oxford: Clarendon Press.

1960b "'Honour' and 'Punishment' in the Homeric Poems." *BICS* 7: 23–32.

1970a *From the Many to the One: A Study of Personality and Views of Human Nature in the Context of Ancient Greek Society, Values, and Beliefs.* London: Constable, and Ithaca, N.Y.: Cornell University Press.

1970b "Clouds, Mysteries, Socrates and Plato." *Antichthon* 4: 13–24.

1972a "Homeric Gods and the Values of Homeric Society." *JHS* 92: 1–19.

1972b *Moral Values and Political Behaviour in Ancient Greece.* London: Chatto and Windus, and Toronto: Clarke, Irwin.

1976 "*Polupragmosune* and 'Minding One's Own Business': A Study in Greek Social and Political Values." *CP* 71: 301–27.

1982 "Laws versus Claims in Early Greek Religion." *HR* 21: 222–39.

D–K

1951 Herman Diels and Walther Kranz, *Die Fragmente der Vorsokratiker.* 6th ed. Berlin: Wiedmannsche Buchhandlung.

Dover, K. J.

1968 *Aristophanes, "The Clouds."* Oxford: Clarendon Press.

Guthrie, W. K. C.

1969 *A History of Greek Philosophy.* Vol. 3. Cambridge: Cambridge University Press.

Heinimann, F. *Nomos und Physis.* Basel: Reinhardt.

LSJ

1940 *A Greek-English Lexicon.* 9th ed. Edited by H. G. Liddell, Robert Scott, and H. Stuart Jones. Oxford: Clarendon Press.

MacIntyre, Alasdair

1981 *After Virtue.* Notre Dame: University of Notre Dame Press.

T Hesiod, *Theogony.*

West

1972 M. L. West. *Iambi et Elegi Graeci ante Alexandrum Cantati.* Oxford: Clarendon Press.

W Hesiod, *Works and Days.*

12 A Confucian Crisis: Mencius' Two Cosmogonies and Their Ethics

Lee H. Yearley

Introduction

Guides for action and general pictures of the world obviously are related. Whether we respond to some devastating reversal of fortune with laughter, equanimity, or tears often rests on whether we see life as a bad farce, an incurable tragedy, or a flawed morality play. Similarly, whether we laugh or cry at our own foolishness, whether we attempt to alleviate or to increase our regret at our failures, whether we worry about or ignore others' pain depends on how we picture the world.

Much can be said about the characteristics of this relationship. I will focus, however, on only one aspect of such general pictures (that represented by cosmogonic myths), on only specific guides for action, and on only those questions that arise when the relationship between the two begins to appear problematic or even to break down. More specifically still, I will examine how those questions appear in Mencius, a fourth-century B.C.E. Confucian, concentrating on the problems that appear when he attempts to correlate the guides for action that arise from the two different cosmogonies that he inherits and utilizes.

A word on terminology before I begin. I will use the word "ethics" for the sake of ease (Mencius often uses words like virtuous, good, or noble), but I mean by it some guides, however clearly or confusedly articulated, for what is and is not appropriate action and attitude.

Many use "ethics" in a much narrower sense than this; for them, my "ethics" is their "ethos." The reason why I prefer the wider sense, especially when dealing with Chinese thinkers, should become clear in time.

Ethics and Cosmogony

The relationships between ethics and cosmogony are, of course, complex. Cosmogonies can be more or less explicit, more or less complicated, more or less guiding, or more or less explanatory. Moreover, a cosmogony affects the general ethos that supports ethics in a variety of ways. It may explain, guide, and comfort, for example, and clearly some functions may remain when other functions have fled. A cosmogony may comfort after it has lost its power to guide and explain. Most important to us, cosmogonies can guide actions in quite specific and powerful ways, as other essays in this volume richly illustrate.

If, however, a strain appears or a breakdown occurs in the relationship between cosmogony and ethics, then a complex and important intellectual process can occur. When the supposedly seamless web of cosmogony and ethics is torn or frayed, then the reinforcing lockstep the web produces develops a significant gimp. In such situations, intellectual ferment occurs, at least among the reflective. Thinkers find themselves having to question what an idea means, how it relates to another idea, and what spheres it may and may not operate in. They often do so in a way that seems to the unsophisticated to be nothing but a picking of nits as when such thinkers differentiate between courage and confidence. Especially important to us, they often have to face striking new problems that demand they disentangle a complicated set of related ideas about how action and attitude should be guided, evaluated, and justified.[1]

Professor Adkins's essay on classical Greece (chapter 11 above) analyzes one kind of problem that can arise in such a situation. Here I will examine another kind of problem: that of relativism, to use a modern formula that does only some violence to the issues at hand. Put abstractly, the problem concerns the relationship or lack of relationship between the natural and the conventional, or the universal and the contingent, or—to use a formulation I will explain later—the claims of morality and of ethos. The problem often arises from reflection on the variety of action and attitude that is both imaginable and evident. But it can also arise if a thinker employs two traditional cosmogonic myths that seem to point in conflicting directions: the one toward the natural, the other toward the conventional. Such is the case, I think, with Mencius, and his attempt to work out the problem presents a striking example of a thinker struggling with issues that arise from the breakdown of a clear relationship between cosmogony and ethics. Moreover, this working out establishes a significant part of the agenda, the key issues and options, with which many later Chinese thinkers will concern themselves.

Mencius: His Two Cosmogonic Myths

A fourth-century B.C.E. thinker (probable dates, 390 to 310–305), Mencius lives in a time when China has been changing substantially, both socially and intellectually, and continues to change. The times are good intellectually and bad socially. The feudal world and all it represented has been replaced by a few large, centralized, and aggressive states, and widespread warfare leads to the usual horrendous violations of people. Not surprisingly, great conflicts of ideas accompany these social changes. With many older ideas and ideals either dead or in decline, numerous professional teachers and counselors and their followers debate about what the world is like and how humans should act, and these debates provide a rich variety of options.

Mencius is such a professional teacher and counselor. He is held by the message articulated by Confucius, that apparent contradiction in terms, a failed sage. Mencius attempts to reinterpret and to revivify that message in order to triumph over the many conflicting views that are current and powerful. He is, then, an apologist in the term's full sense: he attempts to convince both himself and others of the sense and vitality of his tradition. His apologetic, as of course interpreted by others, will have immense power for long periods in later Chinese history; indeed his influence, at times, will resemble that of a combination of Aristotle and St. Paul in the West.

Of special importance to us, however, are the two cosmogonic myths that Mencius works from, the problems that they present for ethics, and the solutions to these problems that Mencius attempts to provide. Put briefly, the two kinds of cosmogonic myth can be called the "heavenly mandate" (*T'ien ming*) myth and the "sagely rule" myth (3A4, 3B9, 5A5).[2] The first kind of myth takes different forms in Mencius, but the crucial theme is that a high God will choose a people or leader to make present his purposes in the world. Heaven employs people to establish a society that will direct all human beings in beneficial ways. Heaven aims, then, to save people, if we use the definition of salvation that often operates in locative rather than open religions: fulfillment through location in a proper culture rather than fulfillment through a state that reaches beyond culture.[3]

The second kind of cosmogonic myth, the "sagely rule" myth, reflects similar locative notions, but it lacks the specific reference to

Heaven. These cosmogonic myths relate how various sages, in the distant past, tamed floodwaters and brought specific elements of civilized order to (or back to) the world. The sages, then, established those social rules and institutions that make possible human flourishing. I will focus here on how these two kinds of myth and the ethics they engender relate to each other and produce a vexing set of issues on which Mencius reflects.[4]

Mencius' Two Cosmogonic Myths and the Problems They Present

Although the two cosmogonies Mencius presents both deal with the character of the social order and appropriate human action within it, they differ substantially. The "sagely rule" myth describes how certain sages saved society by ordering it. The "Heavenly mandate" myth concerns Heaven's actions in the world, especially Heaven's effect on the rulers of society. The relationship between the two is clear: Heaven is responsible for those who correctly order societies. Indeed, one might even say the two myths differ only because one focuses on those who do the ordering and the other on why their ordering is to be seen as correct. The one describes while the other validates.

But important differences between the two do exist, and they become especially clear or even arise because Mencius reinterprets one of the myths. As I shall show more fully later, Mencius recasts the myths about Heaven's action to make Heaven's critical presence in the world the existence of certain general, Heaven-caused characteristics of human nature. This recasting allows him to create a philosophically and religiously powerful framework to undergird ethics. But it also implies that the general characteristics of human nature that are mandated by Heaven may manifest themselves in different ways depending on the precise social context in which they appear and to which they react. Because different manifestations may reflect the same general characteristic, no simple correspondence can be drawn between a characteristic of human nature and a particular social rule.

The myths about sagely rules, however, seem to imply that certain specific rules reflect how the social order must exist if people are to flourish. The "Heavenly mandate" myths can allow for different manifestations of appropriate action, but the "sagely rule" myths imply that certain rules are correct regardless of the social context. Theoretically, of course, the idea of a human nature caused by Heaven could be developed in a way that gives guidelines for validating changes in the original sagely rules. As we shall see, however, that development proves difficult for Mencius to carry off consistently, in part because the theoretical issue of relativism would need to be solved for it to succeed.

The difficulty Mencius faces can be formulated in another way, a way that reveals yet another aspect of the problem. The notion of a Heaven-caused human nature seems to call in question the status of the rules the sages give or perhaps even the need for them. To take seriously the notion that Heaven-given capacities exist in human nature means also to take seriously the possibility that normal people in past times may have discovered these characteristics and extrapolated from them correct rules or social practices. One seemingly must extend, then, the range of Heavenly-mandate myths to include not only the gift of certain characteristics to human nature but also the discovery and application of those characteristics by both sages and normal people in the past. For example, Mencius explains that in ancient times the sight of flies sucking unburied parents and foxes eating them led sons to bury their parents (3A5). Such a story is a cosmogonic myth because it tells how some current social practice started, how people got the rules that now guide them. But it also seems to sharpen still further the conflict between the two kinds of cosmogonic myths.

Put abstractly, the problem is that if all people have access to the characteristics of their nature and can extrapolate guides for action from them (in the case cited, how to deal with dead parents), then the sages do not give to people anything they could not give themselves. Sages are, so to speak, grammarians, not creators, of the good life; they outline the forms and systematize the structures of the language we call natural. Therefore one need not especially honor either the sages or, more important, the sages' insights. Everyone has the same capacity the sages did, and anyone might come to insights or expressions that are as clear as theirs were, or possibly even clearer.

Put in the context of cosmogony and ethics, this means that no rule, no formulation arising from a cosmogony, has a privileged status. Indeed, one might even go farther. To recognize possible extensions of or changes in sagely rules because all people have the capacity to perceive Heaven's directives is to build a significant and perhaps frightening instability into any system of rules. If the rules are in theory expandable and correctable, what, it can be asked, is their present status? We have, then, still another version of the tension in Mencius' thought between the rules themselves and the sources of and the actual justification for the rules.

Such, then, is a general sketch of the difficulties that occur when Mencius attempts to correlate both the guides for action and the justification of those guides that arise from the two different cosmogonies he uses. I need now to turn to a more textured examination of how Mencius faces these difficulties, but in order to do that I must first sketch briefly his most sophisticated general perspective on ethical action. Such a sketch is especially important because this perspective arises from Mencius' reworking of the Heavenly-mandate cosmogony to show that Heaven causes certain characteristics of human nature.

A Sketch of Mencius' General Viewpoint on Ethical Action

Put simply, Mencius' approach resembles that common sort of intellectual readjustment that falls under the general label of "the internalization of the sacred." His genius and situation give this approach a distinctive turn which, put schematically, looks like this: Humans are defined by four Heaven-given potentials (*tuan*, literally shoots or sprouts). The four potentials and their actualizations are: (1) a sensitivity to the sufferings of others, a compassion for others or a suffering with them, that is actualized in the helping of others, in benevolence or co-humanity (*jen*); (2) a yielding to other beings or a deference toward them, a keeping of one's self back to make way for others, that is actualized in ritual (*li*); (3) a tendency to distinguish abstractly between human acts, approving of some and disapproving of others, that is actualized in wisdom (*chih*); and (4) an awareness of falling short of some standard, a repugnance manifested in shame and dislike felt respectively toward the actions of one's self or others, that is actualized in righteousness (*yi*) (2A6, 6A6).[5]

These four capacities are given by Heaven, indeed are to be seen as Heaven's mandate. Heaven does more than just this in Mencius; the old cosmogony has too much vitality left to be so neatly encased. For example, it directly causes things to happen, apparently binds itself to a set timetable for helping the world and gives a special mission to Mencius (1B16, 6B15, 5A5). Ways exist to explain how Mencius deals with the tensions that these various depictions of Heaven produce and they reveal much about his rethinking of the inherited cosmogony, but such an analysis is not directly relevant to our inquiry (see Yearley 1975b).

More important here is how this reformulation of the Heavenly-mandate cosmogony informs his ethical vision. Mencius thinks we have self-validating knowledge of both what these natural capacities are and that Heaven gives them—and connections to Western natural-law theory are not only obvious but more tricky than they may look. But his main interest lies not simply in knowing what they are but in actualizing them so that they can control our action. The key human task, then, is the cultivation of these capacities. Indeed, this cultivation defines both the proper service of Heaven and the incarnation of Heavenly purposes on earth.

The crucial business, then, is self-cultivation, and the key to it is the mastery of what might be called a technique, the technique of extension. The crucial task is to learn to extend one's most basic knowledge and feelings. To extend knowledge is to see that one situation resembles another situation. To extend feelings is to have the feelings that are clearly manifested in one situation break through in another situation. For example, Mencius thinks that anyone who saw a baby about to fall down a well would feel compassion for the baby. One should then extend that feeling to other people in trouble.[6]

Mencius thinks that we have the same freedom to focus our attention on specific movements of the self as we do to focus on specific external objects. If we see a person in psychological or physical pain yet feel no real desire to help him, we can focus attention on the knowing and feeling that we would have either if the person were a close friend or if we saw the endangered baby, and thereby we gain the knowledge and power to act. Mencius thinks, then, that we are predisposed to act in certain benevolent or reverential ways. We must discover or rediscover those predispositions and then apply them to relevant situations.

Mencius uses revealing images to describe what occurs when extension is perfected: a person's actions resemble water breaking through a wall, or running down a hill, or surging up from a spring. Such images reflect an undivided, powerful natural force. In fact, we know whether we are as we should be by the degree of spontaneous force our actions exhibit. If our motivation is full, undivided, and spontaneous, we are properly developed; indeed we have tapped the sacral power in things (*ch'i*) (2A2). Mencius' reformulation of the Heavenly-mandate cosmogony leads him, then, to a picture where the fully developed person's ethical activity flows effortlessly from contact with the sacral constituents of human nature. The perfected person is so finely attuned both to the movements of human nature and to the particular circumstance faced that rules appear to have no obvious place.

And yet Mencius still adheres to a cosmogony which contains rules that are justified by the authority of the sages. Indeed, the sagely-rules cosmogony validates not simply general rules but a complex set of specific rules that govern virtually all aspects of life. These rules form the structure and provide the lifeblood that animates the structure, the saving locative religion, that is the society the sages founded.

We need to look, if briefly, at the character and extent of these rules to appreciate how full and "conventional" is the exfoliation of the sagely-rule cosmogony and thus both how pronounced is the tension that exists with the Heavenly-mandate cosmogony and how vexing is the difficulty that defines this particular version of the problem of relativism. We can best do this by examining briefly the general notion of *li* (a character usually translated as "ritual"), as in it we see a clear exemplification of the nature of the particular rules that should guide attitude and action.

I cannot begin to examine all the problems that swarm around the understanding of *li,* but I do need to note two related points. First, *li* covers two classes of activity that most of us think are distinct. One class of activities is solemn, explicit religious activities such as state or family sacrifices. The other is what we would call etiquette or reasonable learned social conventions such as knowing how to respond to a formal invitation or how to talk to an honored guest. The combination of these two classes within one category reveals how *li* exemplifies a locative religious vision. All activity is said to have a religious quality; indeed, salvation consists in a person functioning appropriately within a complex social order that is thought to be sacred.

My second point concerning *li* is that this locative perspective, especially when reasonable social conventions are the focus, blurs the distinction between what can be called the realm of ethics and the realm of custom in ways that can be productive and perplexing. The potentially productive result of the blurring leads us to consider seriously the idea that we should see our actions as existing on a continuum, with clear moral actions existing on one end and clear customary actions on the other end. Actions that fit on either end of the continuum are clear—conscientious objection to war and fork usage, for example. Arguably, however, most human activity occurs in the middle part of the continuum where distinctions between morality and custom are often difficult to make. We do many things that are inappropriate but that we cannot easily label either immoral or violations of custom. For example, consider my not greeting you with a smile or my being inattentive when you are talking. Most of our actions may fall within a sphere that is neither clearly moral nor clearly customary but that enables us quickly, easily, and humanely to establish and maintain relationships with one another. Such, then, are the possibly productive results of the blurring of custom and morality.

The perplexities generated by such a perspective are also, however, only too obvious. It seems to imply that certain conventional rules provide the only possible means for attaining the desired fulfillment and the needed appropriate action. Put in the form that is most important to us, the idea of *li* focuses the question of how Mencius can maintain that the general moral inclinations implanted in people by Heaven can find correct expression only in the specific rules formulated by the sages. The two cosmogonies seem to point in very different directions. The one points toward Heaven-given characteristics of human nature that may manifest themselves in distinctly different ways depending on the particularities of the social situation. The other points toward complex and particular rules that represent the only way in which the sages, and thus apparently Heaven, think human beings can thrive. Mencius faces here, then, the problem of relativism as that is specified in the relationship between the formulations that arise from and are justified by the Heavenly-mandate cosmogony, on the one hand, and the sagely rule cosmogony, on the other hand.

Mencius' Attempted Solution to the Dilemma Presented by His Two Cosmogonies

Mencius never simply states and faces the issues involved in the dilemma that I have outlined, but some of his most complicated and interesting thought focuses on a version of that dilemma. In it we find a paradigmatic example of a thinker struggling with the issues that arise from the breakdown of a simple relationship between cosmogony and ethics. Before turning to that aspect in Mencius' thought, however, we need to note another strand in his thought that seems barely to recognize the problem. This strand needs to be examined briefly both to give a full picture of Mencius and to illustrate the ways in which thinkers in such breakdown situations often and unsurprisingly can only occasionally focus the new problems they feel.

From this often more dominant perspective, Mencius argues that cosmogonic myths tell us exactly what rules must be followed. These rules are often very specific, reaching down for example to how a prince summons a gamekeeper, and to violate the rules is to violate the fundamental order of things. Therefore people should risk death rather than do so (5B7).

Perhaps even more striking evidence of this aspect of Mencius' work appears in his depiction of human nature's characteristics in terms of what are clearly, to us, culturally determined phenomena. For example, he can depict the constitutive parts of human nature solely in terms of service to one's parents and relationship to one's elder brother (4A27, 7A15). Perhaps most striking—especially because it appears in probably the most philosophically subtle part of the book—is Mencius' argument from the "fact" that all palates have the same preferences in taste, all ears the same ones in music, and all eyes the same ones in beauty to the "fact" that all characteristics of human nature, especially reason and rightness, must be the same (6A7). This side of Mencius manifests, then, so remarkable a naiveté about the distinction between phenomena that are clearly culturally determined and phenomena that may not be so determined that the dilemma sketched here seems to have no affect on him.

In another strand of his work Mencius, however, seems to be aware of the difficulties I have sketched and to work with some considerable sophistication to resolve them. (The existence of these more and less sophisticated strands may point to the existence of esoteric and exoteric teaching in him, but we can reach no firm conclusion on this matter.)[7] In examining this side of Mencius, we see a clear example of a thinker struggling to resolve the problems that arise from the breakdown of a simple relationship between cosmogony and ethics. Let us begin with a brief examination of the abstract resources for resolving the dilemma that exist in Mencius' thought. We can then examine how he actually works through certain aspects of the issue in concrete cases and what implications arise from that process.

At an abstract level one can extrapolate ideas from Mencius' work that can provide the beginnings of an account that might resolve or at least modify the dilemma. For example, one sees in him the inchoate beginnings of a distinction between three ways of analyzing apparently conflicting guides for action within and among cultures: differences in practice that reveal no differences in principle but arise from the differentiating contexts that inform the application of principle; differences in practice that reveal differences best specified as differences between the barbaric and the civilized—the notion that "humans" can be "nonhuman" has a clear meaning for him; and differences in practice that reveal real differences in "civilized" principles.

Only the last kind of differences need really concern Mencius, and the relevant cases here may, to his mind, be very few. From this perspective, then, Mencius could stress the universal characteristics that define human nature and explain the variety of ways in which they might be manifested in a culture by focusing on the specifics of the situation in which those universals are applied. Moreover, such an approach need not be uncritical because one can still query the applications made. Certain ideas in Mencius do point to such a solution: his emphasis on the generality of both the characteristics of human nature and the principles it generates; his stress of the need to attend closely to a situation's distinctive forms when general principles are applied; his notion that a spirit of ritual exists that may be contrary to the actual rituals of a society (2A6, 2A7, 4B6).

Even more revealing, however, than the mere presence of such abstract resources in Mencius' thought are the presence of clear examples of Mencius actually struggling with this issue. Such examples of practical moral reflection cast considerable light both on Mencius and on the issue of relativism. In one case, Mencius is questioned about whether, when faced with a drowning sister-in-law, one ought to take her hand to save her although the ritual forbids that one touch her. Mencius answers quite forcibly that in such cases discretion must always be used and that not to save her is to be a brute (4A17). In another case, Mencius is asked if a particular case of regicide is acceptable, regicide being a clear violation of rules. He responds that in this case only an "outcast"—a mutilator of benevolence and a crippler of righteousness—has been punished. Rather than a king being killed, only an "outcast" has been punished (1B8).[8] These two cases reflect clearly the problems Mencius' position encounters; indeed they may represent stock test-questions by which the interlocuter tries either piously to fathom his teacher's mind or knavishly to trip up his adversary.

Various ways exist to interpret Mencius' responses. In the first case, Mencius could be said to argue that a higher moral principle about care for life should overrule a lesser moral principle about concern for social rules. He might, to use his language, say that the natural movement of sympathy that leads to co-humanity (*jen*) should take precedence over the natural movement of yielding that leads to ritual (*li*). Both movements define human nature, but yielding to ritual should serve the aims of sympathy leading to co-humanity, so that in cases of conflict co-humanity should take precedence. Similarly, in the second case, a higher principle about concern for benevolent government should overrule a lesser principle about concern for social rules, *jen* should take precedence over *li*.

Such an approach surely captures part of what Mencius is doing. An even more productive, if also perhaps more speculative, approach would focus on how in these two cases Mencius employs what could be called the principle of appropriate description. (Such an approach also relates Mencius to the general Confucian notion of the rectification of names, but both that notion and its relationship to Mencius are too complex to examine here.)[9] In the first case, Mencius could be said to argue that of the two relevant descriptions of the woman in this case—sister-in-law and person in mortal danger—the latter must take precedence, and that fact makes one's judgment easy and one's acts spontaneous. In the second case, the same procedure seems clearly to be at work: the person described as a king should in fact be described as an "outcast" and treated accordingly.

Mencius' approach in both cases shows how important to him is the description of an act. Moreover, he implies that once the true description is discovered, the problem of motivating the action disappears. In his terminology, once one decides on what category an event belongs in, the appropriate extension will follow easily if one is truly developed. To see that a friend in distress belongs in the same descriptive category as a baby about to fall down a well is to gain the power to act. The moral judgment is virtually a true description; once that is clear, the proper action will occur spontaneously.

If such an approach captures how Mencius attempts to solve vexing ethical questions, it may cast light on one of Mencius' most important and most cryptic statements. When asked to describe what were his strengths, he replied that he had an insight into words and was good at cultivating his floodlike *ch'i*, a force born of accumulated righteousness that both empowers correct action and brings one into harmony with the universe (2A2). Insight into words may refer, among other things, to his ability to give true descriptions, to see what are the truly relevant aspects of some complex event. Such insight would, then, link up closely with his ability to cultivate *ch'i*. *Ch'i* would be evident in the spontaneous reaction that arose when one clearly described a situation and would be nurtured by the description. Mencius' depiction of his two strengths may further fill in our view of a perspective where the crucial ability is to be able to describe correctly a situation—to see that the person facing you is an "outcast," not a king, or an endangered woman, not a sister-in-law. Once the description is obtained the needed power to act will come, indeed will arise spontaneously with the description.

Such an approach is an interesting and powerful one. Moreover, an interesting variation of it appears in Western thought, notably in the Thomistic tradition, and has even been commended to modern ethicists by at least one notable contemporary philosopher (Anscombe 1958). Mencius, however, never works through either the theoretical or practical subtleties of this position so that we never see him struggle with its most pressing difficulties.

Aquinas, for example, considers a case where a judge, describing an event in one way, wants to kill a criminal, while the criminal's wife, describing the event in another context (that of the criminal's family), wants him released or at least not killed. (S.Th. I, II, 19,10) Such a case is more vexing than the case either of the sister-in-law or the king, and it shows some of the difficulties this approach generates. This case could be "solved," as were the other cases, by establishing a hierarchy among the possible descriptions of the act so that one takes precedence over the other. But the case of the judge and the wife makes evident, as the other two cases do not, that significant disagreements can exist about the appropriateness of the hierarchy of descriptions that is employed. Indeed, Mencius and Aquinas might disagree on this very case because the contexts in which they operate and therefore the principles they use to establish the hierarchy differ so much. It remains unclear, then, that this approach can really solve many of the most difficult problems, although it can sensibly deal with some of them.

That Mencius is able to produce no fully convincing resolution of this problem is unsurprising. The problem of relativism as it is focused in the conflict of Mencius' two cosmogonies represents, I think, one of the most vexing human problems. Indeed, it may be the single most crucial and difficult problem that arises from the breakdown of a close relationship between cosmogony and ethics.

Let us end by investigating that problem as it appears in one particularly relevant and revealing modern formulation. Such an investigation involves a short swim in the, to me, murky waters of Hegelianism. My understanding of Hegel is imperfect (to indulge in a charitable judgment about myself), but the Hegelian point is important enough to make the attempt. It can shed light on Mencius' problem and even clarify the later uses and criticisms of Mencius' ideas in China. Moreover, it can illuminate some important general issues that arise both from cosmogony's relationship to ethics and from the breakdown in any simple relationship between the two.

A Hegelian Swim

Hegel can be said to have spotted a distinctive problem which he focused in two terms, *Sittlichkeit* and *Moralität*, what I will call ethos and morality. Put in terms of what each "commands," the distinction between the two is as follows. Ethos enjoins us to be what we already are, to follow out faithfully the obligations laid on us by our participation in a particular community that in most ways exists apart from us. To follow ethos is to attain an expressive fulfillment of ourselves that joins us with objective standards, with others in our community, and with a whole that is larger and more enduring than we are. Morality enjoins us to follow out those obligations that arise from rational reflection on ourselves as individual, rational wills. To follow morality is to attain a fulfillment of ourselves as free, rational agents. These perceived obligations give birth to needs, to felt desires to realize certain goals that result from our reflection upon our human nature.[10]

For Hegel, the ideal state is a harmonious marriage of the two, a situation where ethos spontaneously produces actions and inclinations that also reflect morality. The modern problem is alienation, a state which arises when the living practices that constitute the ethos are perceived as inadequate reflections of moral norms, when ethos's cultural expression and morality's rational need are at war so that ethos seems to be foreign, irrelevant, or wrong. This problem is particularly taxing, Hegel thinks, because all action must be situated in a particular community and yet must also rest on norms perceived as being more than just the inventions of that particular community. Once the assurance is lost that social practice can be justified by its relationship to a cosmic order, the problem of how to employ both morality and ethos in a way that avoids alienation becomes acute.

Mencius' ideal can be said to resemble Hegel's marriage of morality and ethos. Similarly, one major problem he faces is the alienation that Hegel sees as inevitable when morality and ethos are split. Such alienation seems to exist for many in Mencius' time (3B9). Mencius' job, then, is to justify the ethos of the sages' ritual order by recourse to the morality that arises from the characteristics of the human nature that Heaven caused. The Hegelian formulation does capture, then, the dilemma Mencius faces. Let us look more closely now at how it focuses the particular issues that cosmogony's relationship to ethics raises.

A cosmogony can justify an ethos by declaring that the ethos's "rules" reflect the way things are. A social order with its particular duties and rewards, its differentiated web of obligated actions, is seen as an integral expression because it reflects how the world is or should be. Mencius attempts such a justification when he combines the sagely-rule and the Heavenly-mandate cosmogonies. He employs it when, for example, he uses organic metaphors to describe society and legitimatize particular social rules and social distinctions, such as that between those who labor with their hands and those who labor with their minds (3A4).

But both alienation's presence and problems in validating the sagely-rules cosmogony also lead Mencius to attempt to justify social rules not in terms of what they reflect but in terms of how they fulfill rational human needs. Society's order is to be seen as an instrument that fulfills human needs as they are known from reflection on human nature. Rules that are given, sagely rules for example, must then justify themselves as fulfillers of human needs rather than as mirrors of how things are or must be.

This problem leads Mencius to split off the sagely-rules and Heavenly-mandate cosmogonies, to focus on the latter, and to reformulate Heaven's actions in terms of the causation of certain characteristics of human nature. Human needs can now be discovered by reflection on human nature and people can know they are legitimate because

Heaven is the source of human nature. Sagely rules must then be justified in terms of the needs that reflection makes clear are part of human nature. The sagely-rules cosmogony must now be justified in the terms presented by the reformulated Heavenly-mandate cosmogony.

That situation leads, however, to a major problem. Reflection may well show that an ethos other than the sagely ethos can better fulfill the moral needs and requirements that can be discerned through reflection on one's Heaven-given nature. This situation leads to the "revolutionary" side of Mencius. That side argues reflection on one's nature may show that a particular social order is not justified by morality and that regicide and revolution are valid. In Mencius himself these revolutionary ideas seem to be relevant only to societies that do not incarnate the rules given by the sages. Indeed, he thinks that only the sagely ethos can fulfill moral needs.

But the abstract position entails ideas that are far more radical than just those that Mencius presents. One must test every ethos by morality, and such testing could lead to the validation of an ethos, a social order, fundamentally different from the one the sages designed. Mencius' failure to recognize that implication of his position means that, in Hegelian terms, he has not recognized the full weight of the dialectical situation in which he exists.

Indeed, Hegel might well argue that Mencius' dilemma resembles the situation he thought arose in the last days of classical Greek civilization. A real marriage between morality and ethos had existed, but the marriage was bound to fail because morality had not yet been given its real due. For Hegel, then, Mencius can not possibly solve the dilemma, given the tools he had. Later, Confucians, or Communists for that matter, might be said to have solved it, drawing those conclusions that Mencius' position entails but that he draws back from. That claim is, of course, problematic, but the Hegelian framework does allow us to focus in a productive way the tensions that appear in Mencius' treatment of his two cosmogonies as they relate specifically to questions about ethics.

Such ideas show, I think, that we see in Mencius a particularly striking example of the kind of thinking that arises from the attempt to repair the damage that occurs when a strain, focused in the problem of relativism, develops in the relationship between cosmogony and ethics. Mencius can hardly be said to have solved the problems that result from that strain, even if we stay completely within the framework of his own ideas. But one would, I think, be hard pressed to argue that anyone has solved such problems. Perhaps his hopes as well as his successes, failures, and confusions can tell us something important about those problems. But that is a topic for another time and place.

Notes

1. These ideas relate, of course, to the notion of an axial age or breakthrough that produces a new kind of inquiry. That "theory" is, however, still roughhewn at best and involves a set of hefty claims most of which can be queried and all of which are suspect when they make the "period" too closely resemble our own. For a good collection of articles on the idea, see *Daedalus* 104, 2 (1975). Note especially the papers by Momigliano, Weil, Schwartz, Humphreys, and Dumont.

2. These are representative examples; other interesting references are 1B16; 2B13; 4A1, 17, 28; 4B1, 26, 28; 5A6–7; 6B2, 15; 7A1–4; 7B4. My references to Mencius use the Harvard-Yenching text numbers, e.g., 3A4 equals book 3, part A, section 4. Probably the two most reliable translations of the *Mencius* are Lau and Legge.

3. The notion of open and locative religious visions is developed in Smith (1970), but he applies the notion only to Western religions.

4. It should at this point be noted that the more lavish kind of cosmogonic myths that are evident in many other traditions are strikingly absent in Mencius. This fact has led some to argue that in the tradition Mencius represents we find mytho-poetic restatements of cosmogonic myths or a cosmogonic tradition that just recalls but does not embody fullblooded cosmogonies. (See, for example, Maspero 1978.) Such may be the case, but I find it difficult to know how one could prove the point, with Mencius at least, without presuming the existence of full cosmogonic myths in order to show that they have been transformed or recalled. What we know clearly is that such lavish myths are not obviously central to his thought. The cosmogonic myths that *are* obviously important to his thought are those that describe the origin of the basic social structures and rules that guide human action.

5. Graham (1967) examines the intellectual crisis to which Mencius' idea of human nature responds and that analysis can, I think, be usefully correlated with the general problems treated here.

6. Mencius' use and analysis of extension is very complex. All of 6A relates to it but also see 1A7; 4A10; 4B14, 18–19, 28; 7A15–16, 48; and 7B31, 35, 37, 39.

7. Verification of this distinction is impossible, but see Mencius' comments on *ch'i* and the unmoved mind in 2A2 and the sage in 7B25 as possible examples of esoteric teachings.

8. 4A17 extends the image to the helping of the drowning empire with the Way, and that may either be the story's point or a later addition. The context for 1B8 is a famous historical incident.

9. Some version of the idea of the rectification of names exists at least since the time of Confucius, but it probably receives neither a sophisticated nor a central place until the time of Hsün Tzu.

10. My interpretation of Hegel, particularly on the points discussed here, is much indebted to Taylor's (1975) work. Incidentally, I realize a whole set of issues are finessed by my loose use of the word "needs" when it refers to the results of rational reflection on human nature.

References

Anscombe, G. E. M.
1958 "Modern Moral Philosophy." *Philosophy* 33: 1–19.

Aquinas, Thomas
S.Th *Summa Theologiae*

Girardot, Norman
1976 "The Problem of Creation Mythology in the Study of Chinese Religion." *History of Religion* 15, 4: 289–318.

Graham, A. C.
1958 *Two Chinese Philosophers: Ch'ëng Ming-tao and Ch'ëng Yi-chuan*. London: Lund Humphries.
1967 "The Background of the Mencian Theory of Human Nature." *Tsing Hua Journal of Chinese Studies*, n.s. 6, 1–2: 215–74.

Maspero, Henri
1978 *China in Antiquity*. Trans F. A. Kierman, Jr. Amherst: University of Massachusetts Press. (Original edition: 1927/1965.)

Mencius
1970 Trans. James Legge. New York: Dover Books; reprint of 1895 ed.
1970 Trans. D. C. Lau. Baltimore: Penguin Books.

Smith, Jonathan Z.
1970 "The Influence of Symbols on Social Change: A Place on Which to Stand." *Worship* 44: 457–74.

Taylor, Charles
1975 *Hegel*. Cambridge: Cambridge University Press.

Yearley, Lee H.
1975a "Mencius on Human Nature: The Forms of His Religious Thought." *Journal of the American Academy of Religion* 43: 185–98.

1975b "Toward a Typology of Religious Thought: A Chinese Example." *The Journal of Religion* 55: 426–43.

13 Cosmogony, Contrivance, and Ethical Order

Robin W. Lovin

From an early time in the Christian tradition, ideas of the moral life have been linked to ideas about creation. Actions might be praised as "according to nature" or condemned as "contrary to nature" in phrases borrowed from Greek thought, but the concept of nature was modified in important ways by the Hebrew image of the creator God of Genesis. Indeed, this link between the biblical cosmogony and the Greek ideal of following nature may account for the earliest philosophical formulations of Western natural-law theory. As a pattern of argument for justifying human conduct, the *lex naturalis* owes as much to theology and cosmogony as to jurisprudence (Koester 1968).

The line of argument from created order to moral order, however, is not constant. It varies significantly with changing understandings of nature. These modifications are most dramatic during the rise of modern science and empiricist philosophy, which destroyed the fundamental image of natural order that had prevailed from Aristotle through the Renaissance (Wildiers 1982). For some interpreters, these changes effectively mark the end of natural-law ethics, but the concern for the connection between nature and ethics nevertheless remains a powerful intellectual force. Whether at one extreme the moralists continue to argue for fundamental moral rules that are part of the order of nature, or at the other, positivist pole they argue that nature is so devoid of guidance that all rules must be established by human choice, the arguments that theologians and philosophers offer for one or another system of ethics are determined in part by their ideas about what we know (and what we cannot know) about natural reality.

This essay is a study of the connection between nature and ethics in one quite specific moral system, the theological utilitarianism of William Paley (1743–1805). Paley and his British predecessors of the seventeenth and eighteenth centuries provide a striking study of the reassertion of a natural foundation for ethics within the limits of empirical science. While the turn away from the Aristotelian notion of ends, goals, or perfections inherent in natural objects initially led many to doubt that a purely rational morality was within the grasp of the human intellect, the latitudinarian theologians and the empiricist philosophers agreed on the possibility of a rational morality that did not rest on revealed religion. This made them particularly careful to see that the presuppositions they did use were coherent with the prevailing views of human knowledge.

There was, as we shall see, more than one way to do this, but Paley's system proceeds with striking clarity. A study of the structure of his thought may help us both to understand the alternatives his contemporaries had and to make the structure of the eighteenth century British debate available for comparative study. What Paley wrote in the rectory at Bishop Wearmouth has a distinctive flavor of Anglican rationalism, but if we understand the underlying problem which the moral theologians of his day confronted, we may find that similar things have been said in other times and places.

Cosmogony

Paley begins his *Natural Theology* with a well-known analogy (Paley 1821a, 1–3):

> In crossing a heath, suppose I pitched my foot against a *stone*, and were asked how the stone came to be there: I might possibly answer, that; for any thing I knew to the contrary, it had lain there for ever; nor would it be very easy to show the absurdity of this answer. But suppose I had found a *watch* upon the ground, and it should be inquired how the watch happened to be in that place; I should hardly think of the answer which I had before given,—that for anything I knew, the watch might have always been there.

The reason for this difference, Paley says, is simple and univocal. It is not simply that the watch evidences an intricacy which is missing in the stone. The heath-tramping observer also recognizes that this intricacy serves a purpose.

> Yet why should not this answer serve for the watch as well as for the stone? . . . For this reason, and for no other, viz. that, when we come to inspect the watch, we perceive (what we could not discover in the stone) that its several parts are framed and put together for a purpose. . . .

So the watch exhibits not only order, but an order that serves as means to an end. It is that aspect of the watch that makes it absurd to suppose that it just happened to be there, marking off the hours without any intrusion of human hand or mind.

> The mechanism being observed (it requires indeed an examination of the instrument, and perhaps some previous knowledge of the subject, to perceive and understand it, but being once, as we have said, observed and understood), the inference we think is inevitable, that the watch must have had a maker: that there must have existed, at some time, and at some place or another, an artificer or artificers who formed it for the purpose we find it actually to answer: who comprehended its construction and designed its use.

Now this same evidence of order adjusted to a purpose, Paley suggests, is evident throughout nature, and it warrants the same conclusion. The workings of the eye surely are a more ingenious ordering of materials to the end of vision than even the workings of a watch are to its purpose. If the notion that the watch has no maker is absurd, then the atheist's presumption that the eye and the hand, the intricate adaptation of plants and animals to their environments, and the harmonious order of nature as a whole just happened to be there is even more preposterous, "for every indication of contrivance, every manifestation of design, which existed in the watch, exists in the works of nature; with the difference, on the side of nature, of being greater and more, and that in a degree which exceeds all computation" (Paley 1821a, 14).

It is difficult for a twentieth-century reader, steeped in the heritage of Darwin and the probability calculations of high-energy physics fully to appreciate the impact of Paley's arguments in their own day. While we should no doubt be quite startled if a few billion atoms in a nearby park happened to arrange themselves into a Timex just as we were setting out on our afternoon walk, we would have to admit that with all the atoms there are in the universe, colliding in all the ways they can collide, the thing is just possible. That the order of nature which we do find is in some sense the product of chance, we find not remarkable at all. Given the millions of years of evolution and the mechanisms of natural selection (which we understand rather less than Paley understood his watch), the emergence of this order of squirrels, grass, trees, insects, and Dutch elm disease out of some Precambrian organic soup is perfectly plausible.

For Paley, however, it is not so. While he anticipates Darwin's hypothesis at several points (e.g. 1821a, 51–52) he raises it only to reject it, and we must suspend our reliance on it, too, if we are to understand the point of Paley's system of natural theology and ethics. It is this firm conviction of the evidence of design in nature that transforms Paley's cosmology into a cosmogony and allows him in turn to draw the connection between prescriptions for action that maximize human happiness and the legitimating order of reality which is the basis for his ethics. It is the particular purposes Paley discerns in nature that shape his theological utilitarianism, and not just his admiration for the order he observes. It is not conformity to an existing cosmos that decides moral questions for Paley, but congruity with the purpose which animated the great Watchmaker who created the system. Though the *Natural Theology*, first published in 1802, came last among his published works, it explicates the logical foundation of the rest, as Paley explains in his preface:

> The following discussion alone was wanted to make up my works into a system: in which works, such as they are, the public now have before them, the evidences of Natural Religion, the evidences of revealed religion, and an account of the duties that result from both. It is of small importance that they have been written in an order the very reverse of that in which they ought to be read. (1821a, vii)

For present purposes, then, we will read Paley in the order he proposed. We will examine first the cosmogony he suggests in his *Natural Theology* and then turn to the earlier *Principles of Moral and Political Philosophy* (1785) for an account of the duties that follow from the divine purpose that shapes creation.

The feature of the natural world which Paley suggests ought to remind us of the creative work of an intelligent designer is the way things work together as means to an end, just as human technicians use the material properties of things to solve problems.

> In dioptric telescopes, there is an imperfection of this nature. Pencils of light, passing through glass lenses, are separated into different colors, thereby tinging the object, especially the edges of it, as if it were viewed through a prism. To correct this inconvenience, had long been a desideratum in the art. At last it came to the mind of a sagacious optician, to inquire how this matter was managed in the eye; in which there was exactly the same difficulty to contend with, as in the telescope. His observation taught him, that in the eye, the evil was cured by combining lenses composed of different substances, i.e. of substances which possessed different refracting powers. Our artist borrowed thence his hint; and produced a correction of the defect by imitating, in glasses made from different materials, the effects of the different humours through which the rays of light pass before they reach the bottom of the eye. Could this be in the eye without purpose, which suggested to the optician the only effectual means of attaining that purpose? (Paley 1821a, 17–18)

Much of Paley's *Natural Theology* is taken up with such examples, so that the text is a compendium of natural science, physiology, and astronomy as known at the time, along with the results of Paley's own indefatigable note-taking.

> In the trussing of a fowl, upon bending the legs and thighs up towards the body, the cook finds that the claws close of their own accord. Now let it be remembered, that this is the position of the limbs, in which the bird rests upon his perch. And in this position, it sleeps in safety; for the claws do their office in keeping hold of the support, not by the exertion of voluntary power, which sleep might suspend, but by the traction of the tendons in consequence of the attitude which the legs and thighs take by the bird sitting down, and to which the mere weight of the body gives the force that is necessary. (1821a, 163)

The examples are multiplied to the point of satiation, and perhaps a little beyond, despite Paley's expressed conviction (1821a, 60) that *even one* such observation would be sufficient evidence of contrivance to refute the atheist. The illustrations change, then, but the argument does not: Design implies a designer. The adjustment of means to ends in nature implies an intelligent use of materials with a view to achieving those ends.

The idea of a natural ordering of things by the purposes or ends they serve is not so apparently different from the idea, which we could trace back through Aquinas to Aristotle, that every species in nature has its appropriate perfection, an inherent final cause which determines the course of its development and provides the standard by which we measure the achievements of an individual exemplar. Bacon had declared this teleology unnecessary; Hobbes pronounced it nonexistent, and Hume's skepticism about all causal connections rendered it unknowable. Paley reasserts it, more on commonsense grounds than on the basis of a refutation of his predecessors.[1] It is essential to his argument, of course, but it poses a problem that the more straightforward empiricists did not have to face. What, if any, is the relationship between all these particular systems of means and ends? Are the examples of design which can be multiplied so extensively themselves the instruments of a larger intention, or is the interlocking of designs itself the necessary framework for every particular purpose, the final standard to which every design must sooner or later conform?

As teleology was not a new idea, perhaps this was not a new problem. Paley, at any rate, treats it as the point at issue between the Stoics and the Epicureans, and early in his career it provided the occasion for an apologetic argument to demonstrate the failure of non-Christian, classical thought to solve the problem. Granting that one is to live always "according to nature," does this mean that one ought to develop the virtues of wisdom, courage, temperance, and justice in perfection of particular human capacities? Or is there an overarching intention, such as happiness, that one should pursue according to a well-informed view of how things in the world actually work, using human capacities simply as means to a larger end? Because virtue does not always lead to happiness, while the intelligent pursuit of happiness is not always virtuous, Paley believed he had concrete evidence of the failure of a purely secular, nontheistic view of the world to provide adequate moral guidance. Kant, toward the end of the eighteenth century, would insist that one can resolve the antinomy between Stoicism and Epicureanism only by acknowledging virtue as the necessary condition for a person's worthiness for happiness (Kant 1956, 114–17). Paley, at the beginning of his career, argued that God's judgment on vice provided the necessary link between virtue and happiness that the ancients could not find.

> It was reserved for one greater than Zeno to exalt the dignity of virtue with its utility and by superinducing a future state, to support the paradox of the Stoic on Epicurean principles.[2]

God's power to resolve the antinomy between virtue and happiness rests on the divine power to punish and reward us, especially after this life. Hence, Paley candidly admits, the chief motive to human virtue is "private happiness," though one must sometimes take a rather long view of the matter for that motive to be effective (Paley 1977, 54–57).

What the Christian cosmogony provides that the ancients did not see is the idea of a creator whose comprehensive intention could explain all the diverse designs to which we must conform if we try to live "according to nature." It remains for Paley to establish from the evidences of design he has assembled, first, that the intentions of the creator are still effective in the natural order, and, second, that this comprehensive intention has something to do with human happiness.

Paley's insistence that the natural order continues in an immediate relationship to the explicit, knowable purposes of the original design distinguishes his Anglican natural theology from that of the English Deists, who acknowledged the evidence of design but often denied that God exercises any power over nature's continuing course. (Later Deists especially denied a future state of rewards and punishments.) The extensive defense of miracles in Paley's work, as in Joseph Butler (1692–1752) and Samuel Clarke (1675–1729), is more than a shibboleth to uphold a claim to Christian orthodoxy. These theologians required the possibility of miracles because for them it is the immediate will of God, rather than an initial divine blueprint, which guarantees the coherence of human moral life.

Paley establishes this point in his *Natural Theology* by an elaboration of the example of the watch. Suppose, he suggests, that we find that our newly discovered watch actually has the power to reproduce itself—"the thing is conceivable"—and that we conclude from this that the watch in hand was made, not by a watchmaker, but by another watch (1821a, 6–7). The discovery would not alter the basic point that at some point the series of watches depends on a first creator, and the dependence of each generation in the series on the original designer is no less than that of the first self-reproducing watch itself.

> Contrivance must have had a contriver; design a designer; whether the machine immediately proceeded from another machine or not. . . . No tendency is perceived, no approach towards a diminution of the necessity. It is the same with any and every succession of these machines; a succession of ten, of a hundred of a thousand; with one series as with another; a series which is finite as with a series which is infinite. (1821a: 11)

Miracles are possible because distance from the original creative act does not attenuate nature's dependence on the creator's will. If miracles do not happen very often, that argues only that God's purposes are usually better served by a natural order that is self-regulating and that moves and changes in predictable ways. Samuel Clarke, whose work Paley admired and used regularly during his career as a tutor at Cambridge, expressed it thus:

> The Course of Nature, truly and properly speaking, is nothing else but the *Will of God* producing certain effects in a continual, regular, constant, and uniform manner. (Cited in LeMahieu 1976, 93)

Clarke's position is not quite a doctrine of continuous creation, but it comes closer to that than to the Deist's image of the divine watchmaker who sets the mechanism ticking and walks away from it.

For Clarke and Paley, the very constancy of the natural order on which the Deists so much depended suggested God's continual attention to the system. So understood, both the claims of miraculous intervention and the rarity of verifiable miracles provided evidence for the orthodox conception of God (LeMahieu 1976, 98). Since miracles are defended as occasional and infrequent interruptions of an otherwise constant order, we find in Paley no trace of a periodized or evolutionary view of history. The biblical era is not conceived as an "Age of Miracles" now past, nor is the present time an age of reason in which persons of faith no longer need the extraordinary signs and wonders that formerly were necessary to induce belief. "Paley believed that, in their basic assumptions and fundamental rationality, the early followers of Jesus were not unlike natural theologians of the eighteenth century and that, consequently, extraordinary occurrences such as miracles provided these disciples with convincing proofs of Christ's divinity" (LeMahieu 1976, 111).

Contrivance

So Paley, in company with the Anglican rationalists of his day, offered an interpretation of the design of nature that linked the growing body of scientific thought to the traditional notion of a divine creator, now conceived as a rational artificer who makes good use of his materials and intervenes only rarely to affect directly the course of events. What distinguished Paley from his contemporaries was the force of his conviction that the divine intention behind this stable, well-designed natural order could be reduced to a simply stated and readily recognized concern on God's part to secure human happiness.

Here Paley's optimism and fundamental good humor come to dominate the exposition in an affirmation of the goodness of life and the happiness of the human condition. How one reacts to this probably depends as much on one's disposition as on the course of the argument, but for Paley the evidence of this divine purpose clinches the argument from design. The wanderer who finds a watch on a heath infers a watchmaker because the ordered system of springs and gears serves to keep time. Similarly, Paley treats the natural order as he might some strange device whose purpose he did not immediately understand. That is, he hypothesizes a purpose and asks whether the functions that he can observe are compatible with it. "When God created the human species, either he wished their happiness, or he wished their misery, or he was indifferent and unconcerned about either" (1821a, 364).

If God intended human misery, he might have imposed it much more effectively. "He might have made, for example, every thing we tasted bitter; every thing we saw, loathsome; every thing we touched, a sting; every smell, a stench, and every sound a discord" (1821a, 364). This is clearly not the case. Indeed, the sources of sensual and spiritual gratification are too numerous to be the result of accident. Observation, once again, leads to the happy conclusion that the world has a design and that, indeed, that design has a designer who intends our good.

> The same argument may be proposed in different terms, *thus*: Contrivance proves design, and the predominant tendency of the contrivance indicates the disposition of the designer. The world abounds with contrivances; and all the contrivances which we are acquainted with, are directed to beneficial purposes. Evil, no doubt, exists; but it is never, that we can perceive, the *object* of contrivance. Teeth are contrived to eat, not to ache. . . . This is a distinction which well deserves to be attended to. (1821a, 365)

Paley does not demand that the divine design be foolproof. If teeth occasionally ache, if streams flood and bones break and snakes bite, that is not good, but even an all-wise designer cannot make the whole of nature impervious to misuse and mishap. Our doubts on this point can be largely resolved by taking a somewhat wider view of matters. To say that God intends human happiness is not to say that all things can be judged exclusively from the standpoint of human convenience. From the rattlesnake's perspective, after all, venom is a good thing, and even the reptile's usual prey may be better off quickly poisoned than swallowed alive (1821a, 367–68). Paley's emphasis on happiness is never narrowly anthropocentric. When he exults, "It is a happy world after all!" he counts in his sum of gratifications the bees' enjoyment of flowers, the contentment of a purring cat, and the leaping of fish in the water. (These fish, Paley asserts, "are so happy, that they know not what to do with themselves" [1821a, 358]).

Above all, however, Paley settles doubts by emphasis on his constant theme of intention and purpose. Nature is filled with superfluous pleasure, and the point to the mechanisms that give rise to it must be the creator's intent to provide each species with satisfactions that go well beyond the minimum necessary to sustain life. The accidents and pains that do befall us are not denied, but Paley defies his readers to come up with an instance in which nature devises gratuitous pain. The evidence of superfluous pleasure, however, is all about us, and only the most morose observer can ignore it.

Thus we arrive at a teleology of nature suited to the age of invention. The purposes nature serves are not identified in Aristotelian fashion, by examining the perfection of each thing individually. Rather, nature as a whole is taken to be an engine designed for a total effect. By observation, we can determine what results the creator intended and distinguish these from the accidental outcomes that have no part in the purpose behind the design. Such observations convince us of the goodness of the creator, and that is the intent of Paley's natural theology. They also help us better to operate those parts of the machine that are under our own control, and that is the starting point for Paley's moral thought.

Ethical Order

As a starting point for ethics, the argument that God consistently wills human happiness allows Paley to affirm the orthodox position that moral rightness depends, immediately or indirectly, on the will of God.[3] There is no great difference between the moral obligation to obey God's commandments and other sorts of obligations. All in the end rest on the authority of force. God's power to oblige us in a specially comprehensive way is based on his capacity to punish and reward us, during and even after this life. We have already noted that Paley argued that the concept of divine retribution overcomes the intellectual tension between Stoicism and Epicureanism. He is also quite clear about its role in his own moral system.

> If they be *in fact established*, if the rewards and punishments held forth in the Gospel will actually come to pass, they *must* be considered. Those who reject the Christian religion are to make

> the best shift they can to build up a system, and lay the foundations of morality without it. But it appears to me a great inconsistency in those who receive Christianity, and expect something to come of it, to endeavor to keep all such expectations out of sight in their reasonings concerning human duty. (Paley 1977, 59)

Paley does not linger on this fundamental theological point, however. The important practical question is to determine what God wills, and the principle that guides Paley's cosmogonic speculations is also paramount in the normative inquiry: "Contrivance proves design; and the paramount tendency of the contrivance indicates the disposition of the designer" (Paley 1977, 60). On this basis, we would expect that God who created the immense variety of delights and gratifications that pass in review in the *Natural Theology* would uniformly will that action which, among the available human choices, makes for the greatest happiness. This utilitarian principle, in fact, provides the fundamental normative guidelines in the *Principles of Moral and Political Philosophy*. As a theologian, Paley of course affirms the direct statement of God's will in scripture, but he expects this to correspond exactly with the outcome of a reasoned investigation of the natural order. The normative part of his ethics, then, concentrates on two themes: (1) scriptural injunctions and traditional moral precepts, rightly understood, make for the maximization of human happiness; and (2) when a disputed question arises in morals, it is to be settled by determining which course of action or which interpretation of the rule will lead to the greatest happiness.

> We conclude, therefore, that God wills and wishes the happiness of his creatures. And this conclusion being once established, we are at liberty to go on with the rule built upon it, namely, "that the method of coming at the will of God, concerning any action, by the light of nature, is to inquire into the tendency of that action to promote or diminish the general happiness." (1977, 62)

Because of the close comparison between this interpretation of the will of God and the Benthamite principle of "the greatest happiness for the greatest number," Paley's position is aptly called "theological utilitarianism." This remarkable convergence of ideas appears to be due not to any direct influences between Bentham and Paley, but to extensive borrowing by both from earlier writers who contributed to the growth of utilitarian thought (Albee 1962, 161–82). Both authors assert that actions are to be evaluated according to their tendency to promote the general happiness. For both, happiness consists in the excess of pleasure over pain, and all pleasures are qualitatively equal, differing only in duration and intensity (Paley 1977, 34).[4]

Paley differs from Bentham, however, in his rather conservative application of the greatest-happiness principle. While Bentham proposed it as the foundation for a comprehensive program of social reform, Paley uses it sparingly. Considerations of utility are used to resolve ambiguities in otherwise settled practices. Appeal to the general happiness helps us decide whether to enforce a promise that was made on the basis of mistaken information. "The case of erroneous promises is attended with some difficulty: for to allow every mistake, or change of circumstances, to dissolve the obligation of a promise, would be to allow a latitude, which might evacuate the force of almost all promises; and, on the other hand, to gird the obligation so tight as to make no allowances for manifest and fundamental errors, would, in many instances, be productive of great hardship and absurdity" (1977, 101). The choice between harsh and lenient punishment for debt may be based on a consideration of the general effects of this policy on commerce, as well as on the happiness of the individual debtor. A tradesman who depends on credit for his business "will deem it more eligible, that one out of a thousand should be sent to gaol by his creditor, than that the nine hundred and ninety-nine should be straitened and embarrassed, and many of them lie idle, by the want of credit" (1977: 115).

Paley's respect for established laws and practices makes him clearly what modern parlance calls a "rule-utilitarian." He recognizes the importance of general rules in establishing the framework of expectations that allows rational persons to plan their lives and guide their conduct, and he employs his own version of the utilitarian calculus only to defend a general rule or to settle a dispute concerning its application, never to argue against the idea of having moral rules. Indeed, as the influence of the French Revolution became generally felt, Paley's political writings became a utilitarian apology for the prevailing order in England. He concludes his essay on *Reasons for Contentment, Addressed to the Laboring Part of the British Public,* with the almost Burkean contention that "changes of condition, which are attended with a breaking up and sacrifice of our ancient course and habit of living, never can be productive of happiness" (1821b, 479).

Moral Sentiments

For all their political and temperamental differences, however, Paley and Bentham are united on one point that distinguishes their ethics from the anticipations of utilitarianism in their seventeenth- and eighteenth-century predecessors. The production of human happiness becomes important as a justification for moral choices in British theology and philosophy shortly after the end of the Puritan Commonwealth, when a new attentiveness to the affective side of the moral life began to anticipate the moral-sense theory of the early eighteenth-century. While it is usual for modern scholars to label Jonathan Edwards, Joseph Butler, and other theologians who adopted these themes as "theological utilitarians" (cf. Frankena 1960, vii-viii), it was not until the very end of the eighteenth century that British moralists began to exclude all considerations except anticipated consequences for human happiness from moral decision-making. Bentham and Paley, both writing in the 1780s, offer early formulations of ethics based on a consistent, unbending adherence to the maxim that "Whatever is expedient is right" (Paley 1977, 62).

One of the factors which must be discounted in this utilitarian calculus is, of course, the immediate, affective reaction to a deed which provided the foundation for moral-sense theories of ethics. Bentham treats the appeal to moral sense as a verbal camouflage, designed to obscure the fact that the speaker is really pursuing a subjective preference (1948, 17n). Paley similarly dismisses moral feelings as expressions of prejudice and habit (1977, 26–30). There is no reason why we should ascribe any special moral authority to these habitual responses, and certainly no reason why we should expect to find in them the uniformity that is necessary for moral judgment. By contrast, the estimate of an action's tendency to produce happiness is objective, and if it is not completely reliable, it can at least be debated, while the feelings of different observers cannot. Attentiveness to the principle of utility allows us to look behind immediate, affective moral responses. Once we have grasped the principle, the problem of the moral sense is no longer a key issue on which the validity of our moral system must turn. The reliability and uniformity of our moral responses is now a mere curiosity, an investigation for those "more inquisitive than we are concerned to be, about the natural history and constitution of the human species" (1977, 33).

A consistent utilitarian must, of course, reject appeals to moral sense, instinct, or intuition and adhere rigorously to the greatest-happiness principle. It is precisely the ordinary observer's immediate responsiveness to the special claims of loyalty, gratitude, or indebtedness which play havoc with the attempt to reduce all moral requirements to the single obligation to produce the greatest good for the whole. Later utilitarians have been as steadfast in defending this reductionism as their critics have been in denying it (Ross 1930, 19; Williams 1981, 40–53).

Bernard Williams notes that when a consistent utilitarianism confronts the multiplicity of moral dispositions that people do in fact experience, "there is a very great pressure . . . for it to reject or hopelessly dilute the value of these other dispositions, regressing to that picture of man which early utilitarianism frankly offered, in which he has, ideally, only private or otherwise sacrificable projects, together with the one moral disposition of utilitarian benevolenc" (Williams 1981, 53). In Paley's case, this pressure expresses itself in his readiness to consign moral feelings to the realm of prejudice and habit. To dismiss *this* part of the human constitution, after so many pages and paragraphs on the workings of the eye and the mechanics of the joints, marks a curious departure from his argument from design. It appears that Paley's utilitarianism, when combined with his interest in "the natural history and constitution of the human species," leads him into a dilemma. He can maintain his consistent utilitarianism only by rejecting some evidence of the sort that was decisive for the arguments that established the utilitarian position in the first place.

Moral feelings *are* a part of the human constitution, Paley acknowledges, and no British moralist of his century would disagree. However, in order to exclude these sentiments from any decisive role in the making of moral judgments, Paley is obliged to turn his back rather abruptly on the whole course of his investigations into natural theology. To be a consistent utilitarian, he must either ignore the moral sentiments, or concede that he has here stumbled onto an element in the design that is superfluous.

It was precisely this consideration that led Butler to reject utilitarianism in favor of attention to the moral sense:

> As we are not competent judges, what is upon the whole for the good of the world; there may be other immediate ends appointed us to pursue, besides that one of doing good, or producing happiness. Though the good of the creation may be the only end of the Author of it, yet he may have laid us under particular obligations, which we may discern and feel ourselves under, quite distinct from a perception, that the observance or violation of them is for the happiness or misery of our fellow-creatures. And this is in fact the case. For there are certain dispositions of mind, and certain actions, which are in themselves approved or disapproved by mankind, abstracted from the consideration of their tendency to the happiness or misery of the world. (Butler 1969, 376)

The sharp difference between Butler and Paley on this point marks the distinction between a theological humanism, which Butler surely represented, and true theological utilitarianism, which appears on the scene only with Paley. Paley's readiness to dismiss the affective side of the moral life makes it clear that the design of nature is not, in itself, a source of moral guidance. The design is only a clue to the intentions of the designer, and it is that unifying purpose that decides what is important in the ceaseless activity and the variety of motions that make up the machine.

Contrivance and Comparison

Paley's moral philosophy is just two centuries old, but it seems in large part as distant from our cultural presuppositions as the cosmogonies of ancient Greece or China. The image of an orderly universe, contrived to serve apparent ends, is hopelessly at variance with the world of random genetic changes and subatomic chaos in which the educated person of the late twentieth century lives. Yet there is also in Paley's thought a note that is familiar, and characteristically "modern." When he measures nature by its contribution to human happiness and finely adjusts institutions to serve that supreme end, he seems to belong in the world of genetic engineering and monetary policy, rather than in a culture that stands in awe of "the natural."

The tension we identified in his thought in the preceding section of this essay helps to understand this ambiguity, for Paley both acknowledges the authority of natural order and thinks he has unlocked enough of nature's secrets to take that order in charge. If his philosophy is confused, it is because he stands in a transition period from an Aristotelian world in which people were to note and respect the multiplicity of nature's ends and a technical age in which nature is to serve the ends that people choose. The contrast, however, is not merely between two historic periods. The two conflicting themes in Paley's thought are also recurrent patterns in the human effort to link morality and reality. If the contrast between them becomes particularly clear in the transition to modern ways of thinking, a preliminary identification of analogous patterns in other settings may help us to avoid the easy conclusion that the choice between them is settled by history's "progress."

The first pattern identifies moral choices as those which are responsive to the range of constraints and possibilities reality imposes on us. This pattern of responsiveness treats affectivity as an important element in moral and religious reflection, especially when we seek a method "easily applicable to the several particular relations and circumstances in life" (Butler 1969, 325). In the range of experiences that arouse in us joy, fear, defiance, or remorse, we find ourselves subject to forces and powers that are beyond our control. Religious and moral reflection on these affections need not treat them all as immediately valid, nor need they determine our actions without further examination.

To take affectivity as a starting point does, however, recognize this experience of limitation and finitude as fundamental to human life (Gustafson 1981, 196–209). Because these forces and powers remain beyond our control, religious observances and practices serve to heighten our awareness of these realities and to dispel the illusion that they can be brought under our control or driven out of our experience.

We can identify this pattern in modern thought in the moral-sense theories of the eighteenth century and in contemporary theology which grounds moral responsibility in responsiveness to the demands that our experiences of other persons and external reality make upon us (Niebuhr 1963; Jønsen 1968). It appears also, however, in older forms: in early Christian polemics against magic, for example, where the objection to the manipulation of events for limited human ends is not the "scientific" objection that it does not work, but that it is not an appropriate way to deal with the contingency and riskiness that are pervasive in our human experience (Wildiers 1982).

A theological system which begins with these experiences of contingency and finitude may be monotheistic, unifying the powers we experience in one Power, but moral thinking that begins with affectivity is almost invariably pluralistic. Precisely because we experience the conflict between the claims our different affective responses make upon us, an ethic which takes these responses seriously tends to recognize an irreducible multiplicity of human obligations. Much of the normative reasoning in such a system will then be devoted to resolving these conflicts.

The place of reason in an ethic that begins with multiple demands and powers will, moreover, be limited. Aristotle, who introduces an ethic of many virtues based on our experiences of the good man's character and choices, acknowledges also that in this realm reason does not apply with mathematical precision, and we must not demand more precision than the subject matter will admit. The eighteenth-century theorists of moral sense and moral sentiments—Hutcheson, Hume, and Smith—limited reason even more severely when they insisted, on the basis of their observations, that reason follows the passions, and not vice versa. We are linked to a world of powers beyond our control by affective ties which respond for us before we can reflect and which draw our reasoning processes after them.

That, then, is the first pattern which relates morality to reality in eighteenth-century moral thought: moral choice begins with responsiveness to the variety of constraints and possibilities our experience opens before us. Reason may guide us through the choices, but it cannot arbitrarily overrule the experience. Indeed, the largest part of moral reasoning is finding a satisfactory balance between the multiple tugs and pulls of experienced obligations.

The second pattern of thinking, which comes to dominate Paley's utilitarian ethics, relates morality to reality by evaluating and, if possible, reordering reality in terms of a dominant goal. What marks a morality as rational is its use of instrumental reason to discover the best means for attaining purposes.

The way to identify these comprehensive goals is sometimes unclear in early utilitarianism. Bentham, for example, attaches great weight to democratic decision-making, with "each one to count for one, and no one to count for more than one" (Bentham 1948). It seems, however, that harmony and consensus on these choices is guaranteed by the fundamental fact that human beings all seek happiness and that, when their minds are unclouded by prejudice and superstition, they will understand that happiness in the same way. Paley, of course, is clear that the comprehensive goal of happiness is ultimately God's choice, identifiable in the contrivances arranged in nature to secure it.

However they are determined, the identification of the dominant ends is critical to this pattern of moral thinking, for without knowledge of the ends an ordered system is meant to serve, instrumental reason is powerless. Once a purpose is known, however, nothing limits its pursuit, save conflict with another, more powerful purpose. So, in classical utilitarianism, individual goals succumb to the goals of the greatest number. So, too, in Paley's theological reconciliation of virtue and happiness, God's goal of human happiness has a preemptive claim on any resources available for its fulfillment, precisely because God commands the maximum force that can be put behind any project.

A morality which reorders nature to serve purposes, human or divine, may require us to suppress some immediate affective reactions, but it is not without its compensations. Instrumental reason overcomes superstitions and breaks the hold of worn-out traditions. For both Paley and Bentham, this purging of accumulated prejudices and false ideas was a principal benefit of the new, utilitarian moral system. Instrumental reason provides a clear and objective criterion against which individual claims and preferences can be measured. Above all, it reduces the confusing multiplicity of claims that arise in any affectivity-based ethic and replaces them with a single imperative: pursue the goal.

The contemporary successors of Paley and Bentham look for their goals less in the structures of nature than in the expression of persons' wants and desires. Their political ethics are based on careful attempts to formulate a comprehensive goal which incorporates as many individual aims as possible. That is to say that the characteristic modern uses of instrumental reason in political theory are noncognitivist. Because there is no standpoint from which to judge what persons say they want, the moral task in politics is to find comprehensive goals that satisfy as many of these wants as possible and minimize conflict over the points on which persons cannot agree (Oppenheim 1975). Rational morality is the use of our knowledge of nature to achieve ends that we have chosen for ourselves. Knowing more about reality may be a great help toward those achievements, but what we know about reality will not (the scientists and technologists hasten to assure us) tell us what ends to choose.

Noncognitivism is not, however, the only way to understand the relationship between the goal and the given reality. Paley and the classical utilitarians supposed that created nature dictated the goals as well as the means, or at least that the natural congruence of human wants assured that goals could be formulated that would satisfy everyone. The reordering of reality in accordance with recognized purposes also appears in magical practices and in the "bargaining" with superior powers that is often found in everyday religious life, even where theology stresses affective responsiveness and acceptance of one's finite place in the order of reality. Here the purposes given in reality may be assumed to be indifferent or even hostile to human wants, but there is also some scope for manipulation of one's opportunities. What is to be is not left to chance, fate, or divine decree, but is changed (or at least a change is attempted) according to human purposes. The ordinary religious practices of Buddhism, for example, recognize a host of benevolent and malevolent deities, who may be enlisted to achieve human ends, although *nibbanic* Buddhism insists that these practices are irrelevant to liberation, which comes only from apprehending one's true condition in the fixed order of existence. The cults of saints in Christianity and Islam, especially where these take forms that are disapproved by normative religious authorities, may also exemplify instrumental reason in moral thinking. A complete assessment of these systems and a comparison of them to Paley's theological utilitarianism would take us far beyond the scope of this essay, but brief mention of them should remind us that the use of instrumental reason to achieve a dominant purpose is not exclusively a modern phenomenon.

Nor has the pattern of affective responsiveness to a multiplicity of moral claims entirely disappeared. The formulations of the eighteenth-century theorists, who argued for a distinct "moral sense" by which we know these realities, are no longer adequate, but the underlying idea persists: We make our way in a world that is not designed simply to suit our sensibilities, and our moral life is primarily a response to the varied pressures of this reality, rather than a use of it for determinate ends.

We find this pattern of thinking in recent scientific thought which suggests that humanity is not entirely free to set up its own purposes in the face of nature's silence, but rather that our life is lived in ecological balance with forces whose ordering we have only begun to understand. If these scientific writers enjoy a warm reception among certain theologians, that is perhaps because Christian reformers have usually sought to limit the magical and manipulative uses of religion, with their characteristic instrumental rationality, and to replace manipulation of circumstances with a reverent discernment of one's place in the order of things. They share this not only with the Kantian rejection of goal-oriented ethics, but with the classical Stoic pursuit of virtue without regard to results.

Paley's theology, of course, knows nothing of this. For him, the Stoics were failures, because they could not find the comprehensive purpose that would unite pursuit of virtue and pursuit of individual happiness. His moral world was a world of purposes to be achieved, quite as much as the world of his utilitarian philosophical contemporaries was. The difference is only that, for Paley, the purposes are those of a benevolent deity who has chosen for us the goals that Bentham assumed we would choose for ourselves. Outside of purpose, there is only matter to be used as means. The tensions that we feel between the different systems and powers that claim our attention are the result of our limited knowledge, and the tragedies that ensue when we cannot respond to all of them at once are avoidable by better planning and by having our priorities straight in the first place. Although that position sounds to us most familiar when it accompanies the argument that because reality is devoid of purpose, we must create it for ourselves, it is also the position of theological utilitarians who argue that human purposes echo those imprinted on nature by the Creator, and of magicians, soothsayers, and spellbinders who advance human purposes in the face of the hostility or indifference of suprahuman powers.

Paley's assumption that God created a world of contrivances to advance his purposes through use of the principles of Newton's physics may strike the modern reader initially as quaint, but his continual recourse to the order of reality as he understood it is a reminder that the use of instrumental reason is not a way to do ethics without the trouble of agreeing on fact. The purposes we pursue, no less than the means we employ, depend on our ideas about the reality that is there.

Notes

1. D. L. LeMahieu has noted that Paley largely ignores the epistemological arguments that are the foundation of Hume's skepticism (1976, 29–54). Although he attacks Hume vigorously on secondary points, he leaves the central issue untouched. LeMahieu concludes that Hume's skepticism did not at the time have an intellectual following broad enough to require direct refutation.

2. The quotation is from Paley's English draft of a Latin prize essay which he wrote in 1765 (cited in Clarke 1974, 10).

3. For this purpose, Paley does not need to resolve the familiar question whether something is right because God wills it, or God wills it because it is right. The point is only that we need not fear that God is indifferent to human conduct, which, if it were the case, would necessitate a search for a purely human starting point for ethics.

4. For Bentham's statement of these same points, see Bentham (1948, 2, 22n, 29–32).

References

Albee, Ernest
1962 *A History of English Utilitarianism*. New York: Collier.

Bentham, Jeremy
1948 *The Principles of Morals and Legislation*. New York: Hafner.

Butler, Joseph
1969 "Fifteen Sermons Preached in the Rolls Chapel." In D. D. Raphael, ed., *British Moralists: 1650–1800*. Oxford: Clarendon Press.

Clarke, M. L.
1974 *William Paley: Evidences of The Man*. Toronto: University of Toronto Press.

Frankena, W. K., ed.
1960 *The Nature of True Virtue*, by Jonathan Edwards. Ann Arbor: University of Michigan Press.

Frankena, W. K.
1973 *Ethics*. 2d ed. Englewood Cliffs: Prentice-Hall.

Gustafson, James M.
1981 *Ethics from a Theocentric Perspective*. Chicago: University of Chicago Press.

Jønsen, Albert R.
1968 *Responsibility in Modern Religious Ethics*. Washington, D.C.: Corpus Books.

Kant, Immanuel
1956 *Critique of Practical Reason*. Indianapolis: Bobbs-Merrill. [First published 1788.]

Koester, Helmut
1968 "*Nomos Physeōs*." In Jacob Neusner, ed., *Religions in Antiquity*. Studies in the History of Religions, 14. Leiden: E. J. Brill.

LeMahieu, D. L.
1976 *The Mind of William Paley*. Lincoln: University of Nebraska Press.

Niebuhr, H. Richard
1963 *The Responsible Self*. New York: Harper & Row.

Oppenheim, Felix
1975 *Moral Principles in Political Philosophy*. 2d ed. New York: Random House.

Paley, William
1821a *Natural Theology*. In vol. 4 of *The Works of William Paley*. London: F. C. & J. Rivington. [First published 1802.]

1821b *Reasons for Contentment*. In vol. 4 of *The Works of William Paley*. London: F. C. & J. Rivington. [First published 1791.]

1977 *Principles of Moral and Political Philosophy*. Houston: St. Thomas Press. [First published 1785.]

Ross, W. D.
1930 *The Right and the Good*. Oxford: Clarendon Press.

Smith, Adam
1978 *Theory of the Moral Sentiments*. Oxford: Clarendon Press. [First published 1759.]

Wildiers, Max
1982 *The Theologian and His Universe*. New York: Seabury.

Williams, Bernard
1981 *Moral Luck*. Cambridge: Cambridge University Press.

Part V

Cosmogony, "Science," and Ethics

14 Cosmogony and Ethics in the Marxian Tradition: Premise and Destiny of Nature and History

Douglas Sturm

I

> A modern state does not need a creation myth of the kind ubiquitous among tribal or communal peoples. In a nation-state, such a myth may persist, but it is as part of the ethnic identity of the nation, not in justification of the rational foundations of the state, that this type of ideological weapon is used. Much less, then, does a league of states, striving to form ever larger agglomerations, need such a device to symbolize unity and coordination. The creation myth of the new state, or the new league of states, is a document with articles, clauses, and provisions that are, in principle, amendable as, with every increase in rationality, the partners expect better to manage their incidental irrational consequences. . . . Neither nation states nor the United Nations, nor even the European Community, need the mandate of heaven to legitimize their existence. They stand on the supposedly self-evident rationality of their respective political and economic purposes. (Wilson 1982, 158-59).

Technical rationality is the pride of the modern mind. But technical rationality is not as devoid of presuppositions as its pretends. It propagates its own vision of the world even as it purports neutrality. The genius of the Marxian tradition is to remind us that all activities are manifestations of interest and all interests bear with them a world of meaning. This is as true of academic pursuits as it is of corporate enterprise. The question is what that world of meaning is, what its practical implications are, and how justifiable it might be. Cosmogonies and ethics, the topic of this volume, are efforts to give voice to worlds of meaning. In this essay I shall claim that the Marxian tradition, as a significant world of meaning in the modern epoch, is not without its own cosmogonical understanding and allied ethical orientation. Prior to an examination of selected classical Marxian texts, however, some preliminary considerations are in order.

Bryan Wilson, as quoted above, offers but two forms of statements concerning beginnings and two functions which those statements fulfill. The primitive form ("In the beginning, God") is a myth of legitimation. The modern form ("We, the people of the United States, in order to form a more perfect union") is a calculus of rational control. The former, ideological and presumably superstitious, is characteristic of traditional societies. The latter, functional and presumably clear-headed, is representative of the bourgeois world. Wilson neglects a third possibility which, I contend, is integral to the Marxian tradition. The third possibility is that a statement of beginnings fulfills several functions. It is explanatory: it tells how things came to be what they are. It is predictive: it tells how things are going to be and what the forces of social change are. It is critical: it tells what is deficient in alternative stories about the world. And, finally, it is directive: it tells what can be accomplished (a structure of possibilities) and what should be accomplished (a structure of action).

Furthermore, although sometimes a distinction has been drawn between cosmogony (a theory of the origins of the world) and cosmology (a theory of the character of the world) as though they were radically different in structure and import, I suggest that, at least in some instances, such a distinction is superfluous if not deceptive. There is a double meaning to cosmogony. A cosmogony is a story of the generation of the cosmos, a theory of the beginnings of the world or of some epoch of the world's history. But beginnings may be meant in two senses—chronological and foundational. The two senses may converge, but they may also be distinguished. The words "authority" (*auctoritas*) and "principle" (*principium*) have the same double meaning. An author is one who begins, who initiates, who sets down. But one who is authorized is one whose judgments are fundamental, one who has jurisdiction over a matter. A final authority is one beyond whose word there is no appeal. Similarly, *principium* is a beginning, a start, the point from which matters develop. Yet a principle is also a basic law, a fundamental truth, a central doctrine.

These two meanings converge when there is reference to chronological beginnings as a basis for final appeal or as authorizing specific activities. References to "founding fathers," originating documents, or initial intentions express this dynamic in the political world. Moreover, stories about beginnings and theories about origins, although they have a chronological form, may not be intended to convey anything at all about chronological beginnings or starting times in the literal sense. They may be efforts to convey in narrative form something about foundations or essential meanings or the basic structure of reality. In any case, a foundational cosmogony is expressive of the same dimension of human experience as cosmology or an ontology.

Nevertheless, one confronts a fundamental ambiguity in exploring the topic of cosmogony (whether in the chronological or foundational sense) and ethical order in the Marxian tradition.

On the one hand, the Marxian tradition instructs us to beware of the pretenses of all religions and philosophies, including their stories of origins and their normative systems. Marxism is a hermeneutics of suspicion. It would have us go behind religious myths and reflective arguments to ascertain their roots in forms of human interaction. Myths and arguments are not to be taken at face value. They serve an ideological function. They are both revealing and concealing, but to find out what they reveal one must find out what they conceal. There is little epistemological certainty to be found in religious forms of expression or in ordinary forms of logic. Their meaning does not lie on the surface but must be deciphered by examining the role they play in the practical life of the people who hold them. One might, then, look to Marxism for a method of analysis and interpretation as one investigates the cosmogonies and ethics of, say, the Rig Veda, Mencius, the Navajo, or the ancient Hebrews.

On the other hand, the Marxian tradition may be interpreted as itself promulgating a distinctive story of beginnings and a set of principles for assessing and directing the conduct of individuals and patterns of social life. Marxism is, in its own right, a *Weltanschauung*, a way of viewing the world, and a pragmatics, a way of living in the world. Since its emergence in the nineteenth century, it has had an astounding impact on all civilized peoples, even those who perceive it as a threat. It is one of the more powerful and widespread systems of thought and action in contemporary history. It is possessed of missionary zeal and stands in bold competition with alternative ways of thinking and acting. One might, then, look at Marxism to delineate its own internal cosmogony and ethics, explicit or implicit, and the manner in which its cosmogony and ethics are in concert with each other.

Some critics charge that Marxism as a social movement is incompatible with Marxism as a hermeneutics for, they allege, the hermeneutics relativizes all claims to truth and justice and therefore undercuts the premises of the movement. The argument, however, is trivial. It betrays a failure to comprehend the thick texture of the Marxian tradition. The hermeneutics and the social philosophy are part and parcel of each other. They constitute, respectively, the negative and affirmative sides of a critique. On the negative side, the critique probes into and beyond what is superficially presented; it dispels illusions and throws off the veil of false consciousness. In doing so, on the affirmative side, it reveals, by claim, a deeper reality. As with others, so with Marx and Engels, a critical hermeneutics is employed for the sake of revealing the actual conditions and the eventual destiny of human history.

The Marxian tradition is, in principle and by intention, a totality. Its hermeneutics provides a means of penetrating into the actual meaning of inherited cosmogonies and systems of ethical order. But as a movement, it presents, as an expression of the same method, a constructive way of thinking about beginnings and about categories of human conduct. That, in any case, is the thesis on which I shall proceed. Yet to say that the Marxian tradition is a totality is not to say that it is univocal. It is complex and multidimensional. Like all great traditions of thought and action, it contains strains, even in its originating events, that are not or do not seem to interpreters wholly compatible with each other. It may even be that there were, as Alvin W. Gouldner (1980) argues, "two Marxisms" from the very beginning. Or it may be, as Norman Levine (1975) insists, that, despite the remarkably intense and durable companionship of Marx and Engels, they were pressing in what later proved to be antithetical directions. In any case it is clear that, in subsequent decades, the Marxian tradition, in ways that should not be surprising to those acquainted with developments among the great religions and philosophies of the world, has moved in radically divergent paths (McLellan 1979). A word of caution is thus in order lest one overlook differences at least in emphasis, if not in principle, among the writings of Marx and Engels.

In any case, there is a cosmogonical dimension in the classical texts of the Marxian tradition. It is a dimension with multiple levels and variant forms—cosmological (the motive force of matter), anthropological (the emergence of human history), and historiological (the development of historical epochs). Although, according to Marxism, there is no single moment of absolute beginning, there are moments of creative emergence, of new beginnings and novel formations, even though they constitute the unfolding of potentialities resident in the dynamic structure of prior events. Furthermore, there is an ethical dimension in the texts (cf., however, the controversies reported in Cohen, Nagel, and Scanlon 1980; Buchanan 1982). In keeping with the Marxian cosmogony, the ethical dimension is, in general, naturalistic (its principles derive from the inherent impulse and development of natural processes), social (its focus is on the structural forms and antinomies of human activity), contextual (the appropriateness of kinds of action is contingent on historical conditions), humanistic and teleological (its ultimate concern is with the consummation of specifically human potentialities).

Six texts have been selected for this inquiry: (1) *Difference between the Democritean and Epicurean Philosophy of Nature,* Marx's doctoral dissertation, accepted in 1841 but not published during Marx's lifetime; (2) *The Holy Family or Critique of Critical Criticism: Against Bruno Bauer and Company,* the first collaborative work by Marx and Engels, published in 1845; (3) *The German Ideology,* coauthored by Marx and Engels during the years 1845-47 but published as a whole only in the twentieth century; (4) *Anti-Dühring: Herr Eugen Dühring's Revolution in Science,* first edition 1878, particularly the sections published separately as *Socialism: Utopian and Scientific* in 1880, by Engels; (5) *The Origin of the Family, Private Property and the State,* first edition published by Engels in 1884; and (6) *The Dialectics of Nature,* initiated by Engels in 1873 but never completed, published as a whole only in the twentieth century.

The first three texts, especially *The Holy Family* and *The German Ideology,* are expressions of the initial impulse of Marxism. The last three give testimony to Engels's more mature reflections on the movement and its implications. Some critics make much of the difference between these two moments in the development of classical Marxism, even to the point of suggesting a radical divergence between the younger and the later Marx or of setting Marx and Engels, despite their collaborative work, over against each other. For purposes of investigating the cosmogonical and ethical dimensions of the tradition, however, a commitment on that issue is not needed in advance. Indeed, the continuities of the tradition, even with its inner tensions, may be as significant as, if not more significant than, its discontinuities.

II

Engels, far more than Marx, was concerned to demonstrate the implications of dialectical principles in the natural sciences. Yet Marx was not without interest in the philosophy of nature, as evidenced in his doctoral dissertation in which he defended the integrity and intelligibility of the Epicurean alternative to Democritus's materialism. Although sometimes passed off as of little significance in the development of the Marxian movement, the dissertation presents, through its interpretation of Epicurus, a kind of foundational cosmogony some of whose themes echo throughout Marx's thought during the rest of his life. I would cite three themes in particular: the self-determinative character of the atoms that constitute the universe, the contradictory or tensional character of the relationship between atoms in their abstract essence and in their existential reality, and the practical character of knowledge.

Marx's attention was drawn toward Hellenistic thought through his affiliation with the Young Hegelians to whom Stoicism, Epicureanism, and Skepticism represented divergent forms in the philosophy of self-consciousness. But the twist taken in the dissertation was unique to Marx. Democritus and Epicurus were both materialists. The universe, as each envisioned it, is composed of nothing but atoms and the void. The everyday entities we observe are arrangements of atoms, clustered together for a time. Change is a function of the rearrangement and reformation of the atoms in their positions relative to each other. On this level it would appear that Democritus and Epicurus were proponents of the same understanding of the cosmos.

Yet the strong similarity between Democritus and Epicurus was spoiled, so it seemed, by Epicurus's principle of the declination of the atoms: the atoms swerve. To some interpreters, the swerving of the atoms is an alien principle tacked arbitrarily if not capriciously onto an otherwise thoroughly materialist philosophy in which case Democritus's position would appear to be more consistent than Epicurus's. Yet in Marx's interpretation, the principle of the declination of the atoms is central to Epicurus's philosophy of nature. Epicureanism is an alternative cosmology with its own inner integrity, a cosmology in which the idea of self-relatedness and self-determination is central.

The two philosophies, despite their apparent similarity, were fundamentally different. Democritus was a determinist. The presumption of Democritus was that the atoms move in accordance with laws of strict necessity. Epicurus, on the other hand, was an individualist. It was his presumption that each atom is a being-for-itself and moves in accordance with its own directionality. Furthermore, an atom is not merely a physical entity, but a being with an interest in self-realization through which process it takes on some existential form. It is in the relationship between essence and existence that the tensional character of life is located.

Moreover, in Epicurus's philosophy, the motivation of self-realization underlies forms of reflection. Knowledge is not an end in itself. It is a practical activity whose aim is ataraxia, the tranquility of independence and utter self-relatedness: "Epicurus confesses finally that his method of explaining aims only at the ataraxy of self-consciousness, not at knowledge of nature in and for itself" (Marx 1975a, 45).

Marx, one should note, was critical of the privatism and isolationism of the Epicurean alternative. In his preparatory notebooks, he asserts that the times of Epicurean philosophy were unhappy,

> for their gods have died and the new goddess still reveals the dark aspect of fate, of pure light or of pure darkness. . . . The kernel of the misfortune . . . is that the spirit of the time, the spiritual monad, sated in itself, ideally formed in all aspects in itself, is not allowed to recognize any reality which has come into being without it. . . . Thus . . . the Epicurean . . . philosophy was the boon of its time; thus, when the universal sun has gone down, the moth seeks the lamplight of the private individual. (1975b, 492)

Nonetheless there are central themes in Marx's interpretation of Epicurus's position that constitute prefigurations of his own historical materialism: the declination of the atoms (the principle of freedom), the contradictory or tensional character of the existence of the atoms (the principle of production), and the practical intention of reflection (the principle of praxis). Altogether, these themes are indicative of a foundational cosmogony in which creativity, albeit an individualistic creativity, is a central feature of the universe.

III

The Holy Family, the first joint work by Marx and Engels, is a devastating diatribe against the Young Hegelians, especially Bruno Bauer and his brothers. Marx is its dominant author. Through the treatise, he announces in no uncertain terms his complete break with the idealists. In tone, *The Holy Family* is sarcastic and derisive. In content, it is detailed and diffuse. Overall, it is a harsh attack on the writings of the Hegelian radicals, attending as much to grammatical structure and use of language as to doctrine and principle. Yet in and through the attack, it betrays the outlines of an affirmative position, the position of historical materialism.

The text opens with a declaration:

> *Real humanism* has no more dangerous enemy in Germany than *spiritualism* or *speculative idealism*, which substitutes "*self-consciousness*" or the "*spirit*" for the *real individual man* and with the evangelist teaches: "It is the spirit that quickeneth; the flesh profiteth nothing." Needless to say, this incorporeal spirit is spiritual only in its imagination. (Marx and Engels 1975, 7)

The cosmological-cosmogonical implications of Bauer's idealism are a matter of special note in sections of *The Holy Family* composed by Marx. Responding to Bauer's critique of French materialism, Marx draws out the intention of Bauer's spiritualism:

> the truth of *materialism* is the *opposite* of materialism, absolute, i.e., exclusive unmitigated *idealism*. Self-consciousness, *the Spirit*, is the *Universe*. Outside of it there is nothing. "Self-consciousness," "*the Spirit*," is the almightly creator of the world, of heaven and earth. The *world* is a manifestation of the life of self-consciousness which has to alienate itself and take on *the form of a slave*, but the difference between the world and self-consciousness is only an *apparent difference*. (Marx and Engels 1975, 140)

As the cosmos is a function of and originates in the mind, so also presumably the historical process is a manifestation of spirit. Marx expresses astonishment at this conclusion:

> does Critical Criticism believe that it has reached even the *beginning* of a knowledge of historical reality so long as it excludes *from* the historical movement the theoretical and practical relation of man to nature, i.e., natural science and industry? Or does it think that it actually knows any period without knowing, for example, the industry of that period, the immediate mode of production of life itself? . . . Just as it separates thinking from the senses, the soul from the body and itself from the world, it separates history from natural science and industry and sees the origin of history not in vulgar *material* production on the earth but in vaporous clouds in the heavens. (Marx and Engels 1975, 150)

To Marx, the conditions of history and the possibility of historical change are rooted not in the mind but in the productive process. This is in fact known by the masses, those ridiculed by the idealists as allegedly ignorant of the deeper realities of life.

> But these *mass-minded*, communist workers, employed, for instance, in the Manchester or Lyons workshops, do not believe that by "*pure thinking*" they will be able to argue away their industrial masters and their own practical debasement. They are most painfully aware of the *difference* between *being* and *thinking*, between *consciousness* and *life*. They know that property, capital, money, wage-labour and the like are no ideal figments of the

> brain but very practical, very objective products of their self-estrangement and that therefore they must be abolished in a practical, objective way for man to become man not only in *thinking*, in *consciousness*, but in mass *being*, in life (Marx and Engels 1975, 53)

As Marx notes, Bauer and the idealists were not without ethical sensitivity, but the ethical orientation promoted by them was personalistic and aristocratic (see Hook 1958, 103-8). Those with critical consciousness are under the onus of a kind of noblesse oblige—at the very least, to alleviate suffering and, where possible, to educate the masses into a proper way of thinking about the world; for, ultimately, evil lies in erroneous thinking. Idealism teaches the masses "that they cease in reality to be wage-workers if in thinking they abolish the thought of wage labour . . . . that they abolish real capital by overcoming in *thinking* the category Capital, that they *really* change and transform themselves into real human beings by changing their '*abstract ego*' in consciousness and scorning as an un-Critical operation all *real* change of their real existence, of the real conditions of their existence, that is to say, of their *real ego*" (Marx and Engels 1975, 53).

To Marx, by contrast, the possibility of social change—in particular, the possibility of overcoming the inhumane conditions of capitalist society—is grounded in structural features of the social process itself. Capitalist property, by virtue of its inner character, gives rise to antitheses—wealth versus poverty, the bourgeoisie versus the proletariat. Given its condition, the proletariat is driven to emancipate itself:

> But it cannot abolish the conditions of its own life without abolishing *all* the inhuman conditions of life of society today which are summed up in its own situation . . . . It is not a question of what this or that proletarian, or even the whole proletariat, at the moment *regards* as its aim. It is a question of *what the proletariat is*, and what, in accordance with this *being*, it will historically be compelled to do . . . . There is no need to explain here that a large part of the English and French proletariat is already *conscious* of its historic task and is constantly working to develop that consciousness into complete clarity. (Marx and Engels 1975, 37)

Thus even in this early and largely iconoclastic work, Marx asserts that the emergence of a genuinely humane epoch in history is contingent on the proletariat coming to consciousness of itself as such and adopting its structurally given task, a special ethical vocation, to overcome the institutional conditions of human self-estrangement.

The cosmogonical possibility of a genuinely new historical epoch and the ethical obligation to actualize that possibility are conjoined as rooted in and derived from the social structure of the times.

IV

Following his expulsion from Paris in February 1845, Marx settled in Brussels where he prepared an initial draft of his *Theses on Feuerbach* and developed the outlines of his materialist theory of history. He and Engels then determined to compose a book-length work through which they would refine their critique of left-wing Hegelians and others parading themselves as "true socialists" but, more importantly, in which they would present the materialist alternative. The work, *The German Ideology,* especially the section on Feuerbach, has been called "the first systematic representation of their historical-philosophical conception of the economic development of humankind" (Adoratsky 1934).

Marx and Engels were unsuccessful in finding a publisher for the work. As Marx later remarked, "We abandoned the manuscript to the gnawing criticism of the mice all the more willingly as we had achieved our main purpose—self-clarification" (Marx and Engels 1976, 13). The manuscript, transmitted to the twentieth century, when it finally was published, is in poor condition: pages are missing; some pages are clearly rewritten, others are not; pagination of some sections remains uncertain; marginal notes and corrections are scratched out in places, sometimes by Marx and Engels but other times by later revisionists. Nonetheless there has been a painstaking reconstruction of the text, and there is an inner integrity to the argument which is stated with force and clarity.

The leitmotif of *The German Ideology,* in its negative aspect as a critique of the Young Hegelians, is stated summarily:

> In direct contrast to German philosophy which descends from heaven to earth, here it is a matter of ascending from earth to heaven . . . . The phantoms formed in the brains of men are . . . , necessarily, sublimates of their material life-process, which is empirically verifiable and bound to material premises [*Voraussetzungen*]. Morality, religion, metaphysics, and all the rest of ideology as well as the forms of consciousness corresponding to these, thus no longer retain the semblance of independence. They have no history, no development; but men, developing their material production and their material intercourse, alter, along with this their actual world, also their thinking and the

> products of their thinking. It is not consciousness that determines life but life that determines consciousness. (Marx and Engels 1976, 42)

In contrast to any cosmogony of the form "In the beginning was the word," Marx and Engels present as an alternative, "In the beginning was the deed." In a critical passage, they insist that deeds and words—theory and practice—act reciprocally on each other (Marx and Engels 1976, 61). But the location of the reciprocal interaction between thought and action is the productive process in which specifically human life is grounded. That is, production and reproduction are the necessities of life and the beginnings of history. They constitute the foundation of human reality. As they proceed through various forms, in logical sequence, so the course of historical change develops toward a special time of realization.

> But, from its beginning, human life consists of several aspects or moments. We must begin by stating the first premise of all human existence and, therefore, of all history, the premise, namely, that men must be in a position to live in order to be able to "make history." But life involves before everything else eating and drinking, housing, clothing and various things. The first historical act is thus the production of material life itself. And indeed this is an historical act, a fundamental condition of all history, which today, as thousands of years ago, must daily and hourly be fulfilled merely in order to sustain human life . . . . The second point is that the satisfaction of the first need, the action of satisfying and the instrument of satisfaction which has been acquired, leads to new needs; and this creation of new needs is the first historical act . . . . The third circumstance which, from the outset, enters into historical development, is that men, who daily recreate their own life, begin to make other men, to propagate their kind: the relation between man and woman, parents and children, the *family* . . . . These three aspects of social activity are not of course to be taken as three different stages, but just as three aspects or, to make it clear to the Germans, three "moments," which have existed simultaneously since the dawn of history and the first men, and which still assert themselves in history today. (Marx and Engels 1976, 47-48)

Within the structure of these three moments, there is a twofold relation: a relation with nature (a mode of production) and a social relation (a mode of cooperation). Throughout history, these two modes are inextricably connected. Consciousness is a function of the complex interplay of these two modes; it is, in its roots, of a practical character, it serves the processes of production and cooperation, although it also generates its own possibilities.

Consciousness contains the seed of contradiction (*Widerspruch*), for consciousness bears the prospect of a division of labor between administrators (thinkers) and workers (doers) resulting in

> the *unequal* distribution, both quantitative and qualitative, of labour and its products, hence property, the nucleus, the first form of which lies in the family, where wife and children are the slaves of the husband. This latent slavery in the family, though still very crude, is the first form of property, but even at this stage it corresponds perfectly to the definition of modern economists, who call it the power of disposing of the labour-power of others. Division of labour and private property are, after all, identical expressions; in the one the same thing is affirmed with reference to the product of the activity. (Marx and Engels 1976, 52)

Division of labor gives rise to domination and alienation. In domination, one class, employing the force of the state under the pretense of serving the "common interest," exercises control over the masses. Struggles over forms of political order are in reality struggles among classes. Marx and Engels depict alienation as follows: "as long . . . as activity is not voluntarily, but naturally [i.e. involuntarily/necessarily/coercively], divided, man's own deed becomes an alien power opposed to him, which enslaves him instead of being controlled by him" (1976, 53). One is forced into and cannot escape from one's given sphere of activity. At this point follows one of the more famous passages of the text:

> He is a hunter, a fisherman, a shepherd, or a critical critic, and must remain so if he does not want to lose his means of livelihood; whereas in communist society, where nobody has one exclusive sphere of activity but each can become accomplished in any branch he wishes, society regulates the general production and thus makes it possible for me to do one thing today and another tomorrow, to hunt in the morning, fish in the afternoon, rear cattle in the evening, criticise after dinner, just as I have a mind, without ever becoming hunter, fisherman, shepherd or critic. (Marx and Engels 1976, 53)

Communist society, however, is not, strictly speaking, an abstract ethical ideal, nor is it a categorical imperative, although it does have, as we shall note, ethical significance. It is rather a stage that will eventuate in due course throughout the processes of historical change. It cannot come about by simply thinking or willing, although it will not come about without the emergence of a new form of consciousness and the assumption of a work to be accomplished. Yet for everything there is a season. Communist society is the culminating moment in a sequence of historical periods, each of which is characterized by a particular form of property.

Marx and Engels distinguish four periods of history: tribal ownership in primitive kinship society, the ancient city-state with its communal and state property, medieval society with its feudal property and landed estates, and capitalist society with its bourgeois form of ownership (Marx and Engels 1976, 38-41). The transition from one form to the next is not accidental; it is effected by tensions and contradictions among the productive forces of the prior stage. In effect, the foundations or premises of history contain within themselves the dynamics of change and a directionality that is intelligible. Historiology reflects cosmogony. The patterns of historical process are rooted in the necessities and forms of production.

In this periodization, Marx and Engels devote the bulk of their attention to the transition from medieval to bourgeois society and, within the bourgeois period, to the development through three substages: manufacture, commerce, and large-scale industry (Marx and Engels 1976, 72-82). In each transformation, "Not criticism but revolution is the driving force of history" (Marx and Engels 1976, 61).

Effective revolution, however, is contingent upon the proper conditions. There are two conditions ("practical premises") requisite for the revolutionary transformation of the bourgeois world and the effectuation of a communist society. First, the great mass of humanity must be brought to a condition of propertylessness, that is, must be without any control over the means of production, such that their circumstances are intolerable. Conversely, property and the benefits of property must be concentrated in the hands of an elite, bourgeois class. Second, the development of productive forces—organizationally and technologically—must be so far advanced that they contain the capacity of overcoming poverty and can do so universally, throughout the entire world. In the absence of this capacity, "the struggle for necessities would begin again, and all the old filthy business would necessarily be restored" (Marx and Engels 1976, 54).

However, assuming those conditions and an effective revolution, a new stage of history will emerge, a new society will come into being, and a new structure of human life will be created. In all previous stages, human life has suffered from alienation (*Entfremdung*), dependency (*Abhängikeit*), and a cleavage (*Unterschied*) between private life and social existence. One's life has been governed by material forces and social forms beyond one's control.

The emergence of the new epoch is contingent on action that is not personalistic and aristocratic but structural and proletarian. The proletarian revolution will be an act of liberation (*Befreiung*):

> *All-round* dependence, this primary natural [*naturwüchsige*] form of the *world-historical* co-operation of individuals, will be transformed by this communist revolution into the control and conscious mastery [*bewusste Beherrschung*] of these powers, which, born of action of men on one another, have till now overawed and ruled men as powers completely alien to them. (Marx and Engels 1976, 59)

The new age will be one of freedom (*Freiheit*) and self-activity (*Selbstbetätigung*) wherein individuals will become *complete* individuals (Marx and Engels 1976, 97). But,

> This is not possible without community (*Gemeinschaft*). Only within the community has each individual the means of cultivating his gifts in all directions; hence personal freedom becomes possible only within the community. (Marx and Engels 1976, 86)

The cosmogony of *The German Ideology* thus culminates in a vision of fulfillment, a moment of completion, a time when the human potential comes to fruition. Freedom in community is the end, the telos. Marx and Engels present us with a historical teleology or, if you will, an eschatological ethics. As Kolakowski remarks, "Marx's point of departure is the eschatological question derived from Hegel: how is man to be reconciled with himself and with the world? . . . Marx, like Hegel, looks forward to man's final reconciliation with the world, himself, and others" (1978, 177). Yet, to Marx, unlike Hegel, reconciliation is a matter of proletarian praxis ("The consciousness of the proletariat is not mere passive awareness of the part assigned to it by history, but a free consciousness and a fount of revolutionary initiative" [Kolakowski 1978, 180]) resulting in the formation of a communist society.

> Communism is for us not a *state of affairs* [*Zustand*] which is to be established, an *ideal* [*Ideal*] to which reality [will] have to adjust itself. We call communism the *real* movement [*wirkliche Bewegung*] which abolished the present state of things [*Zustand*]. The conditions of this movement result from the now existing premise [*Voraussetzung*] (Marx and Engels 1976, 57).

V

The last three works in this inquiry into the cosmogonical and ethical aspects of the Marxian tradition were written by Engels during the final third of his life. In Kolakowski's judgment, in relation to Marx,

> Engels, from a more empirical point of view, gave expression to the same vision of a classless communist society, to be brought about by the initiative of the working class activating the natural trend of history. On the other hand, Engels adopted a different standpoint as regards the cognitive and ontological link between man and nature. In his later works the idea of the "philosophy of praxis" . . . gives place to a theory which subjects humanity to the general laws of nature and makes human history a particularization of those laws, thus departing from the conception of man as "the root" (in Marx's phrase) and of the "humanization" of nature. (1978, 181)

Yet, it must be remembered that Marx collaborated directly in the composition of *Anti-Dühring*, that Marx's notes on Lewis Morgan's study of primitive society constituted a primary resource in Engels's composition of *The Origin of the Family, Private Property and the State*, and that Engels corresponded constantly with Marx about his work on *The Dialectics of Nature*. Those who discern a radical divergence between Marx and Engels may be responding more to contemporary political struggles than to the classical texts themselves. Yet it is true that these texts by Engels press the Marxian tradition into new levels and dimensions of cosmogonical and ethical reflection.

The immediate purpose of Engels's *Anti-Dühring: Herr Eugen Dühring's Revolution in Science* was polemical and practical. In the mid-1870s, Dühring, a professor in Berlin, "discovered" socialism and published three books in rapid succession presenting its philosophical grounds and structure. In Engels's typically sardonic style, he remarks,

> As is well known, we Germans are of a terribly ponderous *Gründlichkeit*, radical profundity or profound radicality, whatever you may like to call it. Whenever any one of us expounds what he considers a new doctrine, he has first to elaborate it into an all-encompassing system. He has to prove that both the first principles of logic and the fundamental laws of the universe had existed from all eternity for no other purpose than to ultimately lead to this newly-discovered, crowning theory. (1978, 7)

There was fear lest Dühring's impressive but tendentious and anti-Marxian form of socialism might split the recently unified German

Socialist party. Engels was prevailed upon to prepare a critical response which originally appeared as a series of articles in the house organ of the party during 1877-78 and was published as a book immediately thereafter. Two years later, three chapters, slightly revised, were published separately as *Socialism: Utopian and Scientific*.

In intent, *Anti Dühring* was a polemical work; in import, it was "the first systematic presentation of the philosophy of dialectical materialism" (Bochenski 1972, 38).

In *Anti-Dühring*, the question of the origin of the present state of the universe is posed explicitly for the first time in classical Marxian texts. Engels finds the Kantian hypotheses "of the origin of all existing celestial bodies from rotating nebular masses" (1966, 65) attractive, but only as a hypothesis. Engels's primary concern in this context is to make two points: that even "before the nebular state matter had passed through an infinite series of other forms" (1966, 67) and that "*Motion is the mode of existence of matter*" (1966, 68). To think otherwise is to invoke a deus ex machina in some literal sense of that phrase to explain the beginning of natural and historical processes. Motion is "as uncreatable and indestructible as matter itself" (1966, 68).

There are, however, qualitative changes in the *forms* of motion: "the transition from one form of motion to another always remains a leap, a decisive change" (1966, 75). The emergence of organic life constitutes one radical transition. Others are found in the world of nature, in the world of history, and in the world of thought. But in all spheres, motion proceeds in accordance with the laws of dialectics: the law of contradictoriness (the interpenetration of opposities), the law of transformation (of quantity into quality and vice versa), and the law of negation (the negation of the negation).

Engels contends that, at the present moment in history, we are at a time of radical change, given the contradictory forces of the epoch. We are at a time of new creation, new beginning, a time of fundamental transformation in the structure of human life. Others—Saint-Simon, Fourier, Robert Owen—had had a vision of such a new beginning, but mistakenly thought it would come about simply through the application of reason to human affairs. But reason, by itself, is no agent of social change. Historical transformation occurs when productive forces generated by a particular mode of production are no longer compatible with that mode of production. At the moment, "large-scale industry has developed the contradiction lying dormant in the capitalist mode of production into such crying antagonisms that the imminent collapse of this mode of production is, so to speak, palpable . . . the new productive forces themselves can only be maintained and further developed by the introduction of a new mode of production corresponding to their present stage of development" (1966, 290).

"Scientific socialism" does not invent the means of social change (as "Utopian socialism" had presumed to do); it finds them in the actual contradictory conditions, the material facts, of the productive process. The central contradiction of advanced capitalism is "the incompatibility of social production and capitalist appropriation" (1966, 296). The contradiction is expressed in the class struggle between proletariat and the bourgeois (1966, 297). It is reproduced as well in "the antithesis between the organization of production in the individual factory and the anarchy of production in society as a whole" (1966, 299). The system thus fights against itself, as is manifest in cycles of economic crisis (overproduction, underconsumption, depression, inflation, unemployment). Hence, "the mode of production rebels against the mode of exchange; the productive forces rebel against the mode of production, which they have outgrown" (1966, 302). Over the course of time, various enterprises collapse or merge resulting in an increasing concentration of wealth. The state acts through various means to sustain the system thereby intensifying the growth of giant corporations.

The social forces pressing the historical process in this direction do so "like the forces operating in Nature: blindly, violently, destructively, so long as we do not understand them and fail to take them into account." But when their character is understood, they may be subjected to human will: "once their nature is grasped, in the hands of the producers working in association they can be transformed from demoniac masters into willing servants" (1966, 305).

This is the function, the vocation, of the proletarian revolution: through increased consciousness of the contradictions of the capitalist epoch to bring that epoch to an end. In doing so, it dissolves itself as a proletariat for it brings to an end all class differentiation. Yet division of labor and class stratification were not avoidable realities. For their time, they served needed functions in the development of history. They were required for purposes of economic growth, technological advance, and the increased sophistication of the productive process. Nonetheless, "this does not mean that this division into classes was not established by violence and robbery, by deception and fraud, or that the ruling class, once in the saddle, has ever failed to strengthen its domination at the cost of the working class and to convert its social management into the exploitation of the masses" (1966, 307-8).

At this point in history, a new possibility has emerged: "of securing for every member of society, through social production, an existence which is not only fully sufficient from a material standpoint and becoming richer from day to day, but also guarantees them the completely unrestricted development and exercise of their physical and mental faculties" (1966, 309). Engels in effect is making an eschatological annunciation, for this new possibility means nothing less than the fulfillment of humanity. When that possibility is realized,

> in a certain sense, man finally cuts himself off from the animal world, leaves the conditions of animal existence behind him and enters conditions that are really human. The conditions of existence forming man's environment, which up to now have dominated man, at this point pass under the dominion and control of man, who now for the first time becomes the real conscious master of Nature, because and in so far as he has become master of his own social organisation. The laws of his own social activity, which have hitherto confronted him as external, dominating laws of Nature, will then be applied by man with complete understanding, and hence will be dominated by man. Men's own social organization which has hitherto stood in opposition to them as if arbitrarily decreed by Nature and history, will then become the voluntary act of men themselves. The objective, external forces which have hitherto dominated history, will then pass under the control of men themselves. It is only from this point that men, with full consciousness, will fashion their own history; it is only from this point that the social causes set in motion by men will have, predominantly and in constantly increasing measure, the effect willed by men. It is humanity's leap from the realm of necessity into the realm of freedom (1966, 309-10).

Interpreters of the Marxian tradition have debated at length over whether there is a Marxian ethics and, if so, what the form and the content of that ethics are. In *Anti-Dühring*, there is clearly a teleology-eschatology in the sense of an understanding of the direction of the historical process and its culminating moment. There is also a cosmogony in the sense of a statement of new beginnings and how new beginnings come about, although, as noted, Engels rejects all forms of sheer creationism. There is furthermore an inner connection between cosmogony and teleology-eschatology, for beginnings and ends are conjoined in the dialectical laws of motion. But is there an ethics in the sense of a theory of human conduct and the principles that ought to govern relations among persons?

Engels, rejecting Dühring's argument that ethical truths may be derived analytically through an a priori method, provides an understanding of ethics that reflects his dialectical interpretation of nature and history. In the present epoch, he notes, there are three kinds of morality in contention: a Christian-feudal morality, a bourgeois morality, and a proletarian morality. So "past, present, and future provide three great groups of moral theories which are in force simultaneously." No one of them possesses absolute validity, but "that morality which contains the maximum of durable elements is the one which, in the present, represents the overthrow of the present, represents the future: that is, the proletarian" (1966, 104).

Moral ideas, to Engels, are, consciously or unconsciously, derived from the practical circumstances of a socio-economic class. Where there are moral principles common to several classes, they merely reflect the relatively common historical circumstances in which those classes coexist. Where, for example, there is private property in moveables, a law against stealing might be held by all existing classes. Throughout the history of class antagonisms, however, morality "has either justified the domination and the interests of the ruling class, or, as soon as the oppressed class has become powerful enough, it has represented the revolt against this domination and the future interests of the oppressed" (1966, 105).

In particular, Engels traces the changing fortunes of the principle of equality through Western history, how it was defined and how it functioned variously to support the status quo or to effect social change. In the case of the proletariat, the principle has a double meaning: it is a protest against the "crying inequalities" of feudalism and it is a reaction against the hypocrisies of the bourgeoisie, whose demand for equality is limited and ultimately false. In the latter instance, the proletarian appeal to equality is a means of continued agitation. In both instances, "the real content of the proletarian demand for equality is the demand for the *abolition of classes*" (1966, 118). But this does not mean that the principle of equality is an eternal truth: it is a historical product; it makes sense only within the framework of specific historical conditions.

Yet, Engels asserts, without full explication, that there has been "progress in morality." And he maintains that "A really human morality which transcends class antagonisms and their legacies in thought becomes possible only at a stage of society which has not only overcome class contradictions but has even forgotten them in practical life" (1966, 105).

One additional note. Engels acknowledges that ethics and law presuppose, in some sense, freedom of action, for otherwise responsibility cannot be assigned to agents, and the advance of goals, principles, and duties is absurd. But the meaning of freedom is controversial.

Engels affirms the Hegelian conception, that "freedom is the appreciation of necessity." More precisely, freedom means "the control over ourselves and over external nature which is founded on knowledge of natural necessity; it is therefore necessarily a product of historical development" (1966, 125). Throughout the course of civilization, there has been progress in freedom as there has been increase in knowledge about the laws governing natural events and social intercourse. The resolution of the antagonisms of history through the proletarian revolution constitutes a dramatic step in the progress of freedom, since under conditions of genuinely social production, human action will not be determined by contradictions among productive forces. On what constitutes the form of the dialectical process in history subsequent to the emergence of the communist society, the realm of freedom, the realization of a specifically human life, Engels is silent.

VI

Marx had intended to incorporate into an extended materialist conception of history the findings of Lewis H. Morgan, an American anthropologist, about the development of human society from its most primitive forms. Engels sought to satisfy that intention in *The Origin of the Family, Private Property and the State,* published the year following Marx's death. Using Marx's critical notes on Morgan's work as a primary resource, Engels conceived the book to be "the fulfillment of a bequest" (1962, 170).

Marx and Engels were attracted to Morgan's argument that the most primitive form of human society was communist, that changes in economic relations constituted the catalytic agent in transformations of family structure and political organization, and that human destiny demands a moving beyond civilization in its current form toward a new form of communal life.

As Engels interprets Morgan's work, the emergence of civilization constitutes, in effect, a fall, a movement away from an idyllic form of life. Civilization is degeneration. But it is a kind of degeneration that, dialectically, may be required for movement into a new, higher stage of human existence.

In the most primitive form of society, matriarchy was the rule and group marriage was the practice. The men and women of the tribes were sexually promiscuous. The offspring were associated with their mother, for at least the natural mother was identifiable. Women were responsible for and in charge of the household. Thus "the communistic household implies the supremacy of women in the house, just as the exclusive recognition of a natural mother, because of the impossibility of determining the natural father with certainty, signifies high esteem for the women, that is, for the mothers" (1962, 209). By contrast, "The social status of the lady of civilisation, surrounded by sham homage and estranged from all real work, is socially infinitely lower" (1962, 210).

Furthermore, political life within the primitive gens was thoroughly democratic. Leaders of the gens were elected (and deposed) through democratic process with both men and women voting. The leaders' authority was noncoercive. Basic policies were established in an assembly of all adult members of the gens, with equal voice by all parties, male and female. The principle of mutuality governed social relations: all were obliged to support, protect, and assist each other. Personal property, at the time of one's death, was distributed among all surviving members of the gens (1962, 242-46, 274-76). Within the gens, within the tribe, and even within the confederation of kindred tribes, society existed without the state: "Everything runs smoothly without soldiers, gendarmes or police; without nobles, kings, governors, prefects or judges; without prisons; without trials" (1962, 253).

But such an organization of life was bound for extinction. Wars raged between and among tribes. Forms of production were simple and unsophisticated. Peoples lived under the domination of natural forces that seemed alien and foreign to them. Yet the gradual break with this primitive form of existence was not without its dark side.

> The power of these primordial communities had to be broken, and it was broken. But it was broken by influences which from the outset appear to us as a degradation, a fall from the simple moral grandeur of the ancient gentile society. The lowest interests—base greed, brutal sensuality, sordid avarice, selfish plunder of common possessions—usher in the new civilized society, class society; the most outrageous means—theft, rape, deceit and treachery—undermine and topple the old, classless, gentile society. And the new society, during all the 2,500 years of its existence, has never been anything but the development of the small minority at the expense of the exploited and oppressed great majority; and it is so today more than ever before. (1962, 255)

The stages and substages of evolution/devolution from primitive to civilized society which Engels, following Morgan, outlines are several. But the key factor in the process is the gradual emergence of private property. The dissolution of primitive gentile society is effected by a steady increase in productivity, accumulation of wealth, division of labor, commercial exchange, use of money, and production of commodities (1962, 307-18).

In family structure, the correlative shift is from group marriage to monogamy, which "was the first form of the family based not on natural but on economic conditions, namely, on the victory of private property over original, naturally developed, common ownership" (1962, 224). Monogamy is not, as the bourgeois pretend, a reconciliation of male and female; it is not a relationship of equality. It is instead the subjection of female to the male, who intends control over the household and over accumulated wealth. Monogamy is, in short, a form of patriarchy, and "the overthrow of mother right was the *world-historic defeat of the female sex*. . . . the woman was degraded, enthralled, the slave of the man's lust, a mere instrument for breeding children" (1962, 217). Engels, referring to *The German Ideology*, adds, "The first class antagonism which appears in history coincides with the development of the antagonism between man and woman in monogamian marriage, and the first class oppressions with that of the female sex by the male" (1962, 225).

In political structure, private property engendered the formation of the state. The state, Engels declares, is neither a heteronomous imposition on society nor a final expression of the inner spirit of society.

> Rather, it is a product of society at a certain stage of development; it is the admission that this society has become entangled in an insoluble contradiction with itself, that it is cleft into irreconcilable antagonisms which it is powerless to dispel. But in order that these antagonisms, classes with conflicting economic interests, might not consume themselves and society in sterile struggle, a power seemingly standing above society became necessary for the purpose of moderating the conflict, of keeping it within the bound of "order"; and this power, arisen out of society, but placing itself above it, and increasingly alienating itself from it, is the state (1962, 319).

But the state, as the product of a certain stage in the development of economic structures and as, usually, an agency of the possessing class, is but a temporal phenomenon. The state as a form of political life, like monogamy as a form of family life, will one day cease to be.

> We are now rapidly approaching a stage in the development of production at which the existence of these classes not only will have ceased to be a necessity, but will become a positive hindrance to production. They will fall as inevitably as they arose at an earlier stage. Along with them the state will inevitably fall. The society that will organise production on the basis of a free

> and equal association of the producers will put the whole machinery of state where it will then belong: into the Museum of Antiquities, by the side of the spinning wheel and the bronze axe (1962, 322).

With all the language of inevitability and necessity, Engels conjoins the language of ethical judgment and expectation. The basis of civilization, he complains, is the exploitation of one class by another:

> But this is not as it ought to be. What is good for the ruling class should be good for the whole of the society with which the ruling class identifies itself. Therefore, the more civilisation advances, the more it is compelled to cover the ills it necessarily creates with the cloak of love, to embellish them, or to deny their existence; in short to introduce conventional hypocrisy (1962, 326).

But there is a new time coming: " 'a revival, in a higher form, of the liberty, equality and fraternity of the ancient gentes' " (1962, 327). With that quotation from Morgan, Engels concludes the text.

VII

In his uncompleted but elaborately designed project on *The Dialectics of Nature,* which has come down to us in a set of folders containing notes and unconnected essays, Engels intended to draw the Marxian tradition into a critical relationship with developments in the natural sciences. As Norman Levine interprets the text, Engels

> reduced the physical cosmos to one elemental force, motion. From the laws of motion, Engels believed it possible to derive all other components of the cosmos. . . . Engels was seeking the single origin of all things. . . . he postulated the existence of one pivotal cause of the universe, which existed apart from man, which was eternal and self-generating. . . . his interpretation of the cosmos was deterministic. That is, the processes of the universe could not unfold but in fixed stages, absolutely conditioned series, in the pattern of mechanistic determinism. (Levine 1975, 116; cf. Cornforth 1980, 40, 69)

Levine's interpretation of Engels is, by his own admission, tendentious. He is, however, correct in discerning a central principle of Engels's method, a principle of cosmogonical/cosmological significance. According to Engels, the whole of nature, i.e., the entire universe, is a systematically interconnected totality of bodies, of material existences engaged in constant motion. "Motion, in the most general sense, conceived as the mode of existence, the inherent attribute, of matter, comprehends all changes and processes occurring in the universe, from mere change of place right up to thinking" (1964, 70, cf. 200).

Yet, Engels insists, motion is not mechanistic: "the motion of matter is not merely crude mechanical motion, mere change of place, it is heat and light, electric and magnetic tension, chemical combination and dissociation, life and, finally, consciousness" (1964, 37). There is an inherent directionality to motion, a progressive movement toward a culminating moment of realization. At this point, however, Engels confronts the inexorability of perishing, of finiteness, of death. All systems come to an end. And yet the end is not and cannot be absolute, given the meaning of motion and its potentialities. Engels is thus pressed to his doctrine of the cycle of motion.

> It is an eternal cycle in which matter moves, . . . a cycle in which every finite mode of existence of matter . . . is equally transient, and wherein nothing is eternal but eternally changing, eternally moving matter and the laws according to which it moves and changes. But however often, and however relentlessly, this cycle is completed in time and space; however many millions of suns and earths may arise and pass away; however long it may last before, in one solar system and only on *one* planet, the conditions for organic life develop; however innumerable the organic beings, too, that have to arise and to pass away before animals with a brain capable of thought are developed from their midst, and for a short span of time find conditions suitable for life, only to be exterminated later without mercy—we have the certainty that matter remains eternally the same in all its transformations, that none of its attributes can ever be lost, and therefore, also, that with the same iron necessity that it will exterminate on the earth its highest creation, the thinking mind, it must somewhere else and at another time produce it. (1964, 39-40).

Motion, in short, is not mechanistic; rather it operates according to the laws of dialectics (1964, 63-64, 214-36). There is, on a grand scale, an oppositional interaction between beginnings and endings, life and death, directionality and cyclicity. So also there is an oppositional interaction between freedom and necessity or creativity and determinism which is expressed most acutely in Engels's doctrine of the specific nature and condition of human activity.

One of the more complete sections of *The Dialectics of Nature* is on "The Part Played by Labour in the Transition from Ape to Man" (1964,

172-86). Labor "is the prime basic condition for all human existence, and this to such an extent that, in a sense, we have to say that labour created man himself" (1964, 172). The possibility of labor emerges with "the specialisation of the hand—this implies the *tool*, and the tool implies specific human activity, the transforming reaction of man on nature, production" (1964, 34). To be sure, both human beings and nonhuman animals react to and have an impact on their natural environment. But labor is the specifically human form of action; labor is premeditated, planned, transformative action toward preconceived ends. Labor results in the mastery of nature. Labor gives rise to the human mind which can deliberate about things possible and desirable and which can in interaction with nature transform the world to satisfy its needs and wants: "There is devilishly little left of 'nature' as it was in Germany at the time when the Germanic peoples immigrated into it. The earth's surface, climate, vegetation, fauna, and the human beings themselves have infinitely changed, and all this owing to human activity, while the changes of nature in Germany which have occurred in this period of time without human interference are incalculably small" (1964, 235). Yet the possibilities are not unlimited; for each victory "nature takes its revenge on us" (1964, 182). Engels cites cases of the destruction of forests, the introduction of the potato, the distillation of spirits, the invention of the steam engine to demonstrate the not always acceptable consequences of human labor.

The Dialectics of Nature does not attend to the ethical question in any direct or obvious manner except as it expresses support for the socialist revolution through which workers will wrest control of social production and distribution from the ruling bourgeoisie and thereby, in the formation of a classless society, advance the cause of humanity. That is, labor becomes more truly and fully labor in a thoroughly cooperative and socially planned system of production (1964, 313-14).

VIII

In Kolakowski's interpretation,

> According to Engels . . . the opposition between materialism and idealism is the central question on which philosophy has always turned. In the last analysis it was, in his opinion, a debate concerning the creation of the world. (1978, 378)

The Marxian tradition is materialist. Its cosmogonical dimension is a direct expression of that philosophical alternative. On the anthropological level, materialism means that the principle of labor is the sine qua non of specifically human existence. On the cosmological level, materialism means that the universe is made up of matter in motion, but motion understood in a dialectical, not a mechanistic, manner. On the historiological level, materialism means that the changing epochs of human life are a function of structural crises in the productive process.

The ethical dimension of the Marxian tradition, while not worked out in any detail, is likewise a function of its version of materialism. What is ethically possible at any time depends on the prevailing structure of production. What is ethically promulgated is an expression of social location. What, ultimately, is ethically desirable is the full actualization of the species nature of humanity. "The fundamental human good for Marxism . . . incorporates self-realization in community, freedom as the overcoming of alienation, mastery over nature and the maximization of welfare" (Lukes 1982, 201). In effect, the human good is communism.

Cosmogony and ethics are not, within this tradition, matters of pure theory. They are matters of praxis. As Marx cast this in his eleventh thesis on Feuerbach, "The philosophers have only *interpreted* the world in various ways; the point, however, is to *change* it" (1976, 617). On its critical side, praxis entails a suspicion of theories, whether in the form of mythical narrative or philosophical doctrine. Cosmogonies and ethical doctrines are expressions of social circumstances; they are attempts at self-justification; they may even be self-deceptive in origin and in effect. But on its active side, praxis assumes there is a substantive content to the human species and a direction to human history. Principles of conduct and purposes of action derive from and find their meaning in the real premises and actual destiny of human existence. In the final analysis, however, as Marx remarks, one proves the truth of one's thinking in one's practice (1976, 615).

References

Adoratsky, V.
1934 *Dialectical Materialism: The Theoretical Foundation of Marxism-Leninism*. New York: International Publishers.

Bochenski, Joseph M., ed.
1972 *Guide to Marxist Philosophy: An Introductory Bibliography*. Chicago: Swallow Press.

Buchanan, Allen E.
1982 *Marx and Justice: The Radical Critique of Liberalism*. Totowa, N.J.: Rowman and Littlefield.

Cohen, Marshall; Thomas Nagel; Thomas Scanlon; eds.
1980 *Marx, Justice, and History*. Princeton: Princeton University Press.

Cornforth, Maurice
1980 *Communism and Philosophy: Contemporary Dogmas and Revisions of Marxism*. Humanities Press; London: Lawrence and Wishart.

Engels, Frederick
1962 *The Origin of the Family, Private Property and the State*. In *Karl Marx and Frederick Engels, Selected Works in Two Volumes*. 2:170-327. Moscow: Foreign Languages Publishing House.
1964 *Dialectics of Nature*. Trans. Clemens Dutt. Moscow: Progress Publishers.
1966 *Herr Eugen Dühring's Revolution in Science (Anti-Dühring)*. Trans. Emile Burns. Ed. C. P. Dutt. New York: International Publishers.
1978 *Socialism: Utopian and Scientific*. Trans. Edward Aveling. New York: International Publishers.

Gouldner, Alvin W.
1980 *The Two Marxisms: Contradictions and Anomalies in the Development of Theory*. New York: The Seabury Press.

Hook, Sidney
1958 *From Hegel to Marx: Studies in the Intellectual Development of Karl Marx*. New York: The Humanities Press.

Kolakowski, Leszek
1978 *Main Currents of Marxism: Its Rise, Growth and Dissolution*. Vol. 1. *The Founders*. Trans. P. S. Falla. Oxford: Clarendon Press.

Levine, Norman
1975 *The Tragic Deception: Marx Contra Engels*. Oxford and Santa Barbara: Clio Press.

Lukes, Steven
1982 "Marxism, Morality and Justice." In G. H. R. Parkinson, ed., *Marx and Marxisms*, 177-205. Royal Institute of Philosophy Lecture Series 14. Cambridge: Cambridge University Press.

McLellan, David
1979 *Marxism after Marx: An Introduction*. New York: Harper & Row.

Marx, Karl
1975a *Difference between the Democritean and Epicurean Philosophy of Nature*. In *Karl Marx, Frederick Engels, Collected Works*. Vol. 1. *Karl Marx: 1835-43*, pp. 25-107. New York: International Publishers, Trans. Dirk J. and Sally R. Struik.

1975b *Notebooks on Epicurean Philosophy*. In *Karl Marx, Frederick Engels, Collected Works*. Vol. 1. *Karl Marx: 1835-43*, 403-509. Trans. Richard Dixon. New York: International Publishers.

1976 *Theses on Feuerbach (Original Version.)* In Marx and Engels, *The German Ideology*. Moscow: Progress Publishers.

Marx, Karl, and Frederick Engels
1975 *The Holy Family, or Critique of Critical Criticism: Against Bruno Bauer and Company*. In *Karl Marx, Frederick Engels, Collected Works*. Vol. 4. *Marx and Engels: 1844-45*, Trans. Richard Dixon and Clemens Dutt. New York: International Publishers.

1976 *The German Ideology*. Moscow: Progress Publishers.

Wilson, Bryan
1982 *Religion in Sociological Perspective*. New York: Oxford University Press.

15 Freud as Creator and Critic of Cosmogonies and Their Ethics

Lee H. Yearley

Introduction

If one imagined a parlor game the aim of which was to force the players to make imaginative—or even fanciful—connections between three apparently unrelated words, a leading candidate would be cosmogony, ethics, and Freud. I will argue, however, that inquiry into these three terms not only leads to an interesting perspective on Freud but also says something important about the general issue of cosmogony and ethics, especially as that issue appears in the modern West.

Such an inquiry leads us to consider some of Freud's more bizarre ideas—the death instinct, morality as a pathological phenomenon, and consciousness as a symptom—and thereby to weigh Adorno's comment that in psychoanalysis nothing is true except the exaggerations. It also leads us to reflect on a view of human agency that seems to undermine both normal moral theory and normal models for explicating the relationship between cosmogony and ethics. Moreover, it leads us to think about how a sophisticated modern both criticizes and formulates cosmogonies, a topic that raises various questions about the similarities and differences between cosmogonies and theories and the status of the so-called hermeneutics of suspicion. Finally, the inquiry leads us to consider a distinctively modern notion of cosmogony and ethics, based on ideas of evolution and instinct, that still attracts many. The roads that open out are, then, too many rather than too few. I will try to keep matters manageable by presuming at least some familiarity with Freud's ideas and by sliding over certain difficult issues, especially if those issues threaten to involve us in the arcane world of Freudian scholasticism, the forest inhabited by what Freud called the witch of metapsychology.

Before I proceed, however, a brief comment is needed about certain of the difficulties we encounter in attempting to examine Freud. Freud is difficult to understand for a variety of reasons: the hold that vulgar Freudianism has on all our minds; the fact that resistance is a reality, if not the omnipotent reality that Freud often claims it to be when he uses it as an ax to dismantle his critics; and the extreme interpretative difficulty that confronts any enterprise that attempts to extrapolate to the unknown from symptoms. More important here, however, are the difficulties presented by the sort of thinker, writer, and person that Freud was.

Freud considered himself, luckily he thought, constitutionally incapable of doing philosophy, and he does lack the interests and capacities (he might have said fears and obsessions) that generate tight philosophic formulations or arguments. This fact combines with three probably more important qualities: an admitted sloppiness in formulation, an undervaluing of the importance of coherence in terminology, and a feisty one-sidedness that is exacerbated by his usual tailoring of his approach to specific audiences and occasions. Freud's writings often show the temper of one who sees himself as a conquistador. He opens up new territories through dramatic conquests and then lets lesser bureaucrats administer them and worry about the details (see Freud 1926a, 39-40, 110-12; Jones 1961, 24).[1]

Moreover, and especially with those more daring formulations we are interested in, Freud often has difficulty keeping in focus the most radical aspects of his own ideas. The notion that the master expositer of repression may have repressed certain of his most striking ideas has become an interpretive principle for some. But here I point to a simpler and more understandable phenomenon with which any writer is familiar: the difficulty of keeping in focus an idea that wars with one's normal conceptions and desires.

These difficulties have been a boon to the Freudian scholarly industry and a stimulus to sectarian struggles within psychoanalysis. But they do present real problems and necessitate, I think, an approach to Freud that combines tempered freedom and charitable aim. Such an approach must also recognize, however, that at places Freud is deeply incoherent, sometimes for understandable and possibly curable reasons, sometimes because he tangles with perplexing and significant problems. In either case, and especially with his more radical ideas, some development of the most interesting aspects of the ideas may be necessary. Such developments are always, of course, suspect. But with Freud they often are necessary, particularly in areas such as religion and ethics where his direct analyses of the topics often suffer from glaring faults that are absent from his analyses of related topics.

Given all this, let us begin our examination. It will cover three general areas: Freud's criticism of cosmogonies and their ethics; Freud's creation of a cosmogony based on the life and death instincts; and Freud's account of ethics as a product of that cosmogony. Initially, then, we will examine two sides of Freud (and perhaps, of many moderns). One side is Freud the critic of cosmogonies, the rescuer of people from false gods, the ceaseless deflater of people's narcissistic pretensions. The other side is Freud the constructor of cosmogonies, the prophet of new gods, the proclaimer of a non-narcisstic vision. This side also produces from its cosmogony a distinctive, even bizarre, picture both of ethics in particular and human agency in general. But let us start by analyzing Freud the critic of cosmogonies and their ethics.

Freud as Critic of Cosmogonies and Ethics: Introduction

Seen from one perspective, Freud clearly belongs with that set of modern Western thinkers who attempt to destroy what they see as those damaging mystifications by which humans have previously structured their lives. High on the list of such mystifications, perhaps even first for Freud, are those cosmogonies produced by either the individual neurotic or the religious person. These cosmogonies are especially important targets because they inevitably generate action. They are especially clear targets because virtually all recognize how harmful is the neurotic's cosmogony and virtually all can be led to recognize, Freud thinks, how harmful is the religious person's cosmogony.

The human mind is for Freud a fertile producer of cosmogonies. We all incarnate a process of fictive creation that aims at ordering our cognitive world and guiding our intentional actions (see Meisel 1981). This capacity exists or at least is activated (Freud never clarifies this delicate point) because human beings have been dealt a particularly bad hand by the evolutionary process—and women an even worse hand than men, for whatever cold comfort that may offer to either sex.

Humans face problems—such as establishing appropriate relationships to authority and controlling potentially destructive desires—before they have the capacities to handle them sensibly. The result is that humans are forced into a set of makeshift accommodations that travel under the general name of defenses. These accommodations are necessary because they help us avoid catastrophes. But in time they become somewhat disfunctional in all, and completely disfunctional in some. People both develop their capacity to handle difficulties and meet new situations that present problems about which the formerly saving scenarios of early childhood fail to give adequate guidance. Our interest here is not in Freud's often strikingly incomplete analysis of either the full panoply of possible defenses or the developmental process. But we are interested in those defenses that are cosmogonic in character.

In individual case histories (most notably that of Schreber but also, say, those of the Rat Man and the Wolf Man) Freud presents us with detailed studies both of an individual's cosmogony and of how it guides his actions. (Freud 1963d) These cosmogonies present general pictures of the world's origin and character, have questions asked in terms of them rather than about them, and guide the specific actions of the individual. These cases, of course, deal only with how a particular individual uses specific myths and actions to defend himself. But Freud feels justified in extending the procedures and conclusions that work in examining the individual neurotic (or psychotic, in Schreber's case) to the case of religious cosmogonies that inform whole cultures, in part because of the existence and role of cosmogonies in the two cases.

One can query Freud's procedure of using the abnormal to interpret the normal, although he can marshall formidable arguments to justify that diagnostic model. Moreover, one can legitimately query his fondness for synecdoche, for taking the part for the whole. All too often it informs an analogic method that is only either crudely funny or infuriatingly insensitive as when, for example, he will blithely declare that artists equal hysterics, philosophers equal paranoics, and religious people equal obsessional neurotics.

Most important here, Freud's attempts to examine religious cosmogonies in themselves, and as they guide the actions of individuals in a culture, suffer from numerous specific problems: they lack the rich details of his individual cases; they often suffer from their polemic tone; they generally finesse the question—which Freud himself worries about—of how individual and cultural neuroses differ; they usually fail to deal adequately with important distinctions such as that between primal fantasies which are phylogenetic and cultural expressions which, at the least, may reflect sublimation and individual ontogenetic needs; and they violate Freud's own rule that one-to-one correspondences between images and meanings cannot be made in the abstract but always must be carried out in relation to the subject's own associations.

Freud as interpreter and critic of religious cosmogonies stands, then, on more perilous ground than he does as critic and interpreter of individual cosmogonies—and he often dances over gaping crevices. Nevertheless, it takes a hardier soul than I to fail to read primitive cosmogonies in a different way after one soaks oneself in Freud. Whatever may be the problems and crudities in Freud's general procedures and specific analogies, he gives us an alternative viewpoint that can be revealing and that arises from a perspective which rests both on a theory and on evidence that are strong enough to give his view some plausibility.

I will not attempt here either to examine some specific Freudian interpretations or to examine the more basic and general issue of how to adjudicate among different possible interpretations of a phenomenon such as a cosmogony. Rather I will examine certain distinctively modern principles that underlie Freud's approach to cosmogony and ethics as well as many moderns' unease with the idea of both a cosmogony and a cosmogony's effect on ethics. These principles undergird Freud's distinctive version of the hermeneutics of suspicion. That interpretative method explains phenomena in terms that differ from those the involved agent would use, and may "reduce" those phenomena to categories or entities that the involved agent would reject. These principles do not make clear, however, why the practitioner of the hermeneutics of suspicion turns from interpretation to criticism. That issue finds an especially well-focused form in Freud and is important to the general question of Western modernity's approach to the issue of cosmogony and ethics. But let us turn first to the principles underlying Freud's account.

Freud as Critic: The Underlying Principles

One reason for the power of Freud's interpretation of cosmogonies is that it provides some kind of answer to the questions of why cosmogonic ideas would get started and of why they would be found persuasive. In just asking these questions the interpretation manifests a key Freudian principle: the notion that we do think certain ideas or acts are bizarre, are exaggerated responses to difficulties to which a range of normal responses are understandable. This principle rests on another principle: that we can explain such exaggerated ideas and actions. Both principles reflect deep presumptions in much modern Western thought and raise questions we will return to later. But let us continue on to note the last principle underlying Freud's position.

This principle, or perhaps better presupposition, is that any view that justifies a human being's higher aspirations and values by placing them within a comprehensive picture of the world is simply in error. Freud exemplifies that kind of modern who cannot even entertain seriously the idea that human values may reflect, in however distorted a fashion, the way the world is actually constituted. For Freud the idea that the world's larger pattern does not fit human aspirations is so clear that it is not even painfully clear, as is witnessed by his comment that even to ask the question of life's meaning is to reveal oneself to be ill (Freud 1960, 436). To seek, much less to find, a world congenial to human desires is to reveal the power of one's infantile desire to feel oneself important and to think one's sense of need and of fairness reflects some inviolable standard.

One need not, of course, draw from evolutionary theories the implications Freud did. William James surely took evolution seriously, but he adopted a quite different approach than Freud's. Moreover, one can hardly foreclose as easily as does Freud the question of whether one should ask questions about life's meaning in a way that gives human experience a privileged place (see Nagel 1979, 11-23, 196-213).

Most important to us, one needs to distinguish, in a way Freud did not, among cosmogonies of various sorts. For example, some cosmogonies do reflect human aspirations and values, such as compassion and justice, that are of a noble sort. But others reflect human aspiration and values, such as the acquisition of money and power, that are of a less noble sort. Finally, still others reflect processes that have little relationship to any human aspirations and values. Let us call the first of these cosmogonies a "justice cosmogony," the second a "power cosmogony," and the third an "inhuman cosmogony." Freud might argue that all these cosmogonies are similar because they all attempt both to console one by putting one's problems into a larger picture and to depict an order that justifies or makes understandable some actions. Clearly, however, Freud almost always focuses on justice cosmogonies, a focus that reflects his usual, although not invariable, tendency, when talking about religion, to concentrate not only on Christianity but on Catholic Christianity.

Indeed, it is intriguing to speculate about whether Freud's often confusing approach to cosmogonies may be clarified by seeing that he has very different attitudes to each of these three cosmogonies. Justice cosmogonies are for him religious cosmogonies, and he rejects them. Power cosmogonies are for him what individual neurotics or psychotics create, and they are to be treated by showing the individual that they exact too high a price. Inhuman cosmogonies are for him potentially viable pictures of the world and therefore may be either accepted or even constructed, as one may argue Freud himself does. Were Freud to make these distinctions among cosmogonies, he might go on to claim that this last kind does not really count as a cosmogony, especially if justice cosmogonies are to be seen as paradigmatic. An observer might of course then inquire if the differences between justice and power cosmogonies are so great that Freud's analogy between them is imperiled. Such an inquiry would lead us, however, even further into what is only speculation, and it is probably wiser to return to more solid ground.

For Freud the world has no evident meaning and therefore the existence of cosmogonies and of the guidance they provide is a bizarre fact that demands an explanation. Freud thinks a plausible explanation for these phenomena is provided when we examine how they help to resolve problems that the developmental process produces. Such is his diagnosis, and his motivation for making it seem to resemble that of any careful and dispassionate investigator.

But more is involved because such a portrait of Freud fails to explain both the passion in his inquiry and the complete rationale for it. These arise from the last of our underlying principles: that one is obligated not merely to interpret these cosmogonies and the ethics they produce but also to criticize them in order to destroy them. Examining this principle and its implications at some length is worthwhile because the inquiry tells us much about Freud's perspective on cosmogony and ethics and, by extension, about the perspective of many other practitioners of the hermeneutics of suspicion. Moreover, the inquiry opens up questions of fundamental importance to any attempt to discuss the possible place of either cosmogony or ethics in the modern world. That Freud fails adequately to answer these questions is relatively unimportant. As is often the case with him, his failures can often be more illuminating than the successes lesser figures.

Freud as Critic: The Rationale for the Destructive Criticism

Even if cosmogonies and the actions they guide are explainable as the results of childhood or childlike defenses, the question remains as to why one should criticize them in order to destroy them. One may affirm that one ought to help an individual destroy a cosmogony that leads him, for example, to fear the touch of women and to spend much of each day in elaborate rituals that allow him to eat, especially if he is seeking to rid himself of such problems. But what if a culture has such a cosmogony or, in a more difficult case, has a cosmogony that is, to our eyes, rather quirky and produces odd actions but has little significant impact on people's lives? On what ground can one argue it should be destroyed?

This question fits within the general problem of how to decide about what criteria to use to judge whether humans are flourishing or, in language closer to Freud's, whether they are healthy. The problem is, of course, an old one, but it has a distinctive modern form and bite, as well as many implications for any modern discussion of relativism in general and the relation of cosmogonies and ethics in particular. Moreover, in examining the tangle of issues and claims that arise from Freud's fragmentary attempt to answer it, we discover much about Freud's perspective on cosmogony and ethics.

Freud's desire to criticize cosmogonies rests in significant part on what can be called ethical grounds. He rarely adopts the simple approach, at least in his less positivistic moments, that argues the falsity of an idea necessarily implies one should attempt to destroy it. Not only is he sensitive to the problems involved in falsifying comprehensive notions but also, and more important here, he recognizes how frail a reed mere truthfulness can be if human well-being is one's aim. Rather, he thinks the destruction of cosmogonies and the ethics they produce will help people flourish more freely and fully, will help them obtain more easily and efficiently what they really need and desire.

Nevertheless, a brittle attitude does appear in his criticism of cosmogonies. He will declare, for example, that a life without cosmogonies simply represents a higher stage of the evolutionary process, even if the possession of such cosmogonies was a needed stage in the developmental process. As in many cases where developmental arguments are used, however, Freud never adequately defends, or even explains, the criteria for judging why one stage exceeds another. Generally, he employs criteria such as greater adaptiveness and an enhanced ability to obtain those goods, like food and shelter, that all people really seek. We cannot easily dismiss such arguments but neither can Freud easily accept them. They rest on an anthropology that, for him, sees human desires in too pedestrian a way.

His more complex criticisms rest on two interlocking ideas, both of which represent empirical judgments or claims. These ideas reflect, I think, the reactions of many sensitive and sensible modern observers when they face traditional cosmogonies and the ethics they produce. The first claim is that such cosmogonies have just lost their power, at least for most educated Westerners. They no longer deeply and thoroughly form people's lives by answering important questions and giving significant directives for life. Moreover, their residues are also often destructive. People are haunted rather than consoled by the residues of these traditional cosmogonies; their intentions, actions, and desires are made ambivalent rather than whole. At worst, these residues destroy the possibility of a constructive life, leaving only disabling obsessions. But even at best, these residues disturb people. They generate, for example, a vague sense of guilt for acts or ideas the people rationally accept; the ex-fundamentalist at play presenting a paradigmatic example. If all neurotics are haunted by reminiscences, the residues of a past that remains too powerful, many moderns are haunted by the residues of a cosmogony and its ethics. They have lost the power to guide, explain, and console and kept only the power to disturb, divide, and distort. Many moderns suffer from that kind of neurosis which now does frailly and destructively what previously was done powerfully and curatively.

Closely related to this is Freud's other empirical claim: for an educated modern person no traditional cosmogony can stand up to the alternative explanations provided by either the hard or soft sciences. To understand an alternative explanation—simply to weigh up the benefits and losses each explanation represents—is either to realize the intellectual bankruptcy of a traditional cosmogony and the ethics it produces or to have broken its intellectual hold if not its emotional hold. For Freud, any sensible modern observer will agree on what constitutes the broad outlines of mature belief and action and realize that modern explanations and actions resemble those outlines more clearly than do those that arise from traditional cosmogonies.

Both claims state, then, that educated people live in an era where no choice exists but to reject traditional cosmogonies and ethics. Nostalgia or neurosis may blind people to that fact, but mature self-examination should make them comfortable with the rejection. Such is Freud's position and one would, I think, be hard-pressed simply to dismiss these claims—at least if one desires, for whatever reason, to be a full-fledged participant in modern intellectual culture.

And yet one also has, I think, the sense that this will not do, that something is missing. We cannot examine all the issues that arise here, but we can analyze one aspect of them in terms of a distinction that Freud himself makes: the distinction between illusions and delusions. He fails to develop it or probably even to realize its implications. But the issues this distinction raises help us gain a more textured understanding of Freud's critical stance and foreshadow some problems we will encounter later in examining Freud's own account of cosmogony and ethics.

Freud as Critic: Illusion and Delusion

Freud makes a significant and unfortunately too often overlooked distinction between illusions and delusions (Freud 1927, 30-34; 51-56). Roughly put, illusions are products of wish fulfillments which may or may not both match reality and help the self and others. Delusions are also products of wish fulfillments (at least in an extended sense), but they clearly both fail to match reality and do harm the self and others. Illusion and delusion mark off the two ends of a single spectrum, and fascinating questions arise about how to sort out those activities that do not fit neatly at either end. But let us clarify both the distinction and the spectrum's character by an example.

I am teaching a seminar on Freud. I am in the grip of an illusion if I think that all the students find me articulate, compassionate, and wise and think the seminar is an exciting and valuable intellectual experience. I am in the grip of a delusion if I think the students have concocted a plan to kill me with the guns they have hidden under the table. Moreover, I continue to believe this delusion despite my furtive glances under the table, my knowledge of the students' characters, and my sense of the likelihood of anyone choosing to kill me at that time in that way. No evidence can dislodge my delusion, and I will both live in constant fear and treat the students as the killers I know them to be. Evidence can, however, make me question or even surrender my illusion. Moreover, the illusion is probably harmless and may even make me feel happier and teach a better class. Delusions cripple but illusions may help.

Indeed, Freud thinks that illusions must play a basic role in life and can play a productive role. Illusions must play a basic role because unfulfillable wishes are always a crucial constituent of life. We all continue to carry a full kit bag of childlike desires that we want fulfilled, even when we have realized that they are unfulfillable. Humans remain divided in that wishes, desires to be loved or to operate in a just world, live in one part of the self, while a knowledge of these wishes' obvious futility and archaic character lives in another part of the self.

Such a situation means that illusions can play an often innocent and possibly productive role in life. Indeed, illusions often function for Freud as do what were once called the consolations of civilization. Art, for example, is an immensely beneficial illusion that, in some cases, seemingly can even provide a consoling cosmogony. Humor provides similar benefits, although it is a narcissistic phenomenon that rests on one's childlike sense of invulnerability (Freud 1963a, 263-69).

Given this, the question arises of how to adjudicate among illusions, deciding that some are necessary for human thriving but others are destructive or are acceptable only within limited boundaries. Why, for example, is it acceptable to enjoy and be consoled by Dante's art, but unacceptable to believe in a Dante-like cosmogony? This question, which resembles the issue in utilitarianism of how to adjudicate among various benefits, is complex and vexing. Freud not only fails to answer it satisfactorily, he never even fully faces it.

Freud's failure to investigate these issues may rest on his keen sense of how the analyst's probe can turn on the analyst and cause significant difficulties (Freud 1900, 146). Probably more important, the failure may highlight how crucial for him is the empirical claim that educated modern people simply cannot accept certain illusions once they know they are illusions.

But that claim seems highly suspect, even within Freud's own perspective. One can accept, for example, Freud's interpretations of humor as a narcissistic wish fulfillment and still find oneself laughing. Indeed, one might even accept Freud's position on illusion and because of it decide to cultivate a taste in literature—or perhaps even religion. (Interestingly enough, it is unclear if one could choose to cultivate a taste in humor. Perhaps the grip of humor on someone is more like the grip of a cosmogony than like the grip of literature, possibly because both humor and cosmogony rest in narcissism and reflect comprehensive views about the world.)

Whatever may be the reasons why Freud fails to face these questions, the questions themselves are especially important to us because they focus the difficulties that are involved for Freud (or anyone) in producing and defending criteria that might lead one to reject cosmogonies and the ethics they produce. For example, ethical rules, in Freud's eyes, are usually tied to cosmogonies and therefore are to be seen, at best, as illusions. Yet ethical rules seem to be absolutely necessary to the continuance of civilized life. How can Freud not examine ethics as a necessary and even beneficial illusion, given his commitment, however ambivalent, to the significance of civilized life? The absence of such an analysis seems to fit well with one of the more striking facts about Freud: his remaining unperturbed by all those modern questions that gather under the rubric of "If the gods are dead, is everything allowed?"

The best way to fill in these surprising lacunae in Freud's account of illusion, delusion, and related issues is to turn to Freud the cosmogonist. For this great critic of cosmogonies creates a cosmogony that in ways is as general, as bizarre, and almost as scholastic as some of the cosmogonies he rejected. The form of this cosmogony, why he creates it, how he goes about defending it, and how it links to ethics, tells us much about Freud. Moreover, examining it allows us to understand much about cosmogonies, ethics, and the modern, both in general and as those topics appear in the particular movements that still adhere to some version of the Freudian position.

Freud's Own Cosmogony: Introduction

As we have seen, Freud may be described as a classic critic of cosmogonies. Moreover, he said that psychoanalysis itself lacked even a comprehensive worldview (much less a cosmogony) other than that provided by the procedures of scientific inquiry (Freud 1933, 139, 140). Nevertheless, I think it is clear that Freud has not one but two cosmogonies, both of which generate guides for action. The first, which I will call his historical cosmogony, depicts the origins of the social order in the actions of a primal horde, especially the murder of the father. The second, which I will call his death-instinct cosmogony, depicts life- and death-instincts as cosmic forces that work through everything.

The first cosmogony tells us much about Freud, but I will not concentrate on it. It is an often told story that is especially familiar to those interested in religion. Moreover, much of what is significant for ethics in this cosmogony finds a more focused form in the death-instinct cosmogony. Finally, the significance of this first cosmogony for Freud's mature views can be questioned. This judgment raises complicated technical issues, but for us the main point is that Freud's notion of primal fantasies seems to do all the necessary conceptual work in at least the areas we are interested in. That is, it "shows" people's predispositions to process material in paradigmatic ways, whatever may be the actual historical content of their personal and racial memories. This would make the historical cosmogony's story unnecessary, although the story does help Freud to distance himself from Jung's notion of archetypes, is congenial to his neo-Lamarckian ideas of phylogenetic inheritance, and does resonate with some nineteenth-century views on culture's character and origins. We can, then, by concentrating on the death-instinct cosmogony, avoid certain difficulties and banalities and approach the center of some of Freud's most distinctive and controversial views on the character of ethical action.

Before proceeding to that examination, however, we can briefly note that Freud's historical cosmogony has obvious ethical implications. Some of them are relatively straightforward; for example, his genetic explanation of incest rules. Others are more important, if also more indirect: for example, his propensity, shared by others in his time, to think in terms of conflict rather than cooperation when discussing humans in their "natural state." That propensity, as discussed in Professor Sturm's study of Marx in this volume, defines the human situation in a way that makes necessary artificial constraints on the individual if social order is to exist. Finally, the daring of Freud's approach to ethics is clearly if crudely displayed in his historical cosmogony. For example, he seems to hope that his work can help make taboos as foreign to civilized life as are totems. Such a project may seem unexceptional, but its radical implications become clear when he asserts (in a judgment he will modify elsewhere) that taboos resemble Kant's categorical imperative because both are compulsive and reject any conscious motives (Freud 1912, 22, 66-68, 71-74). In aspiration at least, Freud's historical cosmogony aims to present at least as many moral judgments as forms of pathology or primitivism.

Such an aspiration will come to fruition in the death-instinct cosmogony. Indeed, that cosmogony may be seen as presenting the abstract framework that underlies and helps explain the historical cosmogony in a way that resembles the structural relationship presented in Professor Reynolds' essay in this volume between the Buddhist theory of dependent co-origination and the historical cosmogonies. Moreover, in examining this cosmogony we see a modern who accepts certain fairly sophisticated notions of evidence and explanation, who attacks cosmogonic thought, and yet who tries to defend a comprehensive cosmogony. We glimpse here, then, the power of cosmogony to generate plausibility, stimulate credulity, and cover gaps in evidence or explanation. Finally, this cosmogony links directly with some "quasi-cosmogonic" pictures of ethical action, now evident in sociobiology, that continue to grasp the imagination of certain contemporaries (see Sulloway 1979, 5, 500). I will not pursue these relationships, but they will provide a context for my inquiry. Such pictures contain ideas as speculative as Freud's—selfish genes, for example. Moreover, as with Freud, they both pinpoint such odd phenomena as the fact that ethical concern often seems to be a function of distance. And they try to explain such odd phenomena in terms of evolution and instinct, interpreting them as the product of an adaptive mechanism.

Freud's Own Cosmogony: The Life- and Death–Instincts

Roughly formulated, the death-instinct cosmogony sees literally everything as ruled by two cosmic processes that together generate and control all activity: a drive to greater vitality and more complex structures and a drive to an absence of all vitality and complexity. One causes the productive linkage of things, the building and preserving of more and more complex but still unified entities. The other causes the disintegration of things, the destructive strife or slow decay that reduces and finally annihilates all tension and complexity. The two cosmic forces leave their particular impressions on everything. The quarrels and copulations of dogs, the heroic sacrifices and brutalities of human beings, the disintegration and growth of cells, apparently even the building up and breaking down of mountains and chemical elements all reflect the same forces.[2]

I shall not try here to trace out either the nuances of the idea or the historical permutations of Freud's own development of it. I will focus instead on the idea of instincts that underlies the notion, on the concept of moral agency that emerges from it, on why Freud felt forced to formulate and defend the cosmogony, and on how it relates to his views on ethics. We do need, however, to note something about the origin of the idea and, more important, about one key ambiguity in it.

The early Freud firmly maintained the principle that human processes were governed by a striving after pleasure, although his dualist predilections led him to pair instinctual desires into sexual or libidinal desires and egoistic or self-preservative desires. In 1920, however, Freud published *Beyond the Pleasure Principle*, a book some consider an embarrassment and others his most profound work. In that book Freud tentatively amalgamates all previously divided instincts, such as the libidinal and egoistic instincts, into a life instinct and posits another opposed instinct, the death instinct. The hold this view has on him grows (as he will bemusedly admit) and is clarified theoretically, if modified practically, by his developing sense of the importance of the fusion of the instincts. That idea allows him to declare that the death instinct is mute and will always be found only in fusion with the life instinct.

Most important here, the original description of the death instinct contains within it two significantly different formulations that Freud never, to my mind, successfully either distinguishes or relates. Under one formulation the death instinct is an entropic drive, to use a fashionable phrase. It is a movement to loss of energy and increased internal disorder within a currently stable entity. Under the other formulation, the death instinct is a drive to destructiveness, a drive exemplified when aggressive forces attempt to destroy each other. The two formulations do link loosely because aggression can generate disorder and does expend energy. In this sense, entropy might be said both to underlie and to be served by aggression, assuming, of course, that the animate and inanimate can in fact be closely related. But the relationship is far from clear and Freud never adequately spells it out.

I will focus here on the death instinct as aggression because the entropic formulation diminishes in importance as Freud employs the idea and has little relevance to the issue of cosmogony and ethics. Moreover, it suffers from two major theoretical confusions. Entropy seems better entitled a principle, like gravity, than an instinct, like hunger. Moreover, Freud's entropic formulations are at best confused and at worst in error. He never adequately distinguishes between a movement to reduce all tensions to zero—a version of the constancy principle that is a pure entropic action—and a movement to keep tensions within a tolerable range—a homeostatic principle that nurtures life. These problems are of little importance to us because as Freud develops the idea, especially in relationship to ethics, the focus on aggression becomes very prominent. (Indeed Freud will even at times label it the destructive or aggressive instinct, although he may still consider that manifestation derivative, in part because he is not making our distinctions [see Freud, 1930; 1963c: 252-65]).

My main concern is how this cosmogony relates to ethics, but we do need to note how the cosmogonic idea gripped Freud's mind despite the frailty of the evidence and argument used to support it. That topic uncovers certain important characteristics of Freud's thinking. Moreover, it shows us one interesting instance of the role of cosmogonic speculation in the mind of a sophisticated modern intellectual.

Freud's Own Cosmogony: The Frailty of His Speculation

Freud is aware, of course, of how one's desires can affect one's judgments both in the affairs of Europe and in his own personal life. He realizes that tragic events might make the death instinct congenial to him, and he comments on the general attractiveness of making death a figure of necessity (*Ananke*) (Freud 1920, 39). But he attempts to use rigorous standards of theory building and testing in his creation and defense of the death instinct. Without attempting to probe the technical aspects either of Freud's own theory or of the relationship of myth to overdetermined theory, we can see how Freud tries to use but actually violates such standards. We can do this by examining four phenomena that Freud thought could be explained only by the death instinct. Each phenomenon, however, can be given a plausible and more parsimonious explanation that does not include the death instinct.[3]

One phenomenon is the "negative therapeutic reaction," an umbrella term for diverse matters such as a patient cutting off therapy at an inappropriate time, exhibiting tremendous resistance, or having new symptoms appear. This phenomenon is indeed mysterious, but employing the death instinct to explain it is hardly necessary given both its similarity to other "normal" phenomena (libidinal adherence, narcissistic inaccessibility, and possible gain from illness) and, most important, the evidence that psychoanalysis fails to work for certain kinds of people (see Freud 1963c, 233-71; Madison 1961, 43-71; Laplance and Pontalis 1973, 263-65). The second phenomenon is the presence of widespread, overt aggression, defined here simply as the presence of a conscious or easily identifiable drive to hurt other people. The drive to aggression does demand explanation, but one can explain it in terms of sexual or self-preservative instincts, as have various contemporary ethnographic theories: that is, as drives, say, to fight off threats to one's general well-being, or possession of partners, or occupation of needed space. These two phenomena seem, then, hardly to imply the existence of a death instinct. Positing it from them seems to resemble using a hydrogen bomb to solve a border dispute—a possible action but surely neither a necessary nor even a sensible one.

The next two phenomena, which represent the strongest part of Freud's case, are linked because they manifest forms of self-destruction or masochism. One phenomenon is the compulsion to repeat traumatic incidents in actions, dreams, or even transferences. For example, a person continues to dream about a traumatic injury or, in more complex cases, to pursue benefactors who betray him or authoritative figures who punish him (Freud 1920, 7-11, 12-17, 26-27, 29-32, 50-53). The other phenomenon is an unconscious need for punishment. Because it is unconscious, this need can be seen clearly only in cases where, for example, success brings great psychic stress, or crimes are committed in order to bring retribution, or relationships are destroyed in order to generate suffering (Freud 1963a, 157-81; 1963b, 190-201).

Most would agree, I think, that the engagement of human beings in self-destructive acts is one of the most profoundly puzzling and distressing aspects of human behavior, even if one is not wedded to the idea that humans are simply pleasure-seeking creatures. Nevertheless, other plausible explanations exist besides the death instinct for these phenomena, and on the grounds of parsimony and conservation of beliefs a person would seem best advised to accept them or, at least, to let them make the question moot.

The simple release of energy or gaining of sympathy are two such explanations. But most important is an explanation that focuses on how all these phenomena can be read as attempts to gain mastery over anxiety, as tragically deformed responses that struggle toward health and self-preservation. A repetition compulsion can be seen as a struggle to master a traumatic event by an ill-fated but understandable attempt to relive the event and thus both control its appearance and possibly master the event in its repeated form. Similarly, the phenomenon of a need for punishment can be seen as a natural if tragic attempt to maintain the self under stress by repeating the painful but safe scenario of punishment. All these actions can be seen, then, as defenses against extreme stress and therefore as marks of a healthy ego manifesting strength and seeking pleasure rather than as the death instinct seeking destruction.

I am not arguing that these alternative explanations are obviously true but only that they cast doubt on the need to posit a death instinct. The idea of the death instinct is, then, painfully undetermined. It is neither clearly supported by the evidence nor tested by the usual procedures to which Freud adheres. One might, of course, argue that many Freudian ideas are similarly underdetermined, but this idea differs from most of his because of the combination of the grandeur and implausibility of the speculation and the frailty of the supporting evidence.[4] We see here, I think, how a cosmogony can so grasp the imagination of even a sophisticated modern that questions are asked in terms of it rather than about it, that normal evaluative procedures are displaced, and that differing views are seen as untruths rather than as competing opinions.

Let us now turn to our main topic of inquiry, the relationship of the death-instinct cosmogony to ethics. We can best begin that examination by analyzing briefly Freud's theory of instincts and the portrait of personal agency that emerges from it.

Freud's Own Cosmogony: Instincts and Human Agency

The life and death instincts are, of course, instincts, and in order to understand them we need to say something in general about Freud's theory of instincts. This need is especially pressing because only by examining Freud's theory of instincts can we understand how he "deconstructs" or redescribes the self. Freud portrays the self not as even a potentially unified and intending agent but as a variety of interacting forces that either are simply instinctual in character or are (in a topic we will examine later) personified in the ego, id, superego model.

Such a deconstruction underlies his attempt to make us see that most ethics are a pathological phenomenon, a manifestation of the death instinct in action. Moreover, that deconstruction undermines the understanding of moral agency that normally appears in any interpretation and justification of moral theory. Finally, it also challenges conventional ways of explicating the relationship between cosmogony and ethics because such explications rely on the idea of a unified and conscious agent.

This side of Freud is, then, radical and challenging, questioning as it does the very idea of agency that underlies most moral theories and most interpretations of cosmogony's link to ethics. It is not the only picture of agency that Freud presents. He will, for example, at times portray at least the "actualized" ego in a way that makes it resemble a center to a unified self and even fume against those who have read him (correctly to my mind) as elsewhere portraying a very weak ego (Freud 1926a, 22-23). He will also speak of judgments on what is painful or threatening in a fashion that is reminiscent of more traditional pictures of the self (Freud 1963a, 318).

But such portraits represent only one side of Freud, and I will concentrate here on those radical ideas that Freud himself often has difficulty keeping in focus whether because he represses them or because they differ too much from more common conceptions. I will examine later some implications of these two different sides of Freud, but what needs to be emphasized here is that this radical side of Freud does exist and relates closely to his death-instinct cosmogony or even arises from it. This side too often disappears when Freud is discussed, especially, perhaps, in this country where his ideas have often been domesticated.[5] But the importance of this radical aspect both to him and in general can hardly be overestimated.

The radical side of Freud, especially as it is manifested in his view of agency, rests on his theory of instincts. I can best begin a brief unraveling of that complex topic by noting Freud's awareness of the difficulties involved in understanding instincts. He recognizes that we know instincts only through their mental representations or manifestations in behavior. Moreover, he will use the conception of instincts as an example of how theory guides rather than follows observation (Freud 1963a, 83-84; Ricoeur 1970, 115-16, 134-50, 455-57; Mannoni 1971, 91-97). Indeed, in a late work, where Freud often tries to tidy up his terminology, he will call the theory of the instincts "our mythology" and say instincts are mythical entities, magnificently indefinite, that cannot be disregarded and yet are never, for sure, clearly seen (Freud 1933, 84).

Numerous problems arise, then, when we try to pinpoint the meaning or meanings of Freud's idea of instincts. For example, how does their function in human behavior differ from that of the passions as discussed, say, by Hume? Such questions are too complex and technical to discuss here, although later I will briefly consider some of them. What is important here is seeing how Freud's theory of instincts underlies his deconstruction of the self. That involves noting how human instincts differ from animal instincts, which can be described fairly easily.

The object, aim, and even to some degree impetus of human instincts are far more variable than are their animal counterparts (Freud 1963b, 83-103). Unlike animal instincts that move inexorably to one goal where they find at least momentary satiation, human instincts are too complex to be so pictured. Of course, just as a cigar is sometimes just a cigar, hunger is sometimes just hunger, but Freud aims to make us see the number of times it is not.

Indeed, at places Freud actually contrasts animal instincts (*Instinkt*) and human instincts (*Trieb*). These facts have led some to argue that "drive" is a better translation of *Trieb* than "instinct." I agree, but I do not share their view that such a change helps greatly our understanding of Freud. Both English words are as blessedly or annoyingly vague as are their German counterparts, and I shall follow general practice and translate *Trieb* by "instinct."

Most important here, Freud uses the idea of instincts to take common ideas of the self and intention and "deconstruct" them. To use language I feel more comfortable with, Freud forces or entices us to redescribe our experience in a way that makes instincts and their vicissitudes either "a" or "the" prominent category. These redescriptions are at the heart of Freud's enterprise, especially as that enterprise challenges traditional notions of ethics and reflects his cosmogonic vision.

Given the importance of these redescriptions, let us look at an example of my making that illustrates Freud's procedures and challenge. We may say that Jack is describable as loving Jill. That is, Jack says he loves Jill; he sleeps with and spends a great deal of time with her; and he even is, at present, inattentive to others, losing weight, and constantly thinking of Jill. But Freud says closer investigation may show that the appropriate description is not that Jack loves Jill but that John hates Joan. John is Jack's father—the ego being a precipate of identifications (Freud 1923, 19). Moreover, Joan is his mother—every finding of an object being a refinding of it (Freud 1905, 88). Finally, hate is the only way Jack can, for various reasons, react as John to Joan (Freud 1963b, 97-103). We could, of course, go through similar if more complex procedures (which, incidentally, resemble the transformations one often finds in myths) with "Jack does his duty toward Jill," "Jack behaved justly toward Jill," "Jack lied to Jill," and so forth. In all these cases, Freud redescribes the "same" phenomenon in a way that focuses not on a unified self who intends or even suffers certain actions but on a set of instincts that coalesce into various defenses, move in various ways, and fix on various objects.

I will not try here to fill in all the particulars of this procedure, encased as they are in terms of art, such as identification, or in theories, such as how instincts reverse into their opposites. Rather, I will simply assert that a "plausible" Freudian redescription can be given. Such a redescription will be difficult to invalidate unless one refuses to accept its basic principles. Moreover, it will provide an explanation of certain central and mysterious facts—in the case of love, for example, the selection of one rather than another person to love and the corresponding overestimation of that person by the lover.

Freud's redescriptions illustrate his most dramatic claim: that people are lived by forces whose character and modes of activity most do not understand. He claims that at many crucial times and with many crucial events people do not have direct access to the instincts that express themselves directly or that their defenses cope with these instincts in ways that affect their actions. In his own memorable words, Freud tries to help us "to emancipate ourselves from our sense of the importance of that symptom which consists in 'being conscious'" (Freud 1963b, 139). He asks us to follow him, the archaeologist turned cosmogonist, and describe from their remaining traces the real springs and purposes of our present activity. Using the justification he gives for the notion of the Unconscious (*Unbewusste*, perhaps better as "Unknown"), he asks us to consider what we must presume to explain our behavior, and then builds a new set of descriptions from the presumed operations of the instincts. His explanations or redescriptions are varied: dynamic, economic, ontogenetic, phylogenetic, adaptive, physiological, typological, and structural. But all of them rest on the general idea of instincts.

Obviously a number of complex questions must be answered if we are to understand exactly what Freud means and how valid that meaning is. Unfortunately, Freud often does not address these questions adequately, and we surely cannot explore them here. But the three most important ones are as follows. First, how are we to analyze the distinctions and relationships between a language of force and cause, and a language of intention and reason, especially as that affects how we conceptualize semiautonomous agencies' casual affect on other semiautonomous agencies? (see Ricoeur 1970; Davidson 1982). Second, how are we to judge which is correct: what a skilled observer decides is "a" or even "the" reason for an action or what an agent claims is his or her reason for the action? (see Care and Landesman 1968, 159-78; 214-37). Third, how are we to justify or even explain two apparently contradictory needs: a focus on the abnormal to explain the normal and the presumption that the normal provides the context by which the abnormal is identified? (see Peters 1958, 52-94). Answers to these questions need to be given if we are to evaluate Freud's explanatory apparatus and, perhaps more significantly, his identification of certain apparently mysterious phenomena. More important to us here, however, is the task of taking Freud's theory of instincts, especially as it informs his death-instinct cosmogony, and examining what implications it has for his account of ethical action.

Freud's Cosmogony and Ethics: Introduction

As one would expect, the death-instinct cosmogony has a significant affect on Freud's view of ethics. I will focus on how it affects his picture of an individual's moral action as that appears in his ego/id/superego model. I will begin, however, by briefly noting how it affects his view of social processes. This topic, to which I shall return, is especially important because the late Freud probably turns to the analysis of social processes when he recognizes that they alone present a large enough canvas to reveal clearly the death instinct's workings.

Although Freud's inquiry into social processes appears in various works, the key text is *Civilization and Its Discontents* which can, I think, be interpreted as a cosmogonic romance about how humans were fated to get into the fix they did. The book is, I think, one of the most complex and revealing Freud ever wrote; his categorization and critical analysis of all the normal goals of human life show him at both his most incisive and his most flawed (Freud 1930, 21-32). But most important to us is how his death-instinct cosmogony produces a distinctive picture of the character of ethical action, especially as that action generates discontent within the individual.

The disquiet an individual feels when faced with or acting from moral rules has two sources. The first and most superficial rests on the need for the individual both to abridge private desires and to modify attachment to archiac objects of desire because of the demands made by the larger community. This notion, which resembles ideas in social-contract theory, is an old one in the West, even if it receives a particular character because it rests on Freud's historical cosmogony.

The second source of disquiet arises from the death instinct, however, and here Freud's analysis becomes both more novel and more bizarre. Aggression is the crucial manifestation of the death instinct's action, but "aggression" is not a phenomenon that mere introspection or empirical study can uncover. Only the cosmogonic story can reveal its true character. That story tells us that only by aggression can aggression be mastered and civilization be maintained. That is, only by the death instinct being used against itself, only by instinct taming instinct, can civilized life be nurtured.

Freud explicitly portrays Eros as using Thanatos, the death instinct, to achieve its own end of maintaining a complex entity. He thereby generates a picture whose mythical grandeur he matches in only one other place, interestingly enough in a discussion of the interminable character of psychoanalysis (Freud 1930, 64–92; Freud 1963c, 258–65). Freud asserts, then, that social existence's rules and consequent discontents are understandable only if they are set against the backdrop provided by the cosmogonic story of the character and interaction of Eros and Thanatos. Keeping this general account in mind, we can now examine Freud's account of cosmogony's specific affect on ethics by turning to his so-called second topography (or "personology" in a more accurate if barbaric designation) of the ego, id, and superego.

Cosmogony and Ethics: The Ego/Id/Superego Model

We see in this model of the self both how Freud's theory of instincts assumes a most distinctive form and how his "deconstructionist" or "redescriptionist" approach achieves perhaps its most radical results. In examining this model, however, we need to note again that Freud often cannot keep in focus his more radical ideas. He will, for example, tell us that ego must be where id was right after he has virtually destroyed our ability to think of id and ego as independent beings (Freud 1933, 71). I will concentrate on the most radical thread in his account, for reasons discussed earlier, but we need to remember that other, more normal threads do exist.

Freud sees the three entities in this model as analytic constructs that are to be judged by their ability to further our understanding of the complex flow of life (Freud 1923, 1, 6–17, 28–29; Freud 1926, 17–24). With his new model of the person, he invites us to look at ourselves and others in an unusual but highly suggestive way. To respond adequately to that invitation we need to note that the normal translations of its terms are quite misleading because they imply technical rather than common ideas. "Ego" translates *Ich* ("I"), "superego" translates *Überlch* ("over I," "more than I"), and "id" translates *es* ("it").

Freud asks us, then, to see ourselves as interacting variations on or transformations of the personal and impersonal pronouns, of the grammatical subject. We can, he thinks, productively see ourselves as being our self, more than our self, and other than our self. We are a set of relationships among semiautonomous agencies, one of whom we normally identify as the individual we are, another of whom we normally identify as possessing more capabilities than we as individuals do, and a third of whom we normally identify as being less than or different from our individual selves. All three agencies are constantly intertwined in any action, just as they are intertwined in other ways. Genetically all come from the id, structurally all are largely unconscious, and developmentally all three arise, in part, by contact with or absorption of external objects.

We can, however, describe each agency separately. Such a description can help us grasp ideas that have become part of common speech and suffered accordingly, even if it will necessarily draw most of the blood from this complex body of ideas. The Freudian ego is often painted, sometimes even by Freud, as resembling prudential reason, as being able to guide action by selecting goals, judging the character of external situations, and then controlling or even forming unruly passions (Freud 1926a, 22-23). Such a characterization is misleading because it overlooks various critical ideas; for example, that the ego is fundamentally a body ego (Freud 1923, 15-17). More important, it fails to capture how much of the ego is unconscious, is the agent of those fundamental defenses that form our character and therefore guide much of our action (see Madison 1961).

Similarly, seeing the id as a caldron of seething passions—and again Freud will so describe it—is misleading because it fails to capture how much of both the ego and the superego are conditioned by the id. The id is often perceived by the self as the impersonal, as an alien territory or force within me, as that which I would like to imagine is not me. But it not only constantly asserts itself but also, and most important, informs even those judgments that I think are most clearly and firmly my own, such as my judgments on my unruly passions.

Finally, the superego is neither simply a repository of rules nor just an agent that judges one when those rules are broken. It also is manifested in the sense that one is observed in what one thinks and does, so that not just deeds but also intentions and even fantasies are of ethical import. Even more important, the superego also contains all those ideals that animate me. These ideals, in classical Freudian fashion, neither ennoble my sense of who I am nor guide my acts and aspirations. Rather they induce either a feeling of inferiority and guilt about my inability to realize them or a short-lived and sordid satisfaction about the nobility of my motives. Most important to us, in all these actions of the superego we find the clearest possible manifestation of the death instinct; Freud even calls it a pure culture of the death instinct (Freud 1923, 43). The ideas I aspire to, the moral rules I honor, the observing capacity I employ to purify myself all manifest my aggression against myself (Freud 1933, 52–58). In the superego we find, then, the clearest evidence of Eros using the death instinct to control itself.

In the ego, id, superego model Freud attempts to picture in the most parsimonious yet comprehensive possible way who we really are. He starts from the premise that we are many rather than one—despite the grammatical jokes and theoretical difficulties such a picture presents (see Thalberg 1974). He then attempts to demolish our false picture of ourselves as conscious decision-makers who follow ideals, apply rules, and ajudicate sensibly. Such a view differs radically from the more common Western picture of reason's relation to the passions, whether the Humean view of reason's "slavery" to the passions or the Aristotelian view of reason's "formation" of the passions. Linked directly to his death cosmogony—indeed able to exist because of it—this perspective casts a most distinctive light on the question of what ethics is.

Cosmogony and Ethics: The Freudian Picture

Ethics is understood as that process where Eros uses the death instinct to tame itself, where the superego turns aggression toward others into aggression against the self. The particulars of Freud's genetic account of how the superego develops and the intricacies of his analysis of its actual functioning are too complex to examine here (see Freud 1930, 70-79), but we can examine how such a perspective transforms normal conceptions of the ethical. Duty becomes a form of pathology, morality a form of alienation, and ethical striving the movement to perfect one's slavery to a foreign and inhuman master who asks for sovereignty over both thought and deed. From this perspective, being moral is a masochistic phenomenon, or more precisely satisfies both sadistic and masochistic drives as they are evident in the application and acceptance of moral rules and ethical aspirations. Moreover, all these deformations are seen as necessary and even commendable for at least most people because the alternative—the free reign of the death instinct—is even more frightening than are the deformations.

Such is the bizarre picture that Freud paints, and one is led to ask how this bizarreness can be explained. Two approaches commend themselves. The first emphasizes that the bizarreness of the picture arises in large part because the fully developed death-instinct cosmogony implies such a picture. The bizarreness flows, then, directly from the link between cosmogony and ethics. Given that cosmogony, Freud is naturally if not inevitably led to that picture of ethics.

This link between cosmogony and ethics is most clear in the cosmogonic romance that is *Civilization and Its Discontents*. Behavior that in other contexts is often labeled pathological becomes normal ethical behavior there. That book shows, then, that any depiction which centers on the death-instinct cosmogony must depict ethical activity as pathological. Freud may rarely have carried through an extended treatment of the ethical implication of the death-instinct cosmogony (luckily, some would say) but when he does, as in *Civilization and Its Discontents,* he can only portray ethics as pathological.

The second approach takes a different tack. It declares that Freud's portrait of ethics should be seen as a portrait of the deformations to which ethics is liable. This approach would attempt to do with Freud on morality what some have tried to do with Freud on religion: to see his analysis as a compelling and textured picture of the deformations which religion (or in this case morality) can produce (see Ricoeur 1970, 442-52, 531-57). Freud's picture does capture, without question, some deformations that can easily characterize "the moral life" at least as lived in the modern West. He describes well, for example, the implacable, inhuman severity that conscience can show; the unwillingness to change, adapt, or recognize subtle differences that "moral" judgment can reveal; and either the harshness in the name of compassion or the cruelty in the name of justice that common morality can manifest. Nevertheless, this approach encounters problems that are probably even more striking than those encountered in using Freud to point to the non-narcissistic consolation that should characterize "true" religion.

The major problem with this approach is that Freud never gives us even the beginnings of an adequate account of what morality would be if it were not the pathological phenomenon born of the death instinct that he has described. Fragmentary hints do exist. He will, for example, distinguish between ethics as it concerns the rational relationship of the individual and the community, and as it concerns the relationship of the "deconstructed" individual and the "will of the father" (Freud 1939, 156). Similarly, he will distinguish, if tentatively, between commands arising from taboos and commands arising from morality, on the grounds that the latter arise from reason and aim at universality (Freud 1912, x, 22, 66–68, 71–74). Such hints are intriguing, but they hardly constitute even the beginning of an account of a justifiable morality. This lack is particularly striking because Freud himself was a decent person and at times even a heroically self-sacrificing one. But his work articulates no rationale that explains either why he acted as he did or why anyone should follow one rather than another set of moral directives or, for that matter, follow any moral directives at all.

We can attempt to explain this lacuna in a variety of ways. It may be an understandable if unfortunate extension of the eminently defensible idea that one should suspend the consideration of moral questions in the therapy process. It may rest on an unreflective acceptance by Freud, reinforced by his distrust of philosophical speculation, of an odd amalgam of Hobbesian and utilitarian ideas as the only ideas that provide realistic guides for action. It may arise from a perhaps justifiable commitment to the distinctiveness of the psychoanalytic project that severely limited the areas into which he would inquire.

Whatever may be the explanations for it, the striking absence in Freud of any developed views on nonpathological morality left those who accepted some significant part of the Freudian vision with a job. They needed to "complete" Freud's work either by extrapolating from hints in it or by developing it in the directions that it "should" have gone. The results of this process are of considerable interest and importance, and I shall end by noting three of them. I will concentrate on the one closest to Freud's own ideas, but in no case will I attempt to give an account that is nuanced enough to do justice to the actual historical figures or movements that are involved.

Cosmogony and Ethics: Three Reformulations of the Freudian Picture

Three positions arise from asking the question of how to find appropriate guides for action once one accepts significant aspects of the Freudian vision. They fit into a spectrum with the radical and conservative positions linking ethics with a cosmogony, however loosely, and the middle position attempting to give guides for action that are unsupported by a cosmogony. Because this middle position is both distinctively modern and closest to Freud, I will concentrate on it. We do need to look briefly at the other two positions, however.

Proponents of the radical position argue that Freud failed to distinguish between necessary repression and surplus repression, between the rules needed for the maintenance of a minimal social order and the rules needed to protect the interests, the views, or the neurotic desires of a particular class of people (see Horowitz 1977). They attack both the complexity and severity of traditional rules as examples of surplus repression. Most important to us, this position often rests on a cosmogony that pictures humans, before they are warped by moral codes, as creatures whose spontaneous actions will generate only more complex, vital, and harmonious relationships. Eros's actions alone define the basically human.

The conservative position, which remains relatively undeveloped, sees as implicit in Freud a version of that kind of Western natural-law theory that argues that all humans are defined by natural movements from potentiality to actuality (see Rieff 1979, 358-97). To follow these movements is to reach fulfillment, to act against them is to reach a distorted actuality filled with conflict and pain. In this view, the specific form of fulfillment will always, to some degree, be determined by the society in which people exist, because people can actualize their natural potential only by utilizing the available social forms. Nevertheless, the basic patterns of self-development and social life arise from a noncontingent source and are inescapable in the sense that to violate those patterns is to turn one's life into a sea of pain.

The third position might be entitled a prudential conventionality that is flavored with common decency (see Rieff 1966, 79-107; 1979, 300-357). What proponents of this position reject is most clear and perhaps most important: traditional morality, as that is defined by reliance on a cosmogony and therefore on the superego and its manifestations in guilt, duty, and inflexible rule. They recognize the power of such a morality, resting as it does on habit and the death instinct. Moreover, unlike those who uphold the first two positions, they think people can find no natural directives arising from a cosmogony to guide them. Therefore, they think a person must use considerable energy simply to break the grip of a traditional morality. But reflection on the self and luck—the accidents of social position, native constitution, and early childhood experiences—can allow one to free oneself from at least the more horrendous forms and consequences of traditional morality. (This position will always be haunted by the question of whether all can achieve the hoped-for fulfillment or whether many must continue to be ruled by a pathological morality or face an even worse fate.)

For those who can escape traditional morality, what remains is the recognition that one can live only with the contradictions, conflicts, and uncertainties life presents. Most important is the attempt always to keep both multiple perspectives open and negotiations among them active. One aims only at some prudential accommodation among one's needs, one's opportunities, and one's values, whatever may be the causes or justification of those values. No stable hierarchy of values can be established, and no overarching and unchanging goals can be pursued. Indeed, one must always attempt to tame those hierarchies and ideals that do present themselves. One's moral budget is, so to speak, very limited. One must resign oneself to that fact because overdrafts bring consequences that make living within one's means very attractive.

A critic can argue that such a position is pathetically unheroic, allows one to choose only among choices not worth making, fosters the worst kind of accommodation to regnant social norms, refuses even to ask seriously questions about a healthy morality's character, leaves people prey to the simple pursuit of selfish pleasure, and presents people with no grounds to protest even the most horrible violations of decency. The proponent might admit all of that, might even acknowledge how profoundly unsatisfying is his position. But he can also add that no other options exist save those presented either by pathological morality or by the naive hopes generated by other discredited cosmogonies. People can refuse to endure with resignation, they can indulge themselves in narcissistic wishes about how things should be. But these activities refuse to acknowledge the form of the tragic landscape in which humans live, and they leave a person open to potentially damaging illusions.

The unsatisfying tangle of problems that infests this last position represents, I think, the logical consequence of Freud's analysis. As a critic of cosmogonies, he assumes that cosmogonies must undergird ethics and then attacks the cosmogonies and thus the ethics. As a creator of a cosmogony, he builds a cosmogony that makes all morality into a form of pathology. Each approach represents a side of Freud. The first shows Freud the optimist. He holds to the hope that destruction of the old will bring an undefined better state. The second shows Freud the tragedian. He thinks that resigned incoherence is all we can possess, and he tries simply to replace either uncommon misery with common misery, or obsessionally oppressive guidance with unfulfilling but realistic ambiguity.

The second side of Freud is far sturdier than the first side, but it leads to the thicket of problems and unsatisfying options that we have just examined. The only real remaining questions concern the adequacy of his analysis—and perhaps the need for illusions. Freud, of course, had few doubts about the former and never really examines the latter. We can, of course, differ on both counts.

Notes

1. In making references to Freud's work, I have chosen to refer not to the Standard Edition but to those more readily available editions that contain, except in a few cases, the same translations and notes as does the Standard Edition. Many may have these editions, I assume, while few will have the Standard Edition. Note, however, that Rieff's edition of Freud's papers contains older translations, some of which have been improved considerably in the Standard Edition. In citation's of Freud's work, the date given is that of original publication, except in the case of the collected papers.

2. For particularly important texts in Freud, see 1920, 6–17, 28–56; 1921, 32–36; 1923, 30–49, 1930, 48–50, 58–92; 1940, 5–7, 54–55; 1963b, 190–201; 1963c, 252–65. Also note other Freudian statements on death (Freud 1963a, 121–33, 148–51), including one where he links his and Einstein's pacifistic tendencies to organic causes (Freud 1963a, 146). Good secondary studies of the whole cosmogony but especially of the death instinct are found in Brown 1959, 77–134; Laplanche and Pontalis 1973, 97–103, 153–54, 241–42, 447; Mannoni 1971, 145–59; Ricoeur 1970, 228–29, 257–58, 281–320, 451–52; Rieff's Introduction in Freud, 1963b, 9–17; and Sulloway 1979, 393–415. On this issue, as well as on most others, Brown and Sulloway represent well the two contrasting positions on Freud that can be taken. To oversimplify, Brown often employs categories drawn from Hegel and focuses on Freud the humanist who sees the importance of biology. Sulloway focuses on Freud the scientist and employs categories drawn from the history of science. Ricoeur attempts, sometimes quite successfully, to bring together these two positions.

3. Many technical issues arise when one discusses the question of how logical or even sensible is Freud's postulation of a death instinct. I cannot discuss them here, but for two influential positions that differ from mine, see Brown (1959, 87-109) and Sulloway (1979, 393-415). Brown argues that of the three sets of phenomena Freud points to as implying the death instinct only one in fact is even partially adequate to the task. Sulloway presents the death instinct as the logical result of Freud's biological perspective. I am not, incidentally, arguing that the four phenomena discussed here are the only

ones to which Freud refers when he discusses the death instinct; regression, for example, is also very important.

4. The question of the validity of Freud's interpretative method has generated much controversy. For, to my mind, Freud's most mature statement on the question of interpretation within the therapeutic context, see Freud (1963c, 273-86). For a discussion of recent philosophic criticisms of Freud's interpretive method, see Erwin (1981). For attempts to defend Freud by relying on, among other things, the concept of narrative, see Sherwood (1969) and Ricoeur (1978, 184-210).

5. See, for example, the very different picture of Freud that arises from the work of Lacan and those influenced by him: Lacan (1968); Lemarie (1979); Jameson (1977).

References

Brown, Norman O.

1959 *Life Against Death: The Psychoanalytic Meaning of History*. Middleton: Wesleyan University Press.

Care, Norman and Charles Landesman, eds.

1968 *Readings in the Theory of Action*. Bloomington, Indiana: Indiana University Press.

Davidson, Donald

1982 "Paradoxes of Irrationality." In Richard Wollheim and James Hopkins, eds., *Philosophical Essays on Freud*, pp. 289-305. Cambridge: Cambridge University Press.

Erwin, Edward

1981 "The Truth about Psychoanalysis." *The Journal of Philosophy* 78: 10 (October): 540-60.

Freud, Sigmund

1900 *The Interpretation of Dreams*. Trans. J. Strachey. New York: Basic Books, 1965.

1905 *Three Essays on the Theory of Sexuality*. Trans. J. Strachey. New York: Basic Books, 1962.

1912 *Totem and Taboo*. Trans. J. Strachey. New York: W. W. Norton, 1950.

1920 *Beyond the Pleasure Principle*. Trans. J. Strachey. New York: W. W. Norton, 1961.

1921 *Group Psychology and the Analysis of the Ego*. Trans. J. Strachey, New York: W. W. Norton, 1959.

1923 *The Ego and the Id*. Trans. J. Riviere, rev., J. Strachey. New York: W. W. Norton, 1962.

1926a *The Problem of Anxiety*. Trans. H. Bunker. New York: W. W. Norton, 1936.

1926b *The Question of Lay Analysis with Freud's 1927 Postscript*. Trans. J. Strachey. New York: W. W. Norton, 1959.

1927 *The Future of an Illusion*. Trans. J. Strachey. New York: W. W. Norton, 1961.

1930 *Civilization and Its Discontents*. Trans. J. Strachey. New York: W. W. Norton, 1930.

1933 *New Introductory Lectures on Psychoanalysis*. Trans. J. Strachey. New York: W. W. Norton, 1964.

1939 Moses and Monotheism. Trans. K. Jones. New York: Random House, 1967.

1940 *An Outline of Psychoanalysis*. Trans. J. Strachey. New York: W. W. Norton, 1949.

1960 *Letters of Sigmund Freud*. Ed. Ernst Jones. Trans. T. and J. Stern. New York: Basic Books.

1963a *Character and Culture. The Collected Papers of Sigmund Freud*. Ed. Phillip Rieff. New York: Collier Books.

1963b *General Psychoanalytic Theory. The Collected Papers of Sigmund Freud*. Ed. Phillip Rieff. New York: Collier Books.

1963c *Therapy and Technique. The Collected Papers of Sigmund Freud*. Ed. Phillip Rieff. New York: Collier Books.

1963d *Three Case Histories. The Collected Papers of Sigmund Freud*. Ed. Phillip Rieff. New York: Collier Books.

Horowitz, Gad

1977 *Repression: Basic and Surplus Repression in Psychoanalytic Theory: Freud, Reich, and Marcuse*. Toronto: University of Toronto Press.

Jameson, Fredric

1977 "Imaginary and Symbolic in Lacan: Marxism, Psychoanalytic Criticism, and the Problem of the Subject." In Shoshana Felman, ed., *Literature and Psychoanalysis: The Question of Reading Otherwise*, pp. 338-95. Yale French Studies 55/56.

Jones, Ernest
1961 *The Life and Work of Sigmund Freud*. Ed. L. Trilling and S. Marcus. London: Hogarth Press.

Lacan, Jacques
1968 *The Language of the Self: The Function of Language in Psychoanalysis*. Trans., with notes and commentary, A. Widen. Baltimore: The John Hopkins University Press.

Laplanche, J. and J. B. Pontalis
1973 The Language of Psychoanalysis. Trans. D. Nicholson-Smith. New York: W. W. Norton.

Lemarie, Anike
1979 *Jacques Lacan*. Trans. D. Macey. London: Routledge and Kegan Paul.

Madison, Peter
1961 *Freud's Concept of Repression and Defense: Its Theoretical and Observational Language*. Minneapolis: University of Minnesota Press.

Mannoni, O.
1971 *Freud*. Trans. R. Bruce. New York: Random House.

Meisel, Perry, ed.
1981 *Freud: A Collection of Critical Essays*. Englewood Cliffs, N.J.: Prentice-Hall.

Nagel, Thomas
1979 *Mortal Questions*. New York: Cambridge University Press.

Peters, R. S.
1958 *The Concept of Motivation*. New York: Humanities Press.

Ricoeur, Paul
1970 *Freud and Philosophy: An Essay on Interpretation*. Trans. D. Savage. New Haven: Yale University Press.
1978 *The Philosophy of Paul Ricoeur: An Anthology of His Work*. Ed. C. Regan and D. Stewart. Boston: Beacon Press.

Rieff, Philip
1979 *Freud: The Mind of the Moralist*. Chicago: University of Chicago Press (reprint of 1959).
1966 *The Triumph of the Therapeutic Uses of Faith after Freud*. New York: Harper and Row.

Sherwood, M.
1969 *The Logic of Explanation in Psychoanalysis*. New York: Academic Press.

Sulloway, Frank
1979 *Freud, Biologist of the Mind*. New York: Basic Books.

Thalberg, Irving
1974 "Freud's Anatomies of the Self." In Richard Wollheim, ed., *Freud: A Collection of Critical Essays*, pp. 147–71. Garden City, N.Y.: Doubleday.

Contributors

Arthur W. H. Adkins
Professor of Greek, Philosophy, and New Testament,
the University of Chicago

Hans Dieter Betz
Professor of New Testament,
the University of Chicago

Sheryl L. Burkhalter
Graduate student in History of Religions,
the University of Chicago

Norman J. Girardot
Associate Professor of Religion Studies,
Lehigh University

Douglas A. Knight
Associate Professor of Old Testament,
Vanderbilt University

Robin W. Lovin
Associate Professor of Ethics and Society,
the University of Chicago

Wendy Doniger O'Flaherty
Professor of History of Religions and Indian Studies, the University of Chicago

Frank E. Reynolds
Professor of History of Religions and Buddhist Studies, the University of Chicago

Douglas Sturm
Professor of Religion and Political Science, Bucknell University

Lawrence E. Sullivan
Associate Professor of History of Religions, the University of Chicago

Kay Barbara Warren
Associate Professor of Anthropology, Princeton University

Lee H. Yearley
Associate Professor of Religious Studies, Stanford University